Springer Handbook of Auditory Research

Volume 54

Series editors

Richard R. Fay, Woods Hole, MA, USA
Arthur N. Popper, College Park, MD, USA

More information about this series at http://www.springer.com/series/2506

The ASA Press

The ASA Press imprint represents a collaboration between the Acoustical Society of America and Springer dedicated to encouraging the publication of important new books in acoustics. Published titles are intended to reflect the full range of research in acoustics. ASA Press books can include all types of books published by Springer and may appear in any appropriate Springer book series.

ASA Press

M. Brock Fenton • Alan D. Grinnell
Arthur N. Popper • Richard R. Fay
Editors

Bat Bioacoustics

With 62 Illustrations

ASA Press

Springer

Editors
M. Brock Fenton
Department of Biology
University of Western Ontario
London, ON, Canada

Alan D. Grinnell
Department of Physiology
David Geffen School of Medicine
University of California
Los Angeles, CA, USA

Arthur N. Popper
Department of Biology
University of Maryland
College Park, MD, USA

Richard R. Fay
Marine Biological Laboratory
Woods Hole, MA, USA

ISSN 0947-2657 ISSN 2197-1897 (electronic)
Springer Handbook of Auditory Research
ISBN 978-1-4939-3525-3 ISBN 978-1-4939-3527-7 (eBook)
DOI 10.1007/978-1-4939-3527-7

Library of Congress Control Number: 2016936785

Printed on acid-free paper

This Springer imprint is published by Springer Nature
The registered company is Springer Science+Business Media LLC New York

This book is dedicated to the memory of Annemarie Surlykke (1955–2015), A brilliant scholar, delightful colleague, and good friend.

Acoustical Society of America

The mission of the **Acoustical Society of America** (www.acousticalsociety.org) is to increase and diffuse the knowledge of acoustics and promote its practical applications. The ASA is recognized as the world's premier international scientific society in acoustics, and counts among its more than 7,000 members, professionals in the fields of bioacoustics, engineering, architecture, speech, music, oceanography, signal processing, sound and vibration, and noise control.

Since its first meeting in 1929, The Acoustical Society of America has enjoyed a healthy growth in membership and in stature. The present membership of approximately 7,500 includes leaders in acoustics in the United States of America and other countries. The Society has attracted members from various fields related to sound including engineering, physics, oceanography, life sciences, noise and noise control, architectural acoustics; psychological and physiological acoustics; applied acoustics; music and musical instruments; speech communication; ultrasonics, radiation, and scattering; mechanical vibrations and shock; underwater sound; aeroacoustics; macrosonics; acoustical signal processing; bioacoustics; and many more topics.

To assure adequate attention to these separate fields and to new ones that may develop, the Society establishes technical committees and technical groups charged with keeping abreast of developments and needs of the membership in their specialized fields. This diversity and the opportunity it provides for interchange of knowledge and points of view has become one of the strengths of the Society.

The Society's publishing program has historically included the *Journal of the Acoustical Society of America*, the magazine *Acoustics Today*, a newsletter, and various books authored by its members across the many topical areas of acoustics. In addition, ASA members are involved in the development of acoustical standards concerned with terminology, measurement procedures, and criteria for determining the effects of noise and vibration.

Series Preface

The following preface is the one that we published in Volume 1 of the *Springer Handbook of Auditory Research* (SHAR) back in 1992. As anyone reading the original preface, or the many users of the series, will note, we have far exceeded our original expectation of 8 volumes. Indeed, with books published to date and those in the pipeline, we are now set for over 50 volumes in SHAR, and we are still open to new and exciting ideas for additional books.

We are very proud that there seems to be consensus, at least among our friends and colleagues, that SHAR has become an important and influential part of the auditory literature. While we have worked hard to develop and maintain the quality and value of SHAR, the real value of the books is very much because of the numerous authors who have given their time to write outstanding chapters and to our many coeditors who have provided the intellectual leadership to the individual volumes. We have worked with a remarkable and wonderful group of people, many of whom have become great personal friends of both of us. We also continue to work with a spectacular group of editors at Springer. Indeed, several of our past editors have moved on in the publishing world to become senior executives. To our delight, this includes the current president of Springer USA, Dr. William Curtis.

But the truth is that the series would and could not be possible without the support of our families, and we want to take this opportunity to dedicate all of the SHAR books, past and future, to them. Our wives, Catherine Fay and Helen Popper, and our children, Michelle Popper Levit, Melissa Popper Levinsohn, Christian Fay, and Amanda Fay, have been immensely patient as we developed and worked on this series. We thank them and state, without doubt, that this series could not have happened without them. We also dedicate the future of SHAR to our next generation of (potential) auditory researchers—our grandchildren—Ethan and Sophie Levinsohn, Emma Levit, and Nathaniel, Evan, and Stella Fay.

Preface 1992

The *Springer Handbook of Auditory Research* presents a series of comprehensive and synthetic reviews of the fundamental topics in modern auditory research. The volumes are aimed at all individuals with interests in hearing research including advanced graduate students, postdoctoral researchers, and clinical investigators. The volumes are intended to introduce new investigators to important aspects of hearing science and to help established investigators to better understand the fundamental theories and data in fields of hearing that they may not normally follow closely.

Each volume presents a particular topic comprehensively, and each serves as a synthetic overview and guide to the literature. As such, the chapters present neither exhaustive data reviews nor original research that has not yet appeared in peer-reviewed journals. The volumes focus on topics that have developed a solid data and conceptual foundation rather than on those for which a literature is only beginning to develop. New research areas will be covered on a timely basis in the series as they begin to mature.

Each volume in the series consists of a few substantial chapters on a particular topic. In some cases, the topics will be ones of traditional interest for which there is a substantial body of data and theory, such as auditory neuroanatomy (Vol. 1) and neurophysiology (Vol. 2). Other volumes in the series deal with topics that have begun to mature more recently, such as development, plasticity, and computational models of neural processing. In many cases, the series editors are joined by a coeditor having special expertise in the topic of the volume.

Richard R. Fay
Woods Hole, MA, USA

Arthur N. Popper
College Park, MD, USA

Volume Preface

Bats are one of the most vociferous of all mammalian groups. They not only use sound and hearing for communication but also for echolocation in ways that are rivaled only by toothed whales. While most papers, including previous SHAR volumes, focus on the echolocation behavior of bats, this volume extends the story to include overall bioacoustics, including communication and behavior.

Chapter 1, by Alan Grinnell, Ed Gould, and M. Brock Fenton, goes into the history of the study of bat bioacoustics, including a description of the work by some of the pioneers in the field (with photographs). This is followed by Chapter 2 in which Emma Teeling, Steve Rossiter, and Gareth Jones review the contributions of genetic and genomic databases in advancing our understanding of the phylogeny of bats and their classification among the mammals.

In Chapter 3, Walter Metzner and Rolf Müller review the diversity of laryngeal and facial structural specializations for production of both echolocation pulses and communication signals in echolocating bats. This is followed by Chapter 4 in which Patricia Jones, Rachel Page, and John Ratcliffe discuss the use of laryngeal echolocation when bats are foraging, including the differences and similarities among aerial hawkers, substrate gleaners, and bats that alternate between those two strategies. In Chapter 5, Erin Gillam and M. Brock Fenton consider the many situations in which bats communicate with acoustic signals, from behaviors underlying social systems to interactions between mothers and their young.

In Chapter 6, Annette Denzinger, the late Eli Kalko, Marco Tschapka, Alan Grinnell, and Uli Schnitzler review the evidence supporting the view that communities of bats consist of guilds of species. They focus on the overlaps among sympatric species in terms of echolocation strategies, foraging behavior, habitat use, and diet, and they define seven different guilds that suffice to include all species.

In the first of two chapters that deal with central processing of sound, Paul A. Faure and Uwe Firzlaff (Chapter 7) discuss the importance of temporal resolution in the hearing of echolocating bats. They consider the neural mechanisms that code signal duration and the functional significance of this information in achieving selectivity of the responses to communication sounds and echolocation signals. In Chapter 8, George Pollak considers the diversity of communication and echolocation sounds used by different bats and the diversity of neural adaptations for processing the sounds, especially at the level of the inferior colliculus.

Chapter 9, by Shizuko Hiryu, Emanuel Mora, and Hiroshi Riquimaroux, examines how bats use Doppler shift compensation (DSC) in the context of high-duty cycle echolocation. Chapter 10, by Annemarie Surlykke, Jim Simmons, and Cynthia Moss, considers the question of "scene analysis" via echolocation. They compare the information available to the auditory system, through echoes, with information processing in the visual system. In Chapter 11, Lutz Wiegrebe, Brock Fenton, and Alan Grinnell identify some of the major challenges to understanding mechanisms of echolocation and the technological advances that make it feasible to address them.

This volume in the *Springer Handbook of Auditory Research* has close ties to several earlier volumes that discuss aspects of bat bioacoustics. Hearing is covered extensively in *Hearing by Bats* (Vol. 5, edited by Arthur N. Popper and Richard R. Fay), while biosonar is covered in a chapter in *Auditory Computation* (Vol. 6, edited by Harold L. Hawkins, Teresa A. McMullen, Arthur N. Popper, and Richard R. Fay) and in multiple chapters comparing bats and toothed whales in *Biosonar* (Vol. 51 edited by Annemarie Surlykke, Paul Nachtigal, Richard R. Fay, and Arthur N. Popper).

M. Brock Fenton
Toronto, ON, Canada

Alan D. Grinnell
Los Angeles, CA, USA

Arthur N. Popper
College Park, MD, USA

Richard R. Fay
Woods Hole, MA, USA

Contents

Contributors

Annette Denzinger Animal Physiology, Institute for Neurobiology, University of Tübingen, Tübingen, Germany

Paul A. Faure Department of Psychology, Neuroscience and Behaviour, McMaster University, Hamilton, ON, Canada

M. Brock Fenton Department of Biology, University of Western Ontario, London, ON, Canada

Uwe Firzlaff Lehrstuhl für Zoologie, Technische Universität München, Freising-Weihenstephan, Germany

Erin Gillam Department of Biological Sciences, North Dakota State University, Fargo, ND, USA

Edwin Gould Columbia, MD, USA

Alan D. Grinnell Department of Physiology, David Geffen School of Medicine, University of California, Los Angeles, CA, USA

Shizuko Hiryu Faculty of Life and Medical Sciences, Doshisha University, Kyotanabe, Japan

Patricia L. Jones Department of Ecology and Evolutionary Biology, Cornell University, Ithaca, NY, USA

Gareth Jones School of Biological Sciences, University of Bristol, Bristol, UK

Elisabeth K.V. Kalko Institute of Experimental Ecology, University of Ulm, Ulm, Germany

Smithsonian Tropical Research Institute, Panamá, Republica de Panama

Walter Metzner Department of Integrative Biology and Physiology, University of California, Los Angeles, CA, USA

Emanuel C. Mora Research Group in Bioacoustics and Neuroethology, Havana University, Havana, CP, Cuba

Cynthia F. Moss Department of Psychological and Brain Sciences, The Johns Hopkins University, Baltimore, MD, USA

Rolf Müller Department of Mechanical Engineering, Virginia Tech, Blacksburg, VA, USA

Shandong University—Virginia Tech International Laboratory, Shandong University, Jinan, China

Rachel A. Page Smithsonian Tropical Research Institute, Balboa, Panama

George D. Pollak Department of Neuroscience and Center for Perceptual Systems, The University of Texas at Austin, Austin, TX, USA

John M. Ratcliffe Department of Ecology and Evolutionary Biology, University of Toronto, Toronto, ON, Canada

Hiroshi Riquimaroux Faculty of Life and Medical Sciences, Doshisha University, Kyotanabe, Japan

Stephen J. Rossiter School of Biological and Chemical Sciences, Queen Mary University, London, UK

Hans-Ulrich Schnitzler Animal Physiology, Institute for Neurobiology, University of Tübingen, Tübingen, Germany

James A. Simmons Department of Neuroscience, Brown University, Providence, RI, USA

Annemarie Surlykke Department of Biology, University of Southern Denmark, Odense, Denmark

Emma C. Teeling School of Biology and Environmental Science, University College Dublin, Dublin 4, Ireland

Marco Tschapka Institute for Evolutionary Ecology and Conservation Genomics, Ulm, Germany

Smithsonian Tropical Research Institute, Panamá, Republica de Panama

Lutz Wiegrebe Division of Neurobiology, Department Biology II, Ludwig-Maximilians-Universität München, Planegg-Martinsried, Germany

Chapter 1
A History of the Study of Echolocation

Alan D. Grinnell, Edwin Gould, and M. Brock Fenton

Keywords Bats • Don Griffin • Doppler shift compensation • Electrophysiology • Field studies • Galambos • Lazaro Spallanzani • Louis Jurine • Technological developments

1.1 Spallanzani to Griffin

Although separated by 150 years, Lazaro Spallanzani (Figure 1.1a) and Donald Griffin (Figure 1.1c) are the two individuals most associated with the discovery of echolocation. Both were audacious thinkers with an extraordinary, wide-ranging curiosity about nature. Spallanzani was a professor at the University of Pavia, Italy. He is famous for pioneering work on digestion, artificial fertilization, and volcanism. His disproof of spontaneous generation predated Pasteur's similar work by 50 years. Spallanzani's interest in bats stemmed from his observation, in the 1790s, that a pet owl could not orient in total darkness. When he tested bats, he found that even when blinded they could survive, feed, and find their way back to a home roost. In an extensive series of ear-plugging experiments performed in a long-distance collaboration, he and the Swiss physician/naturalist Louis Jurine established that bats depend on hearing for orientation (MacArthur 2000). Spallanzani and Jurine could not detect sound emissions, however, and they did not arrive at the idea of echolocation. Their findings were not accepted by leading nineteenth-century zoologists.

A.D. Grinnell (✉)
Department of Physiology, David Geffen School of Medicine, University of California, Los Angeles, CA, USA
e-mail: adg@ucla.edu

E. Gould
6505 Carlinda Avenue, Columbia, MD 21046-1019, USA
e-mail: pottergould@gmail.com

M.B. Fenton
Department of Biology, University of Western Ontario, London, ON, Canada N6A 5B7
e-mail: bfenton@uwo.ca

M.B. Fenton et al. (eds.), *Bat Bioacoustics*, Springer Handbook of Auditory Research 54, DOI 10.1007/978-1-4939-3527-7_1

Fig. 1.1 (**a**) Lazaro Spallanzani. (**b**) Robert Galambos presenting a bat to Pierce's Sonic Detector in 1940. (**c**) Donald R. Griffin netting echolocating swiftlets in Chillagoe, Australia, in 1980. (**d**) Sven Dijkgraaf. (**e**) Gerhard Neuweiler. (**f**) Nobuo Suga. (**g**) Jim Simmons, Phil Jen, and Don Griffin at Chillagoe in Australia, 1980. (**h**) Ed Gould

After the Titanic disaster, Maxim (1912) proposed detecting icebergs by mounting low-frequency sound generators on the hulls of transatlantic vessels based on the principles by which bats avoided obstacles. Responding to this suggestion, Hartridge (1920) pointed out that low frequencies would not work for the small obstacles bats must avoid and suggested the idea of echolocation using high-frequency sounds. Proof of this did not come until 20 years later, when Donald Griffin entered the field.

Griffin was born in 1915 and grew up in New York and Cape Cod. Before entering Harvard in 1934, he was largely homeschooled by his father, Henry F. Griffin, a newspaper reporter, advertising agent, historian, and novelist. At 12, Don was already trapping and skinning small mammals; he was a "naturalist in the making" (Gross 2005). At 15, with the encouragement of his maternal uncle Alfred Redfield, a professor at the Harvard Medical School, Griffin joined the American Society of Mammalogists, which provided him with a subscription to the *Journal of Mammalogy*. Griffin had published five papers in the *Journal of Mammalogy* before graduating from Harvard (e.g., Griffin 1934).

In 1932, after helping birdbanders Oliver Austin and Maurice Braun with their projects, he developed an ambition to band bats. "My small mammal collecting had included only one or two bats that I either found dead or captured away from their roosts. But, again through my uncle Alfred, I learned of a nursery colony of little brown myotis (*Myotis lucifugus*) between Barnstable and Woods Hole. I improvised crude methods of capturing these bats as they emerged from crevices in an abandoned frame house, and furtively placed standard bird bands on the legs of a few"….. "Having recaptured some the following summer, and doubtless with encouragement from the Austins, I applied to the U.S. Bureau of Biological Survey for permission to use bird bands on bats" (Griffin 1983, p. 124). His studies of banded migrating bats as a Harvard undergraduate were his transition from natural history to future years as an experimental zoologist. Griffin (personal observation to EG), like his forebearer. Spallanzani (Barbellion 1920), displayed distain for taxonomy. When he was a junior at Harvard, two of Griffin's classmates, Talbot Waterman and Jim Fisk, told him of a Rochelle salt-crystal microphone developed by the Harvard physicist George W. Pierce to listen to high-frequency insect sounds. They urged Griffin to use Pierce's microphone to test his idea that bats use high-frequency sounds to detect and avoid obstacles. Pierce was enthusiastic about the idea, and they soon determined that, indeed, caged bats emitted a constant stream of ultrasonic pulses (Pierce and Griffin 1938). But they could not detect any of these sounds when bats were flying because the parabolic horn on the microphone made it very directional.

The following year Griffin was joined by Robert Galambos (Figure 1.1b), a graduate student working with Hallowell Davis of the Harvard Medical School. Together they found that when the parabolic horn was pointed in the right direction, they detected streams of high-frequency pulses produced by flying bats. They also found that bats prevented from hearing echoes or with their mouths covered could not detect and avoid obstacles (Griffin and Galambos 1940, 1941; Galambos and Griffin 1942). This was the first of many breakthroughs in understanding echolocation that can be attributed to the development of new instrumentation.

Subsequent improvements in microphones and the introduction of the cathode ray oscilloscope allowed one to view and photograph bat sounds and measure their duration and frequency structure. Then, the advent of increasingly "portable" equipment (battery-powered FM tape recorders and, later, digital instrumentation and computers) made these measurements ever more accurate, even under field conditions. These continuous improvements in technology have played a major role in the many subsequent discoveries in the field.

At nearly the same time as Griffin and Galambos, Dijkgraaf (1943) (Figure 1.1d) came to similar conclusions about bat sounds. He reported hearing the faint "ticklaut" accompanying each pulse produced by flying bats in his native Netherlands during the Nazi occupation. The negative impact of the Nazi occupation surely restricted Dijkgraaf's ability to share his findings and discuss them with others. Dijkgraaf was well-known for a variety of studies (Görner et al. 1996).

After the initial discovery of echolocation, Griffin turned his attention to bird migration, the subject of his Ph.D. thesis, while Galambos studied bat auditory microphonics for his Ph.D. He found that little brown bats could hear sounds up to almost 100 kHz. Only after Griffin accepted his first faculty position at Cornell in 1946 did he begin to record sounds of big brown bats (*Eptesicus fuscus*) in the field (1953). He described the significance of those early experiments:

> Because it was impossible to aim the parabola at a bat flying 50 ft above the pond and at the same time watch the oscilloscope, I threw together a crude audible detector by feeding the ultrasonic signal into the second detector stage of a portable radio. The results were unexpectedly spectacular (Griffin 1953, 1958). Although Galambos and I had learned a decade earlier that bats increase the repetition rate of their orientation sounds when approaching obstacles, the increases in pulse rate during insect pursuit were very much greater. The "put-put-put" from the portable radio as a bat approached speeded up as it closed in on a moth or beetle to a buzz of up to 200 pulses per second. Neither I nor anyone else had previously suspected that bats might catch small flying insects by echolocation. We had always thought of echolocation as a collision warning system, and it seemed out of the question that small insects could return strong enough echoes to be audible to a rapidly moving bat. It is difficult to realize three decades later how much of a change in viewpoint was necessitated by this evidence that bats use echolocation not only for locating and avoiding stationary obstacles but for their hunting of small rapidly moving insect prey. Echolocation of stationary obstacles had seemed remarkable enough, but our scientific imaginations had simply failed to consider, even speculatively, this other possibility with such far reaching ramifications. (Griffin 1983, p. 138)

He went on to add:

> I consider that this was the most important scientific discovery which I can claim to have made entirely on my own. (Griffin, pers. comm. to ADG)

It is not easy for students of bat echolocation using twenty-first century instrumentation to appreciate fully the complexity and difficulty of making field recordings of bats in the 1950s. Griffin's early studies, supported by the Office of Naval Research, were conducted in local farmland where feeding bats were common. He drove about 10 miles to Freeville, then over bumpy meadows in an oversized, wooden-bodied Navy-surplus station wagon. The tailgate was prominently labeled "MYOTIS." The vehicle was fitted with shelving to accommodate the necessary equipment. This included electronic filters, wave generator, old radios, amplifiers, extra vacuum tubes, a heavy dual-beam oscilloscope fitted with a fast 35 mm film camera, parabolic reflectors, microphones, multiple tripods, a very heavy gasoline generator, 200 ft of heavy extension cord, tanks of gasoline, and repair supplies.

Electronic equipment was very heavy in those days because of iron core transformers and steel chassis. Most items of equipment required two people for lifting. Arrival at the site early in the afternoon was essential to prepare for 15–20 min of recording time scheduled to begin about 8:30 PM. Oscilloscope data were obtained directly with fast-moving 35 mm film or paper to be developed in a small tank the following day.

In 1953, Griffin returned to Harvard and expanded his interests to other bats, especially Neotropical species. Georg von Békésy, the Nobel laureate expert on hearing, told Griffin that it would be a waste of time to study other types of bats because the sounds are simply noise bursts and a bat is a bat. Nevertheless, Griffin went to Panama and recorded the sounds of tropical bats (with oscilloscope and camera). For this purpose, he built a wooden box approximately 2×2 ft. square housing a perch for a single bat. Perplexed by his inability to detect sounds from Seba's short-tailed fruit bat (*Carollia perspicillata*), a common phyllostomid, Griffin sat in a lotus position inside the box (nicknamed Griffin's Orgone Box). He held a microphone immediately in front of a bat's mouth. In this way he recorded the calls of what he called "whispering bats." At about the same time, Möhres (1951) described long pulses emitted by horseshoe bats that were quite different from those recorded by Griffin from vespertilionids and phyllostomids.

Those early attempts marked the beginning of what has become a major area of echolocation research. Biologists could characterize the vocalizations of different species of bats and correlate them with different ecological niches and behavior. In 1958, Griffin published "*Listening in the Dark*," which summarized the field to that date. This book won several awards and has become a classic in the field of animal behavior.

The next major step was taken by Alvin Novick, who gave up his medical career to join Griffin's lab and pursue his new-found interest in bat biology. Novick (1958a, b, 1962, 1963) traveled to Mexico, Panama, Sri Lanka, the Philippines, and the Congo recording the sounds of numerous bat species. In 1977, he summarized the diversity of bats in which some aspects of echolocation had been studied up to that time. They numbered more than 130 species, studied by 30 scientists (Novick 1977). A major leap in understanding echolocation came with the study of insect capture and other bat behaviors under laboratory conditions. This coincided with the expansion of our knowledge through studies of the neural correlates of echolocation. Griffin described the latter development in his 1998 autobiography:

> Neurophysiology of brain mechanisms of bat echolocation began when Alan Grinnell (Figure 1.2a, k) first exposed the dorsal surface of the brain of a little brown bats under nembutal anesthesia and placed a simple wire electrode to the surface of the inferior colliculus. The beautiful evoked potentials so easily recorded formed the basis of his extensive studies of the neurophysiology of bat echolocation (Grinnell 1963). These studies were extended later by Nobuo Suga and several others, so that in many ways the neural bases of auditory discriminations in bat brains are better understood than those of any other nonhuman animal. (Griffin 1998, p. 83).

Fig. 1.2 (**a**) Alan Grinnell, Eli Kalko, Annemarie Surlykke, 2005. (**b**) Annette Denzinger, Uli Schnitzler, and Bjoern Siemers in Belize, 2009. (**c**) Rod Suthers and Jeff Wenstrup in Zimbabwe, 1982. (**d**) George Pollack, Jersey Biosonar Conference in 1978. (**e**) Donald W Thomas and James Fullard in Zimbabwe, 1979. (**f**) Cindy Moss. Photo by Will Kirk. (**g**) Gareth Jones in the Negev Desert in Israel, 2003. (**h**) Jens Rydell in the Mexican Yucatan, 1999. (**i**) Lee Miller, Paul Faure, Dody Dunning, Eli Kalko, Asher Treat (with family friend), and Annemarie Surlykke, 1992. (**j**) Brock Fenton in Cuba, 2012. (**k**) Alan Grinnell in Panama, 1970s. (**l**) Marc Holderied in Israel, 2003

Fig 1.2 (continued)

1.2 Early Lab and Field Experiments

1.2.1 Experiments in the Griffin Lab (1956–1965)

In 1955 Grinnell joined the Griffin lab as a Harvard sophomore at a time when the lab was addressing three main questions in bat behavior:

1. What is the absolute sensitivity of echolocation—at what distance can objects of different sizes be detected, and how does echolocation behavior differ for targets of different sizes?
2. How easy is it to "jam" echolocation?
3. What is the correlation between flight behavior and echolocation during insect capture in flight?

Assessing the sensitivity of echolocation was done with little brown myotis in a large flight room with one or more rows of moveable wires of different gauges arranged a few feet from one end of the room. The improvement on earlier wire avoidance experiments was in recording the bat's emitted sounds while simultaneously filming the flights and—with the help of grid lines on the wall—determining accurately the bat's position throughout the flight. These experiments helped establish the terminology of search, approach (or pursuit), and terminal buzz. The work revealed that pulses during the approach phase were commonly emitted in pairs (later to be called "strobe groups"), showed that the bats must have detected echoes from large-diameter wires for a considerable time/distance before they reacted by increasing their repetition rate, and confirmed the ability of little brown bats to detect and avoid wires down to a diameter of 0.12 cm (Grinnell and Griffin 1958).

Griffin was already thinking in terms of how bats overcome interference, so the wire avoidance study above morphed into a study of what it took to "jam" bats with white noise. For these experiments, Jerry McCue from the Lincoln Labs of MIT joined the team. Fifty-two custom-made loudspeakers were mounted at the two ends of the flight room, driven non-coherently by five powerful amplifiers. The bats, Rafinesque's big-eared bat (*Corynorhinus rafinesquii*—then classified as a *Plecotus*) increased the intensity of their emitted cries but could be faced with so much noise that they could no longer avoid smaller wires. On the other hand, the ratio of the calculated returning echoes to the noise power per cycle bandwidth (E/N_o), which information theory says must be >1 for something to be detectable (by a single detector), was more like −10 dB. The solution to this conundrum, presumably, was that bats use two ears, which are directional, and the bats didn't always fly directly at the bank of loudspeakers. So it probably wasn't a disproof of information theory, merely another seemingly impossible ability of bats (Griffin and Grinnell 1958; Griffin et al. 1963).

The third major research thrust was the first detailed laboratory study of bat flight and insect capture on the wing. This was made possible by the participation of Fred Webster, one of the Forbes family, and David Cahlander from the Harold Edgerton lab at MIT. Webster became fascinated with bats and converted a large Quonset hut in his back yard from a trampoline studio into a bat flight facility. Several bat species, including big brown bats and little brown bats, learned first to catch fruit flies around a mesh bag of overripe fruit, then transferred their skills to catching mealworms projected into the air by a "mealworm gun." With a high-speed camera and superfast strobe from the Edgerton lab, superb footage of insect captures, with simultaneous records of emitted sounds, showed that the bats not only localized targets accurately enough to catch them with an outstretched wing-tip (in contrast to the usual tail membrane catch) but that they could discriminate and catch a mealworm in the presence of multiple other targets of similar size but different shapes. Included among the targets were spheres, disks, washers, rolled-up adhesive tape, and a variety of other items (Griffin et al. 1960; Webster and Brazier 1965; Webster 1967). The technology was far before its time, and the extraordinary abilities demonstrated by the bats in these experiments have driven research and modeling of echolocation mechanisms ever since.

1.2.2 Other Advances in the Mid-1960s

The work of David and Ade Pye, in London, also boosted our understanding of echolocation. David brought sophisticated acoustic and engineering expertise to early attempts to model echolocation mechanisms, in both frequency-modulated (FM) and constant-frequency (CF) bats, and, equally important, helped to develop a new, state-of-the-art heterodyning bat detector. Although not as sensitive as the old Lincoln Labs detector, it had major advantages for field identification of species (Pye 1960; Pye and Flinn 1964). Ade contributed important early comparative studies of cochlear morphology in a variety of bat species (Pye 1966a, b; 1967). Her work was extended throughout the auditory nervous system in subsequent years, especially by Marianne Vater and Manfred Kössl in Germany and by John Zook, John ("Pete") Cassaday, and Ellen Covey in the U.S. (see Popper and Fay 1995).

In the mid-1960s, Bill Henson succeeded in recording cochlear microphonics and auditory evoked potentials in flying Brazilian free-tailed bats (*Tadarida brasiliensis*) and Parnell's mustached bats (*Pteronotus parnellii*). His studies established first, that this kind of recording was possible, and, second, that middle ear muscles began contracting well before pulse emission and relaxed quickly enough to suppress responses to emitted pulses while having a much reduced effect on echo responses, at least at moderate pulse-echo separations (Henson 1965). This was a real *tour de force* at the time.

1.2.3 Uli Schnitzler and Doppler Shift Compensation

Thanks to research in the lab of Professor Möhres in Tubingen, Germany, and to the field recordings done by Griffin and by A. Novick worldwide, the tremendous variety in the types of echolocation signals used by different species of bats was recognized. After Dijkgraaf, Möhres (1953) had shown that greater horseshoe bats (*Rhinolophus ferrumequinum*) used relatively long, constant-frequency (CF) signals ending in a frequency-modulated (FM) tail. In the mid- to late 1960s, as his students Gerhard Neuweiler (Figure 1.1e) and Hans-Ulrich Schnitzler were coming into their own, the "German group" became an important center for the study of echolocation.

The whole field was given a boost by the first NATO International Bat Research Conference held at Frascati, Italy, in 1967. Shortly after that, Schnitzler published an important paper revealing that when flying, the greater horseshoe bat lowered its emitted CF just enough to compensate for the Doppler shift of the echo, which was thereby maintained at an approximately constant frequency (Schnitzler 1967, 1968, 1973). The precision of maintaining a constant frequency in that component of the emitted sound within 0.06 % (Schuller et al. 1974), combined with the great precision of Doppler shift compensation (DSC), makes this one of the most accurately regulated motor behaviors in the animal kingdom. There are many potential uses for Doppler shift information. It now appears that one of the principal uses is in detection of wing-beat flutter of potential prey, as well as recognizing species by wing-beat signatures (Suga and Jen 1975, 1977; see Figure 1.1f; Neuweiler et al., 1980) (Schnitzler and Denzinger 2011; see Figure 1.2b).

This phenomenon also explained the sharp tuning of the auditory system in CF/FM bats to a narrowband around the emitted CF, since slight changes in the echo frequency were clearly driving this behavior. This specialization of the auditory system had been shown in the Neotropical bat, Parnell's mustached bat (then known as *Chilonycteris rubiginosa*) (Grinnell 1970) and soon after was shown to be even more extreme in greater horseshoe bats (Neuweiler 1970; Neuweiler et al. 1971; see Figure 1.1e). The brilliant neurophysiological studies of Nobuo Suga (Figure 1.1f) showed that DSC was a central organizing feature of the auditory system (see below) of Parnell's mustached bat.

1.2.4 Enter Jim Simmons

Griffin and his students had attempted operant conditioning with bats without notable success. In the late 1960s, however, one of E.G. Wever's students at Princeton, James Simmons (Figure 1.1g), found that he could use positive reinforcement to persuade big brown bats to fly to a target. He then embarked on a brilliant series of experiments testing the ability of these bats to discriminate target distances. Soon he had them crawling to one side of a platform or the other instead of flying to it. He demonstrated that big brown bats could discriminate distance differences down to about 12 mm (at an absolute distance of ~34 cm) and, in 1973, he published a report showing that bats seamlessly transferred from distinguishing differences in the distances of real targets to distinguishing differences in the timing of "echoes" from virtual targets. The virtual targets were represented by a microphone-loudspeaker circuit that allowed him to electronically delay one echo more than the other. This was proof that pulse-echo delay was the criterion bats used to determine target distance (Simmons 1973).

Subsequently, Simmons (1979) designed a clever variant on this experiment in which the delay of an echo from one virtual target was held constant, while the echo delay from the other was "jittered" back and forth for successive echoes. The difference in delay of the jittered echo was reduced until the bat could not distinguish between the two options. The smallest detectable "jitter" is still controversial but appears to be in the range of a few microseconds, if not sub-microsecond! A major reason for suspending disbelief in this remarkable result is the ability of bats to discriminate target shapes on the basis of glints differing in arrival time by a few microseconds. Additional research by Shelley Kick (1982) in Simmons' lab showed that big brown bats could detect a 2-cm-diameter sphere at a maximum absolute distance of about 5 m (roughly 30 ms absolute pulse-echo delay). At the time, this appeared to be an effective working range for echolocation of small targets, but more recent work suggests ranges of up to 30 m for detection of insect-sized targets (e.g., Surlykke and Kalko 2008; see Figure 1.2a).

Many other laboratories have adopted forced choice behavioral conditioning to study the abilities of bat species adapted to different behavioral niches to discriminate small range differences; to pinpoint target direction in both the horizontal and vertical dimensions; to discriminate target shapes, dimensions, and surface textures; and to do all of these things in the presence of differing amounts of backward and forward acoustic "clutter." With improvements in instrumentation, these studies have been expanded to encompass bats flying in the laboratory, even carrying miniaturized microphones and radio transmitters during quasi-natural behaviors. The results of these experiments, complemented by field observations of natural behavior in an increasingly wide variety of species, occupying a diversity of habitats, and using different echolocation strategies, have guided and continue to guide attempts to explain the neural adaptations that enable bats to use echoes so successfully as a substitute for vision.

1.2.5 1957–1980: Studies on Adaptations of the Auditory Nervous System for Echolocation

In his junior year, fascinated by the results of all of the behavioral experiments going on in the Griffin Lab, Grinnell proposed using electrophysiological techniques to look for neural adaptations that could help to account for the bats' abilities. These began on a trial basis at the Woods Hole Marine Biological Laboratory during the summer of 1957 where he served as an assistant in the physiology course taught by Griffin. Fine silver wire electrodes were used initially, and later etched tungsten electrodes of the type developed by David Hubel were used to record evoked potentials and clusters of single unit responses from the inferior colliculus (IC) of anesthetized little brown bats. The work required production of tone pips of controlled duration and rise time as well as physiological amplifiers and Peltier elements for keeping the anesthetized bats warm.

Happily, the experiments went well, and Griffin gave his blessing to continuing them during the school year at Harvard. More than that, he arranged that Grinnell spend his spring vacation working in the laboratory of Bob Galambos at Walter Reed Army Hospital in Washington, D.C. At the time, Galambos was one of the premier experts on central auditory processing, with a state-of-the-art stimulating-recording setup. Although Galambos' equipment was intended for cats, it worked beautifully with bats. The results convinced Griffin that electrophysiology should go ahead full speed, which it did, first as Grinnell's Ph.D. work and subsequently in the hands of Nobuo Suga, who came from Katsuki's lab in Tokyo.

There were nine principal findings from the initial electrophysiology experiments:

1. Bats were most sensitive in the range of their emitted sounds, even though the frequencies of greatest sensitivity were higher than in other mammals.
2. Temporal resolution exhibited clear adaptations for echolocation. The auditory pathways of bats exaggerated the response to echoes, especially between the levels of the auditory nerve and the inputs to the inferior colliculus (IC). This contrasts with humans or other experimental animals in which responses to echoes are suppressed by neural mechanisms.
3. Most single units in the IC responded only once to a brief tone pip at any intensity.
4. Many single units in the IC responded to the first of a pair of stimuli and to the second stimulus only within a narrow range of intervals (corresponding to echo delays), or only once to a pair of stimuli within a narrow range of intervals (delay-tuned neurons, although not yet given that name).
5. The threshold for response to the second of two sounds (an "echo") could be as much as 30 dB lower than to a single tone pip of the same frequency.
6. Input to the ipsilateral ear suppressed the response to contralateral input, thereby sharpening directional sensitivity of the auditory system in general.
7. Some individual units in the IC were extremely directional in their sensitivity, with threshold shifts as great as 60–80 dB over a change of 30° in the horizontal axis (Grinnell 1963).
8. Evoked potentials in the IC were strongly directional, especially at high frequencies, and suggested the hypothesis that identification of the source of a broadband echo could be determined accurately by binaural comparison of intensity differences at many frequencies in the echo. Moreover, deflecting the pinna greatly reduced directionality, consistent with the effect in degrading obstacle avoidance, and deflection of the tragus also had a significant effect (Grinnell and Grinnell 1965).
9. A series of experiments revealed that many single units in the IC and other nuclei responded better to downward-sweeping FM sounds than to tone pips of any frequency—sometimes only to FM signals (Grinnell and McCue 1963). These data were obtained with the help of Jerry McCue of the MIT Lincoln Labs, who

provided a large Lincoln Lab-constructed box capable of producing FM signals. Suga confirmed and greatly extended these findings in a systematic analysis of response properties at different neural levels in the auditory system of FM bats (Suga 1964; Suga and Schlegel 1973).

In the mid-1960s, Grinnell studied a number of species of Neotropical bats in Panama, showing clear correlations between neural adaptations and the types of emitted pulses and the frequencies used. In particular, he found that the long CF/FM bat, Parnell's mustached bat, had an extraordinary evoked potential audiogram, with very sharp tuning in the region of the CF second harmonic (Grinnell 1966, 1970). Short CF bats, such as bulldog bats (*Noctilio* spp.) and the smaller species of mustached bats (*Pteronotus* spp.), also showed major emphasis on the frequencies of their CF components, but tuning was not as sharp.

George Pollak (Figure 1.2d), who with his students was to contribute many important findings about neural mechanisms of auditory processing in bats, entered the field in a collaboration with Henson (Pollak et al. 1972). They showed that the phenomenally sharp tuning and resonance are already present in the cochlear microphonics of Parnell's mustached bat near its CF. The long CF/FM bat (the greater horseshoe bat) was shown by Gerhard Neuweiler's group (Neuweiler 1970; Neuweiler et al. 1971) to have equivalent emphasis on its CF but without the resonance that is so conspicuous in mustached bats (Schnitzler et al. 1976). This sharp tuning, of the auditory system to frequencies around the echo CF, later described as an "acoustic fovea" (Schuller and Pollak 1979), has led to a large and still growing literature on the long CF/FM bats.

Subsequently, in a beautiful series of electrophysiological experiments, Nobuo Suga and his colleagues showed that the CNS of Parnell's mustached bat is organized to emphasize combination-sensitive neurons. Some of these neurons compare the CF of the echo's second harmonic with the CF of the emitted first harmonic to detect and quantify echo Doppler shift, while others compare the delays of the echo's FM components in the second, third, and fourth harmonics with the time of the emitted fundamental FM to accurately measure target range. These Doppler shift and FM delay-tuned populations are mapped beautifully in the cortex (e.g., Suga et al. 1975a, b; Suga and O'Neill 1979; Suga 1984). The auditory nervous system of greater horseshoe bats shows similar emphasis on the CF but with a different organization in the higher auditory centers that is consistent with the lack of multiple harmonics in the emitted sounds.

Much of the understanding of how echolocation is achieved was coming into focus by the mid-1970s, but the field has exploded since then as the students of Grinnell, Simmons, Suga, Pollak, Neuweiler, and Schnitzler established their own labs and began training the next generation of bat experts. At the same time, field studies of bats and the correlation of pulse structure and echolocation strategies with ecological niches have tremendously increased our appreciation for the sophistication of this sensorimotor system. The growth of the field is dramatically illustrated in Figure 1.3.

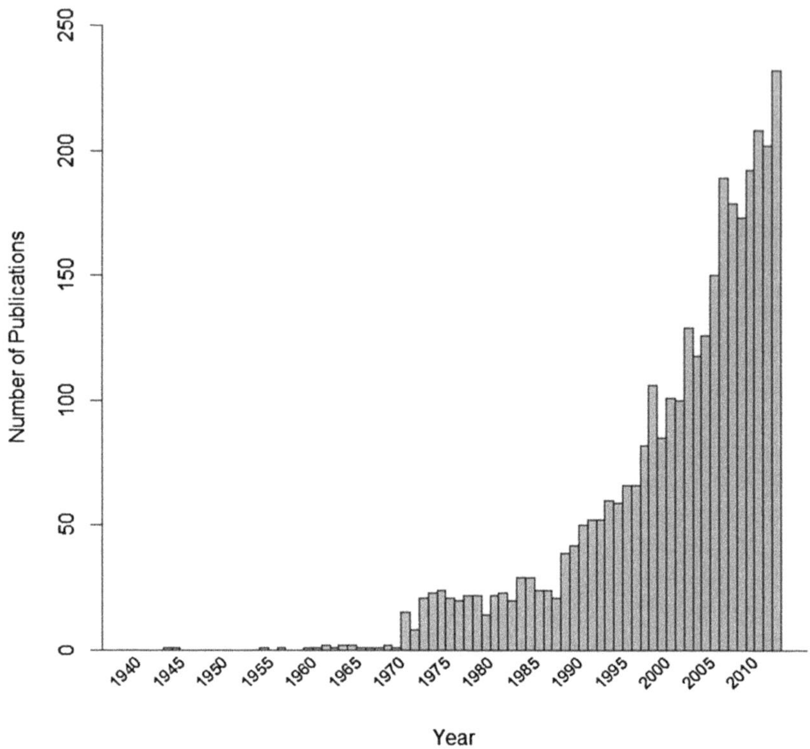

Fig. 1.3 Numbers of papers published about echolocation by year from 1940 to 2013 (Courtesy of Liam McGuire)

1.3 Taking Up the Tradition of Studying Bats in the Field

Ever since Spallanzani's discovery that blinded bats released into the wild returned to their roosts having fed en route, there has been a link between what was discovered in the laboratory and what happened in the field. Griffin continued these endeavors even while working in laboratory settings. The appearance of commercially available 'bat detectors,' instruments sensitive to the echolocation calls of bats, was an extension of Griffin's crude audible bat detector (Section 1.1). John Hooper (1964) was one of the first to publish details of habitat use obtained by monitoring echolocation calls and others followed (e.g., Fenton 1970; see Figure 1.2j). A next step was identifying species of bats by their echolocation calls (Ahlén 1981; Fenton and Bell 1981; Jung et al. 2007), a development that has resulted in many publications (e.g., Erickson and West 2002; Lacki et al. 2009), many discussions of accuracy (e.g., Barclay 1999), and assessments of variation (e.g., Chiu et al. 2010). This approach to using echolocation calls as proxies for measuring bat activity (e.g., Patriquin and Barclay 2003; Dzal et al. 2011; Randall et al. 2011) and vulnerability, for example, to wind turbines, has opened up new opportunities for bat biologists (Barclay 2009; Baerwald and Barclay 2011).

By about 1976, it was possible to use a reel-to-reel tape recorder operated at 76 cm s^{-1} to record the echolocation calls of bats in the field. The equipment Griffin used to record bat calls in the field 20 years earlier filled the back of a pickup truck. There was a movement to recording and playing back bat echolocation calls in the field, expanding our knowledge of the dynamics and details of echolocation behavior. By the 1980 biosonar meetings on the Isle of Jersey (UK), several presentations were based on field work. At that time there was a sharp contrast between work on bat and odontocete echolocation. Both had laboratory components, but there was little field work on odontocetes. This discrepancy persisted at the 1986 biosonar meetings in Denmark but had vanished by the 1998 meeting in Portugal. In both venues, previously unimagined technologies were being used to advantage to study echolocation, and this has continued. The diversity of approaches to research on echolocation is proving to be almost as diverse as echolocation itself.

1.4 The Echolocation Calls of Bats

It was clear by 1958 that features of echolocation calls of largyngeally echolocating bats varied considerably (Griffin 1958). Obvious features that vary include intensity (Surlykke and Kalko 2008; Brinkløv et al. 2010, 2011), patterns of frequency change over time (Gillam and McCracken 2007; Guillén-Servent and Ibáñez 2007; Mora et al. 2011) as well as the presence of harmonics (Fenton et al. 2011). While some bats use high-intensity echolocation calls, others use much less intense calls, for example, the "whispering" phyllostomids. As more information has accumulated, it is clear that these calls span a spectrum from about 130 dB SPL at 10 cm to about 60 dB SPL at 10 cm, and that many phyllostomids, for example, are not whispering according to the intensity of their calls (Mora and Macias 2007; Brinkløv et al. 2009).

As Griffin et al. (1960) and then others (e.g., Kalko 1995; Schnitzler and Kalko, 2001) have demonstrated, echolocation calls vary in duration from <1 to >50 ms (Figure 1.4), often reflecting the situation in which the bats are operating (searching for prey as opposed to tracking and closing with it). Simmons and Stein (1980) proposed how changes in call design affect a bat's ability to localize a target. More recently, others have demonstrated how prey (Goertlitz et al. 2010; Holderied et al. 2011) and clutter (Hiryu et al. 2010) influence the design of a bat's echolocation calls, including the incidence of harmonics (e.g., Bates and Simmons 2011) and production of multiple calls (Moss et al. 2006, 2011) (see Figure 1.2f).

Early studies showed that frequency modulated (FM) calls were common and ranged from narrowband to broadband, with some narrowband signals approaching bandwidths of 100 kHz. There was a tendency to distinguish between FM and CF signals and to use shorthand to describe calls that combined CF and FM components (e.g., CF-FM calls; Figs. 1.4 and 1.5). Furthermore, Griffin (1953), Griffin

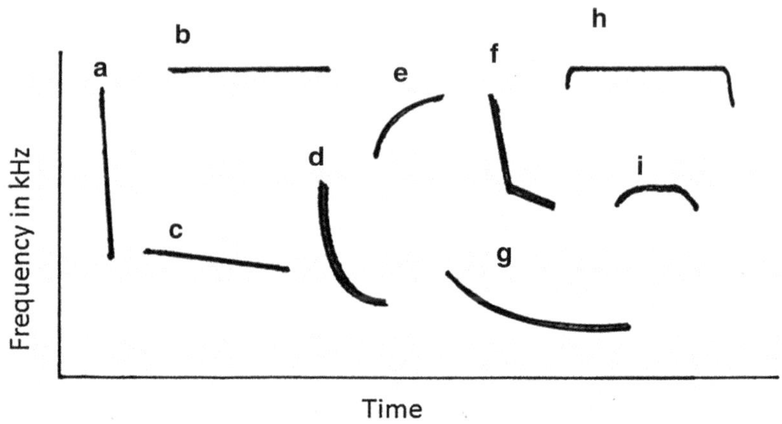

Fig. 1.4 An indication of the variations in designs of search phase echolocation calls used by largyngeally echolocating bats. Calls range from steep (**a**) to shallow (**c**, **g**) frequency-modulated (FM) sweeps that vary in details (**d, f**). Narrowband calls range from almost pure tone (**b**) to long constant frequency signals (**h**), beginning with upward and ending with downward FM sweeps. A few species use short calls that combine upward and downward sweeps separated by a narrowband component (**i**), and much less often, upward FM sweeps (**e**)

et al. (1960), and others (e.g., Kalko 1995; Siemers and Schnitzler 2000; Siemers et al. 2001) demonstrated how individual bats changed the details of their calls according to the task at hand, namely, searching for, tracking, or closing with a flying insect. Eventually it was obvious that while terms such as CF, FM, or some combinations (FM-CF-FM) were useful when describing calls and their components, they fell short of describing the echolocation behavior of bats (Fenton 1999). The variety of echolocation call types produced by laryngeally echolocating bats searching for prey (e.g., Figure 1.4) illustrates the diversity that can be involved. The calls presented in Figure 1.4 do not show variations in the use of harmonics by echolocating bats. Some molossids, such as dwarf dog-faced bats (*Molossops temminckii*) or little goblin bats (*Mormopterus minutus*), are particularly impressive (Guillén-Servent and Ibáñez 2007; Mora et al. 2011, respectively), but the sequence leading to a feeding buzz by black mastiff bats (*Molossus rufus*) makes the point about variation quite clearly (Figure 1.5).

The duty cycles of call production by echolocating animals, particularly bats, are important (Fenton et al. 2012). Most echolocators (and most echolocating bats) emit calls at low duty cycles (<10 %), which separates the pulse and the echo in time. A few bats emit calls at high duty cycles, and, by Doppler shift and Doppler shift compensation (DSC), separate pulse and echo in frequency. Included are species in the families Rhinolophidae and Hipposideridae (Old World) as well as a few (but not all) species of Mormoopidae (*Pteronotus parnellii*, *Pteronotus mesoamericanus*, *Pteronotus paraguensis*, New World). High duty cycle echolocators produce long calls separated by short periods of silence. Their calls are dominated by a single frequency (the CF component) which is central to DSC. High duty

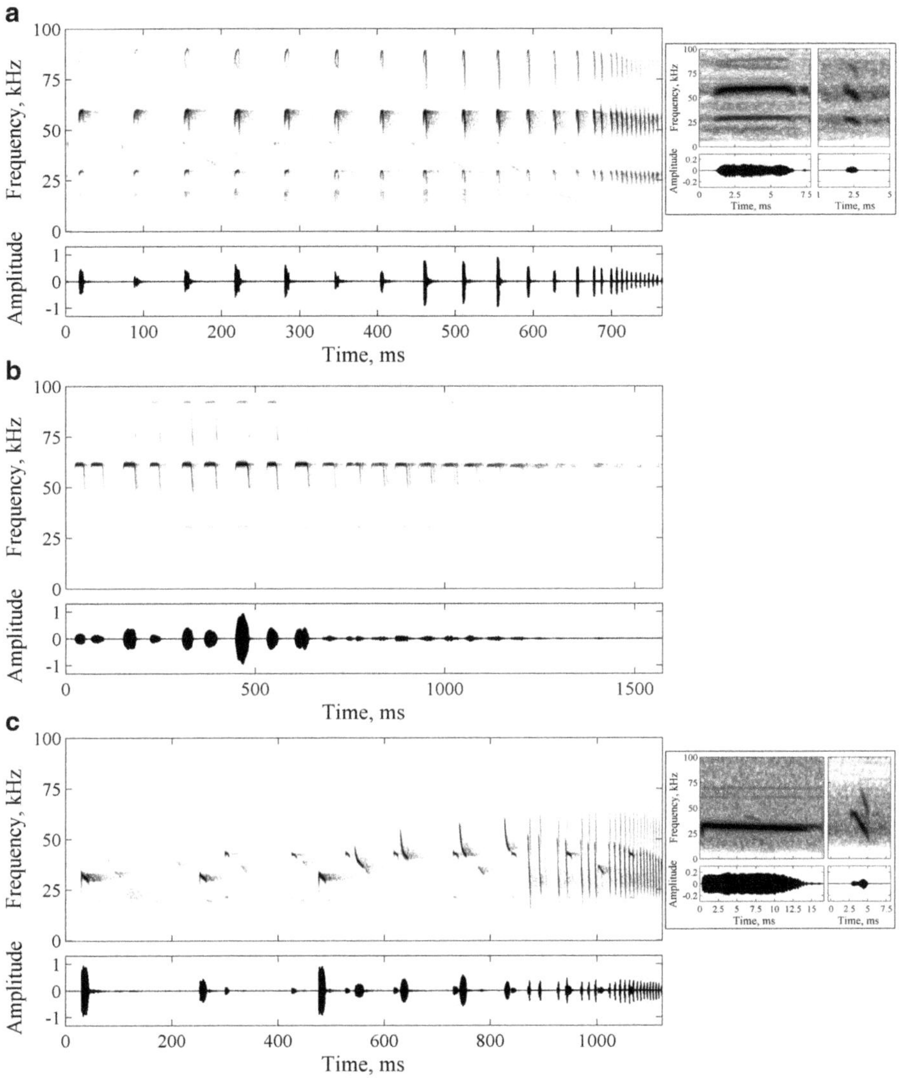

Fig. 1.5 During attacks on flying prey (feeding buzzes), some bats produce calls of variable design (**c**) while others are more consistent (**a**, **b**). These feeding buzzes were produced by an emballonurid (**a**), a Parnell's mustached bat (**b**), and a black mastiff bat (**c**). There are calls of a *Saccopteryx* (most energy about 47 kHz) in the background in (**c**)

cycle echolocation is one way to detect fluttering targets in clutter (Lazure and Fenton 2011), but some low duty cycle echolocators achieve this in different ways. To date we know of one hipposiderid, the East Asian tailless leaf-nosed bat (*Coelops frithii*), that uses low duty cycle signals to detect fluttering prey (Ho et al. 2013).

1.5 Overview of This Volume

The chapters that follow illustrate how studies of bat bioacoustics have proliferated and diversified (see Figure 1.3). Although this book focuses on the bioacoustics of bats, it repeatedly connects to topics in basic biology, functional morphology, evolution, and diversification. Although interactions between bats and prey, notably insects, are briefly discussed, this intersection of disciplines remains an active area of study. The same pattern is beginning to emerge with studies of how bats interact with plants.

In Chapter 2, Emma Teeling, Steve Rossiter, and Gareth Jones (Figure 1.2g) review the contributions of genetic and genomic databases in advancing our understanding of the phylogeny of bats and their classification among the mammals. They address the question about whether echolocation in bats is an ancestral or a derived trait. They also discuss the contributions of 'hearing genes' to our knowledge of echolocation in animals.

In Chapter 3, Walter Metzner and Rolf Mueller summarize the diversity of laryngeal and facial structural specializations for production of both echolocation pulses and communication signals in echolocating bats. They analyze the various levels of neural control of vocalizations, including sensory feedback and circuitry controlling Doppler shift compensation in long CF-FM bats. They also present a sophisticated analysis of both the static and dynamic roles of facial features and pinna structure in echolocation.

In Chapter 4, Patricia Jones, Rachel Page, and John Ratcliffe present information about the use of laryngeal echolocation when foraging, noting the differences and similarities among aerial hawkers, substrate gleaners, and bats that alternate between these two strategies. They focus, in part, on the arms race between bats and their animal prey, particularly examples of hearing-based defenses that have evolved several times in insects. They feature species of bats that appear to use a variety of approaches to find their prey, stepping outside the behavior expected from animal-eating bats.

In Chapter 5, Annette Denzinger, the late Eli Kalko, Marco Tschapka, Alan Grinnell, and Uli Schnitzler review the evidence supporting the view that communities of bats consist of guilds of species. They focus on overlaps among sympatric species in terms of echolocation strategies, foraging behavior, habitat use, and diet and define seven different guilds that suffice to include all species. They point out that sympatric species do not appear to compete for resources and that niche dimensions tend to vary and overlap among species.

In Chapter 6, Erin Gillam and Brock Fenton consider the many situations in which bats communicate with acoustic signals. These range from behaviors underlying social systems to interactions between mothers and their young, including the signals conspecifics use for finding their roosts. They also explore the obvious continuum between echolocation and communication, reporting the repeated demonstration that echolocation calls serve roles in communication.

In Chapter 7, Paul Faure (Figure 1.2i) and Uwe Firzlaff discuss the importance of temporal resolution in the hearing of echolocating bats. They focus on the neural mechanisms of coding signal duration and the need for this information in achieving selectivity of the responses to communication sounds and echolocation signals. They then describe the neural mechanisms, mainly at the cortical level, that can account for the ability of bats to recognize target features through echolocation. These "acoustic images" depend on intensity, spectral interference patterns, "roughness" of the echo, spectral directionality, and a variety of subtle temporal features of the echoes. They also discuss neural processing of species-specific social calls and other vocalizations.

In Chapter 8, George Pollak considers the diversity of communication and echolocation sounds used by different bat species and the diversity of neural adaptations for processing the sounds, especially at the level of the inferior colliculus. He discusses the dominant role of inhibition in shaping selective responses to different components of the calls, and directs attention to the multiple neural mechanisms for extracting any given kind of information. He also argues that many of these neural mechanisms were adaptations first for analysis of communication sounds, which were then refined for echolocation.

In Chapter 9, Shizuko Hiryu, Emanuel Mora, and Hiroshi Riquimaroux examine how bats use Doppler shift compensation (DSC) in the context of high duty cycle echolocation. Although most echolocators separate pulse and echo in time (low duty cycle echolocation), horseshoe bats, old world leaf-nosed bats, and at least three species of mustached bats use high duty cycle echolocation, separating pulse and echo in frequency. Doppler shift compensation is central to this and has evolved independently in at least two lineages of bats. An acoustic fovea and other specializations are central to DSC.

In Chapter 10, Cynthia Moss, James Simmons, and the late Annemarie Surlykke consider the question of "scene analysis" via echolocation and then compare the information available to the auditory system, through echoes, with information processing conducted by the visual system. Using state-of-the-art instrumentation and clever experimentation, they report a variety of findings that document the richness as well as the limitations of information that can be obtained by echolocation.

In Chapter 11, Lutz Wiegrebe, Alan Grinnell, and Brock Fenton reflect on some of the opportunities and challenges facing those who study the bioacoustics of bats. They also discuss the new technologies that will make it easier to take on these challenges.

This chapter has provided highlights of the contributions from the first four decades of research on echolocation in bats. The field already has lost many of the pioneers: Griffin and Galambos, Dijkgraaf and Möhres, and K. D. Roeder, who pioneered the studies of sound-based interactions between bats and insect prey. We also have lost some of the second generation, including Novick and Neuweiler. Annemarie Surlykke, to whom this book is dedicated, died on 28 July 2015. She is the most recent loss among these colleagues who entered the field in the 1980s: James H. Fullard (1952–2010), Donald W. Thomas (1953–2009), Elisabeth K. V. Kalko (1962–2011), and Bjorn Siemers (1972–2012), whose creativity and personalities enormously enriched the field. We salute all of these strong contributors to the field of bat bioacoustics.

References

Ahlén, L. (1981). Identification of Scandinavian bats by their sounds. *Sveriges Lantbruksuniversitet, Institutionen för Viltekologi*, 6, Uppsala.

Baerwald, E. F., & Barclay, R. M. R. (2011). Patterns of activity and fatality of migratory bats at a wind energy facility in Alberta, Canada. *Journal of Wildlife Management*, doi:10.1002/jwmg.147

Barbellion, W. N. P. (1920). Enjoying life and other literary remains of W. N. P. Barbellion. New York: George H Coran Co. [Barbellion was the *nom-de-plume* for Bruce Frederick Cummings]

Barclay, R. M. R. (1999). Bats are not birds—a cautionary note on using echolocation calls to identify bats: A comment. *Journal of Mammalogy*, 80, 290–296.

Barclay, R. M. R. (2009). Causes of bat fatalities at wind turbines: Hypotheses and predictions. *Journal of Mammalogy*, 90, 1330–1340.

Bates, M. E., & Simmons, J. A. (2011). Perception of echo delay is disrupted by small temporal misalignment of echo harmonics in bat sonar. *Journal of Experimental Biology*, 214, 394–401.

Brinkløv, S., Kalko, E. K. V., & Surlykke, A. (2009). Intense echolocation calls from two 'whispering' bats, *Artibeus jamaicensis* and *Macrophyllum macrophyllum* (Phyllostomidae). *Journal of Experimental Biology*, 212, 11–20.

Brinkløv, S., Kalko, E. K. V., & Surlykke, A. (2010). Dynamic adjustment of biosonar intensity to habitat clutter in the bat *Macrophyllum macrophyllum* (Phyllostomidae). *Behavioral Ecology and Sociobiology*, 64, 1867–187.

Brinkløv, S., Jakobsen, L., Ratcliffe, J. M., Kalko, E. K. V., & Surlykke, A. (2011). Echolocation call intensity and directionality in flying short-tailed fruit bats, *Carollia perspicillata* (Phyllostomidae). *The Journal of the Acoustical Society of America*, 129, 427–435.

Chiu, C., Reddy P. V., Xian, X., Krishnaprasad, P. S., & Moss, C.F. (2010). Effects of competitive prey capture on flight behavior and sonar beam pattern in paired big brown bats, *Eptesicus fuscus*. *Journal of Experimental Biology*, 213, 3348–3356.

Dijkgraaf, S. (1943). Over een merkwaardige functie van den gehoorzin bij vleermuizen. *Verslagen Nederlandsche Akademie van Wetenschappen. Afd. Naturkunde*, 52, 622–7. (Cited in Griffin, 1958)

Dzal, Y. L., McGuire, L. P., Veselka, N., & Fenton, M. B. (2011). Going, going, gone: The impact of white-nose syndrome on the summer activity of the little brown bat (*Myotis lucifugus*). *Biology Letters*, 7, 392–394.

Erickson, J. L., & West, S. D. (2002). Associations of bats with local structure and landscape features of forested stands in western Oregon and Washington. *Biological Conservation*, 109, 95–102.

Fenton, M. B. (1970). A technique for monitoring bat activity with results obtained from different environments in southern Ontario. *Canadian Journal of Zoology*, 48, 847–851.

Fenton, M. B. (1999). Describing the echolocation calls and behavior of bats. *Acta Chiropterologica*, 1, 127–136.

Fenton, M. B., & Bell, G. P. (1981). Recognition of species of insectivorous bats by their echolocation calls. *Journal of Mammalogy*, 62, 233–243.

Fenton, M. B., Skowronski, M. D., McGuire, L. P., & Faure, P. A. (2011). Variation in the use of harmonics in the calls of laryngeally echolocating bats. *Acta Chiropterologica*, 13, 169–178.

Fenton, M. B., Faure, P. A., & Ratcliffe, J. R. (2012). Evolution of high duty cycle echolocation in bats. *Journal of Experimental Biology*, 215, 2935–2944.

Galambos, R., & Griffin, D. R. (1942). Obstacle avoidance by flying bats: The cries of bats. *Journal of Experimental Zoology*, 89, 475–490.

Gillam, E. H., & McCracken, G. F. (2007). Variability in the echolocation of *Tadarida brasiliensis*: Effects of geography and local acoustic environment. *Animal Behaviour*, 74, 277–286.

Goertlitz, H. R., ter Hofstede, H. M., Zeale, M. R. K., Jones, G., & Holderied, M. W. (2010). An aerial hawking bat uses stealth echolocation to counter moth hearing. *Current Biology*, 20, 1568–1572.

Görner, P., Coombs, S., & Atema, J. (1996). Sven Dijkgraaf 1908–1995. *Brain, Behavior and Evolution*, 48, 350–358.

Griffin, D. R. (1934). Marking bats. *Journal of Mammalogy*, 15, 202–207.

Griffin, D. R. (1953). Bat sounds under natural conditions, with evidence for echolocation of insect prey. *Journal of Experimental Zoology*, 123, 435–66.

Griffin D. R. (1958). *Listening in the dark: The acoustic orientation of bats and men*. New Haven, CT: Yale University Press.

Griffin, D. R. (1983). Scientific recollections. In D. A. Dewsbury (Ed.), *Autobiographies in animal behavior* (pp. 120–142). Bucknell: Bucknell University Press (reprinted by U. Chicago Press, 1985, *Studying animal behavior*)

Griffin D. R. (1998). Donald R. Griffin. In L.R. Squire (Ed.), Volume 2 of *The history of neuroscience in autobiography* (pp. 68–93). New York: Academic Press.

Griffin, D. R., & Galambos, R. (1940). Obstacle avoidance by flying bats. *Anatomical Record*, 78, 95.

Griffin, D. R., & Galambos, R. (1941). The sensory basis of obstacle avoidance by flying bats. *Journal of Experimental Zoology*, 86: 481–506.

Griffin, D. R., & Grinnell, A. D. (1958). Ability of bats to discriminate echoes from louder noise. *Science*, 128, 145–147.

Griffin, D. R., McCue, J. J. G., & Grinnell, A. D. (1963). The resistance of bats to jamming. *Journal of Experimental Zoo*logy, 152, 229–250. (With added discussion published as Technical Report #285, MIT Lincoln Laboratory, October 1962)

Griffin, D. R., Webster, F. A., & Michael, C. R. (1960). The echolocation of flying insects by *bats. Animal Behaviour*, 8, 141–154.

Grinnell, A. D. (1963). The neurophysiology of audition in bats. *Journal of Physiology*, 167, 38–113.

Grinnell, A. D. (1966). Mechanisms of overcoming interference in echolocating animals. In R.-G. Busnel (Ed.), *Animal sonar systems* (pp. 451–481). Jouy-en-Josas, France: Imprimerie Louis Jean (Gap).

Grinnell, A. D. (1970). Comparative auditory neurophysiology of neotrophical bats employing different echolocation signals. *Zeitschrift fur Vergleithende Physiologie*, 68, 117–153.

Grinnell, A. D., & Griffin, D. R. (1958). The sensitivity of echolocation in bats. *Biological Bulletin*, 114, 10–22.

Grinnell, A.D., & Grinnell, V.S. (1965). Neural correlates of vertical localization by echolocating bats. *Journal of Physiology*, London, 181: 830–851.

Grinnell, A. D., & McCue, J. J. G. (1963). Neurophysiological investigations of the bat, *Myotis lucifugus*, stimulated by frequency modulated acoustical pulses. *Nature*, 198, 453–455.

Gross, C. G. (2005). *Biographical memoirs*, Volume 86 (pp. 188–207). Washington, DC: National Academies Press.

Guillén-Servent, A., & Ibáñez, C. (2007). Unusual echolocation behaviour in a small molossid bat, *Molossops temminckii*, that forages near background clutter. *Behavioral Ecology and Sociobiology*, 61, 1599–1613.

Hartridge, H. (1920). The avoidance of objects by bats in their flight. *Nature*, 54, 54–57.

Henson, O. W. (1965). Activity and function of middle-ear muscles in echolocating bats. *Journal of Physiology, London*, 180, 871–887.

Hiryu, S., Bates, M. E., Simmons, J. A., & Riquimaroux, H. (2010). FM broadcasting bats shift frequencies to avoid broadcast-echo ambiguity in clutter. *Proceedings of the National Academy of Sciences of the USA*, 107, 7048–7053.

Ho, Y-Y., Fang, Y-P., Chou, C-H., Cheng, H-S., & Chang, H-W. (2013). High duty cycle to low duty cycle: Echolocation behaviour of the hipposiderid bat *Coelops frithii. PLoS ONE*, 8(5), e62938.

Holderied, M., Korin, C., & Moritz, T. (2011). Hemprich's long-eared bat (*Otonycteris hemprichii*) as a predator of scorpions: Whispering echolocation, passive gleaning and prey selection. *Journal of Comparative Physiology A: Neuroethology, Sensory, Neural, and Behavioral Physiology*, 197, 425–433.

Hooper, J. D. H. (1964). Bats and the amateur naturalist. *Studies in Speleology*, 1, 9–15.

Jung, K., Kalko, E. K. V., & von Helversen, O. (2007). Echolocation calls in Central American emballonurid bats: Signal design and call frequency alternation. *Journal of Zoology*, 272, 125–137.

Kalko, E. K. V. (1995). Insect pursuit, prey capture and echolocation in Pipistrelle bats (Microchiroptera). *Animal Behaviour*, 50, 861–880.

Kick, S. A. (1982). Target-detection by the echolocating bat, *Eptesicus fuscus*. *Journal of Comparative Physiology A: Neuroethology, Sensory, Neural, and Behavioral Physiology*, 145, 431–435.

Lacki, M. J., Cox, D. R., Dodd, L. E. & Dickinson, M. B. (2009). Response of northern long-eared bats (*Myotis septentrionalis*) to prescribed fires in eastern Kentucky forests. *Journal of Mammalogy*, 90, 1165–1175.

Lazure, L., & Fenton, M. B. (2011). High duty cycle echolocation and prey detection by bats. *Journal of Experimental Biology*, 214, 1131–1137.

MacArthur, C. W. P. (2000). Louis Jurine's papers on the hearing of bats and on their parasites: A retrospective with new information. *Le Rhinolophe*, 14, 1–35.

Maxim, H. (1912). The sixth sense of the bat. Sir Hiram's contention. The possible prevention of sea collisions. *Scientific American*, Suppl. Sept. 7, 148–150.

Möhres, F. P. (1951). Uber eine neue Art von Ultraschall-Orientierung bein Fledermausen. Verhandlungen der deutschen zoologischen Gesselschaft in Wilhelmshaven (pp. 179–86). (cited in Griffin, 1958)

Möhres, F. P. (1953). Über die Ultraschallorientierung der Hufeisennasen (Chiroptera-Rhinolophinae). *Zeitschrift fur Vergleichende Physiologie*, 34, 547–588.

Mora, E. C., & Macias, S. (2007). Echolocation calls of Poey's flower bat (*Phyllonycteris poeyi*) unlike those of other phyllostomids. *Naturwissenschaften*, 94, 380–383.

Mora, E. C., Ibanez, C., Macias, S., Juste, J., Lopez, I., & Torres, L. (2011). Plasticity in the echolocation inventory of *Mormopterus minutus* (Chiroptera, Molossidae). *Acta Chiropterologica*, 13, 179–187.

Moss, C.F., Bohn, K., Gilkenson, H., & Surlykke, A. (2006). Active listening for spatial orientation in a complex auditory scene. *PLoS ONE*, 4, 1–12.

Moss, C. F., Chiu, C., & Surlykke, A. (2011). Adaptive vocal behavior drives perception by echolocation in bats. *Current Opinion in Neurobiology*, 21, 645–652.

Neuweiler, G. (1970). Neurophysiologische Untersuchungen zum Echo-ortungssystem der Grossen Hufeisennase Rhinoloophus ferrumequinum. *Zeitschrift vergleichende Physiologie*, 67, 273–306.

Neuweiler, G., Bruns, V., & Schuller, G. (1980). Ears adapted for the detection of motion, or how echolocating bats have exploited the capacities of the mammalian auditory system. *The Journal of the Acoustical Society of America*, 68, 741–753.

Neuweiler, G., Schuller, G., & Schnitzler, H.-U. (1971). On- and off- responses in the inferior colliculus of the greater horseshoe bats to pure tones. *Zeitschrift Vergleichende Physiologie*, 74, 57–63.

Novick, A. (1977). Acoustic orientation. In W.A. Wimsatt (Ed.), *Biology of bats*. Vol. 3 (pp. 73–287). New York: Academic Press.

Novick, A. (1958a). Orientation in Paleotropical bats. II. Megachiroptera. *Journal of Experimental Zoology*, 137, 443–462.

Novick, A. (1958b). Orientation in Paleotropical bats. I. Microchiroptera. *Journal of Experimental Zoology*, 138, 81–154.

Novick, A. (1962). Orientation in Neotropical bats. I. Natalidae and Emballonuridae. *Journal of Mammalogy*, 43, 449–455.

Novick, A., (1963). Orientation in Neotropical bats. II. Phyllostomatidae and Desmodontidae. *Journal of Mammalogy*, 44, 44–56.

Patriquin, K. J., & Barclay, R. M. R. (2003). Foraging by bats in cleared, thinned and unharvested boreal forest. *Journal of Applied Ecology*, 40, 646–657.

Pierce, G. W., & Griffin, D. R. (1938). Experimental determination of supersonic notes emitted by bats. *Journal of Mammalogy*, 19, 454–455.
Pollak, G. D., Henson, O. W., Jr., & Novick, A. (1972). Cochlear microphonic audiograms in the 'pure tone' bat, *Chilonycteris parnellii parnellii. Science*, 176, 66–68.
Popper, A. N., & Fay, R. R. (Eds). (1995). *Hearing by bats*. New York: Springer-Verlag.
Pye, A. (1966a). The structure of the cochlea in Chiroptera. Microchiroptera. *Journal of Morphology*, 118, 495–510.
Pye, A. (1966b) The Megachiroptera and Vespertilionidae of the Microchiroptera. *Journal of Morphology*, 119, 101–120.
Pye, A. (1967). The structure of the cochlea in Chiroptera. Phyllostomatoidea. *Journal of Morphology*, 119, 241–254.
Pye, J. D. (1960). A theory of echolocation by bats. *Journal of Laryngology and Otology*, 74, 718–729.
Pye, J. D., & Flinn, M. (1964). Equipment for detecting animal ultrasound. *Ultrasonics*, 2, 23–28.
Randall, L.A., Barclay, R. M. R., Reid, M. L. & Jung, T.S. (2011). Recent infestation of forest stands by spruce beetles does not predict habitat use by little brown bats (*Myotis lucifugus*) in southwestern Yukon, Canada. *Forest Ecology and Management*, 261, 1950–1956.
Schuller, G., & Pollak, G. (1979). Disproportionate frequency representation in the inferior colliculus of horseshoe bats: Evidence for an "acoustic fovea". *Journal of Comparative Physiology A: Neuroethology, Sensory, Neural, and Behavioral Physiology*, 132, 47–54.
Schnitzler, H.-U. (1967). Discrimination of thin wires by flying horseshoe bats (Rhiolophidae). In R.-G. Busnel (Ed.), *Animal sonar systems*, Volume I (pp. 69–87). Juoy-en-Josas, France: Laboratoire de Physiologie Acoustique.
Schnitzler, H.-U. (1968). Die Ultraschall-Ortungslaute der Hufeisen-Fledermause (Chiroptera-Rhinolophidae) und der Mehanismus der Bildhorens. *Zeitschrift fur Vergleichende Physiologie*, 57, 376–408.
Schnitzler, H.-U. (1973). Control of Doppler shift compensation in the greater horseshoe bat, *Rhinolophus ferrumequinum. Journal of Comparative Physiology A: Neuroethology, Sensory, Neural, and Behavioral Physiology*, 82, 79–92.
Schnitzler, H.-U., & Kalko, E. K. V. (2001). Echolocation by insect-eating bats. *BioScience*, 51, 557–569.
Schnitzler, H.-U., & Denzinger, A. (2011). Auditory fovea and Doppler shift compensation: Adaptations for flutter detection in echolocating bats using CF-FM signals. *Journal of Comparative Physiology A: Neuroethology, Sensory, Neural, and Behavioral Physiology*, 197, 541–559.
Schnitzler H-U., Suga, N., & Simmons, J. A. (1976). Peripheral auditory tuning for fine frequency analysis by the CF-FM bat, *Rhinolophus ferrumequinum. Journal of Comparative Physiology*, 106, 99–110.
Schuller, G., Beutner, K. & Schnitzler, H.-U. (1974). Response to frequency-shifted artificial echoes in the bat, *Rhinlophus ferrumequinum. Journal of Comparative Physiology A: Neuroethology, Sensory, Neural, and Behavioral Physiology*, 89, 275–286.
Schuller, G., & Pollak, G.D. (1979). Disproportionate frequency representation in the inferior colliculus of horseshoe bats: Evidence for an "acoustic fovea". *Journal of Comparative Physiology*, 132,121–128.
Siemers, B. M., & Schnitzler, H.-U. (2000). Natterer's bat (*Myotis nattereri* Kuhl 1818) hawks for prey close to vegetation using echolocation signals of very broad bandwidth. *Behavioral Ecology and Sociobiology*, 47, 400–412.
Siemers, B. M., Kalko, E. K. V., & Schnitzler, H.-U. (2001). Echolocation behavior and signal plasticity in the Neotropical bat *Myotis nigricans* (Schinz, 1821) (Vespertilionidae): A convergent case with European species of *Pipistrellus*? *Behavioral Ecology and Sociobiology*, 50, 317–328.
Simmons, J.A. (1973). The resolution of target range by echolocating bats. *Journal of the Acoustical Society of America*, 54, 157–173.

Simmons, J.A. (1979). Perception of echo phase in bat sonar. *Science,* 204, 1336–1338.
Simmons, J. A., & Stein, R. A. (1980). Acoustic imaging in bat sonar: Echolocation signals and the evolution of echolocation. *Journal of Comparative Physiology A: Neuroethology, Sensory, Neural, and Behavioral Physiology*, 135, 61–84.
Suga, N. (1964). Recovery cycles and responses to frequency-modulated tone pulses in auditory neurons of echolocating bats. *Journal of Physiology*, London, 175, 50–80.
Suga, N. (1984). The extent to which biosonar information is represented in the bat auditory cortex. In G. M. Edelman, W. E. Gall, & W. M. Cowan (Eds.), *Dynamic aspects of neocortical function* (pp. 315–373). New York: John Wiley & Sons, Inc.
Suga, N., & Jen, PH- S. (1975). Peripheral control of acoustic-signals in auditory system of echolocating bats. *Journal of Experimental Biology*, 62, 277–311.
Suga, N., & Jen, PH- S. (1977). Further studies on the peripheral auditory system of 'CF-FM' bats specialized for fine frequency analysis of Doppler-shifted echoes. *Journal of Experimental Biology*, 69, 207–232.
Suga, N., & O'Neill, W. E. (1979). Neural axis representing target range in the auditory cortex of the mustached bat. *Science*, 177, 82–84.
Suga, N., & Schlegel, F. (1973). Coding and processing in the auditory systems of FM-signal producing bats. *Journal of the Acoustical Society of America*, 54, 174–190.
Suga, N., Simmons, J. A. & Jen, P. H-S. (1975). Peripheral specialization for fine analysis of Doppler shifted echoes in the auditory system of the CF-FM bat *Pteronotus parnellii. Journal of Experimental Biology*, 63, 161–192.
Surlykke, A., & Kalko, E. K. V. (2008). Echolocating bats cry out loud to detect their prey. *PLoS ONE*, 3(4), e2036.
Webster, F. A. (1967). Performance of echolocating bats in the presence of interference. In R.-G. Busnel (Ed.) *Animal sonar systems: Biology and bionics* (pp. 673–713). Jouy-en-Josas-78, France: Laboratoire de Physiologie Acoustique.
Webster, F. A., & Brazier, O. G. (1965). *Experimental studies on echolocation mechanisms in bats*. Aerospace Medical Research Laboratories, Wright-Patterson Air Force Base, Ohio, AD 673373.

Chapter 2
Phylogeny, Genes, and Hearing: Implications for the Evolution of Echolocation in Bats

Emma C. Teeling, Gareth Jones, and Stephen J. Rossiter

Keywords Bat hearing • Bat vision • Cetacea • Convergence • Laurasiatheria • Microchiroptera • Megachiroptera • Mammalian phylogeny • Molecular evolution • Olfaction • Sensory perception • Taste • Whales • Pseudogenes • Yangochiroptera • Yinpterochiroptera

2.1 Introduction

Of all the mammals, bats are arguably the most unusual, uniquely able to fly and also distinctively capable of laryngeal echolocation, enabling them to orient and move in complete darkness (Kunz and Fenton 2003). Being highly specialized mammals with unique adaptations, it is not surprising that the phylogenetic position and evolutionary history of the order Chiroptera has been argued and debated since they were first named by Blumenbach in 1779 (Simmons and Geisler 1998). Indeed, of all the mammalian orders, Chiroptera has been considered the most contentious in terms of its phylogenetic controversies. This has resulted in provocative conclusions about the evolution of echolocation and flight in mammals (see extensive reviews in Teeling et al. 2012; Jones et al. 2013).

To understand how echolocation has evolved, as well as the molecular basis of this spectacular trait, bat inter- and intra-ordinal relationships must first be resolved (see Figs. 2.1 and 2.2 for consensus relationships from molecular phylogenies). This has proven difficult over the past century because of conflicting phylogenies,

E.C. Teeling (✉)
School of Biology and Environmental Science, University College Dublin, Dublin 4, Ireland
e-mail: emma.teeling@ucd.ie

G. Jones
School of Biological Sciences, University of Bristol, Bristol, UK
e-mail: gareth.jones@bristol.ac.uk

S.J. Rossiter
School of Biological and Chemical Sciences, Queen Mary University, London, UK
e-mail: s.j.rossiter@qmul.ac.uk

M.B. Fenton et al. (eds.), *Bat Bioacoustics*, Springer Handbook of Auditory Research 54, DOI 10.1007/978-1-4939-3527-7_2

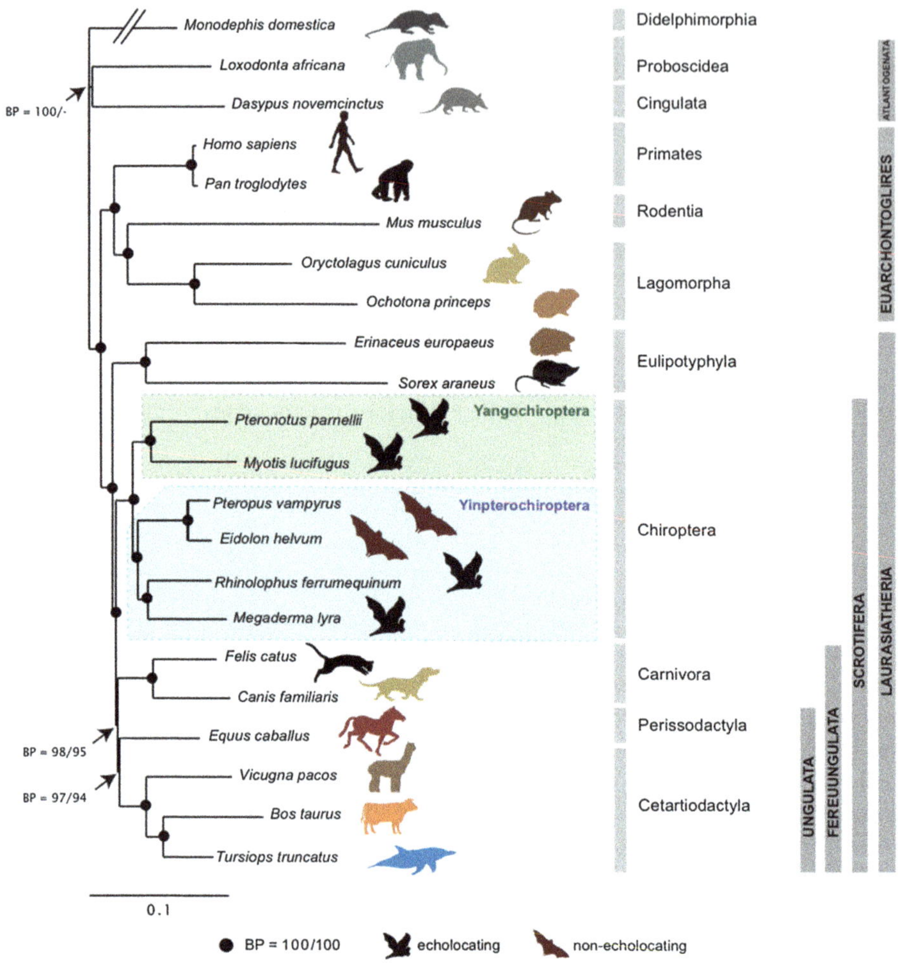

Fig. 2.1 The evolutionary relationships of Chiroptera among other mammalian lineages based on a large phylogenomic study with basal bat representatives (Taken from Tsagkogeorga et al. 2013 with permission)

arguably due to different data types, limited taxonomic sampling, and a poor fossil record (Teeling et al. 2012). However, the dawn of the "genomic and genetic era," in which molecular data have been used to build evolutionary trees, has seen great advances and paradigm shifts in our understanding of the evolutionary history of bats and other mammals (Figs. 2.1 and 2.2) (Springer et al. 2004; Jones and Teeling 2006). Within the past decade, we have approached a consensus stemming from analyses of large molecular and genomic data sets (Figure 2.1) (Meredith et al. 2011; Tsagkogeorga et al. 2013), and we have started to uncover the molecular basis of echolocation (Liu et al. 2010a, b; Parker et al. 2013).

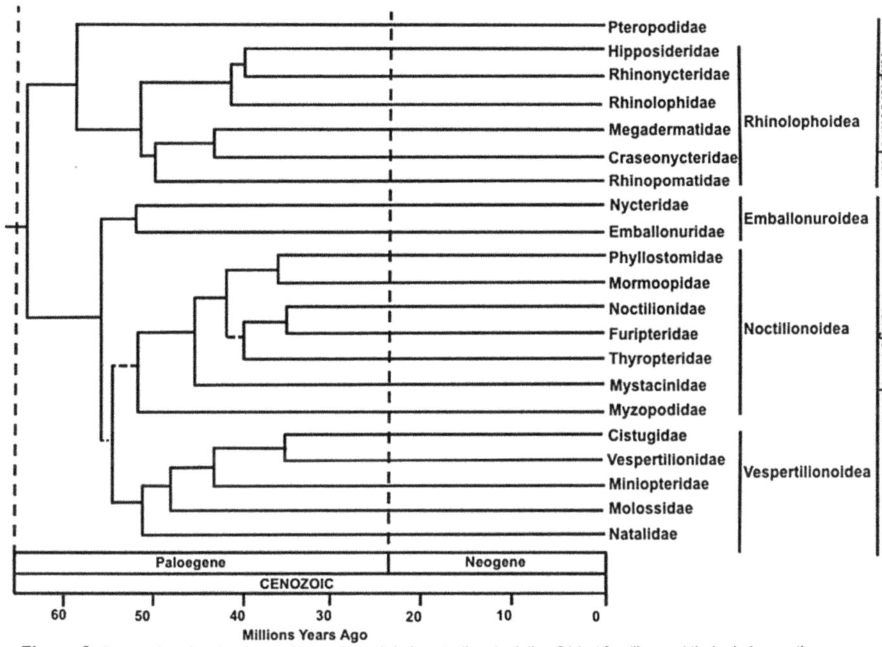

Figure 2. Composite of molecular phylogenetic and dating studies depicting 21 bat families and their phylogenetic relationships. Dotted branches indicates ambiguous clades with conflicting support.

Fig. 2.2 A composite figure summarizing the consensus divergence times and family relationships among bats (Based on: Teeling 2009b; Lack et al. 2010; Meredith et al. 2011; Teeling et al. 2012 and references therein; Foley et al. 2014)

Currently within Chiroptera, there is agreement regarding the majority of interfamilial relationships based on large nuclear data sets (Teeling et al. 2005; Miller-Butterworth et al. 2007), mitochondrial, and nuclear whole genome studies (Meganthan et al. 2012; Tsagkogeorga et al. 2013). However, the position of the Pteropodidae, or Old World flying foxes, has recently been questioned by a large phenomic (morphological) data set (O'Leary et al. 2013), which contrasted with all published molecular-based phylogenies (Springer et al. 2013) and, therefore, questioned all current molecular-based conclusions regarding the evolution of laryngeal echolocation in bats.

Below, the consensus and conflict regarding bat evolutionary relationships is explored based on key phylogenetic studies over the past 15 years. The consequential conclusions regarding the evolution of bats' unique sensory perception are detailed. Using these consensus phylogenies as an evolutionary framework, the search for the genomic signatures of echolocation is described, and the spectacular unprecedented sequence convergence observed between echolocating bats and whales is highlighted. What is currently known about the molecular basis of echolocation is detailed, and the future steps required to link this trait with its genomic bases are explored. It is often argued that the evolution of one specialization will have consequences for other traits, and, in the case of echolocation, a "trade-off" between the senses must occur (Jones et al. 2013). Here, the potential molecular sensory "trade-offs" between echolocation, vision, olfaction, and taste are explored. Finally, the next steps required to ultimately reach a universal consensus regarding bat phylogenetic relationships are discussed together with future directions for elucidating the genomic basis of echolocation in mammals and the extent of molecular trade-offs that occur in these remarkable species.

2.1.1 The Molecular Phylogenetic Position of Chiroptera Within Eutheria

Three key studies published in *Nature* and *Science* in 2001 provided the first robust support for four superordinal clades of mammals (Afrotheria, Laurasiatheria, Euarchontoglires, and Xenarthra; Table 2.1) and represented the largest genic and taxonomic data sets at that time (Madsen et al. 2001; Murphy et al. 2001a, b), ultimately changing the landscape of mammalian biology (Springer et al. 2004). The order Chiroptera was placed in the superordinal group Laurasiatheria along with carnivores (e.g., cats, dogs, seals), pangolins, cetartiodactyls (e.g., whales, cows), perrisodactyls (e.g., horses, rhinos), and eulipotyphylan insectivores (e.g., hedgehogs, shrews). This disassociated Chiroptera from its traditional morphology-based position in the superordinal clade Archonta, along with primates, tree shrews, and as the sister group to the flying lemurs, Dermoptera (Springer et al. 2004). Since these seminal papers, further molecular-based support for the four superordinal groups of placental mammals (Table 2.1), including Laurasiatheria, has come from many large phylogenomic data sets (e.g., Song et al. 2012; McCormack et al. 2012). Yet despite much congruence among studies, questions still remain regarding the position of the root of placental mammals (Morgan et al. 2013; Romiguier et al. 2013) and the branching patterns within Laurasiatheria (Meredith et al. 2011; Tsagkogeorga et al. 2013).

One of the largest studies in terms of taxonomic representation included 164 mammals representing up to 99 % of all recognized mammalian families for ~35,000 aligned nucleotide positions from 26 gene fragments (Meredith et al. 2011). Phylogenetic analyses and divergence time estimates provided high support for the four superordinal clades of mammals and estimated that Laurasiatheria originated approximately 85 million years ago (MYA) and that crown-group bats (i.e., all modern bats and their close fossil relatives) started to diverge approximately 66 MYA (Meredith et al. 2011). However, despite this comprehensive data set, it was still not possible to fully resolve the branching pattern within Laurasiatheria, arguably because of incomplete lineage sorting resulting from the rapid radiation and divergence of the laurasiatherian lineages (Springer et al. 2003; Romiguier et al. 2013). This has made it difficult to conclude which laurasiatherian family is the sister group to the bats and, therefore, has direct implications for interpreting how and when flight and echolocation originated in mammals.

Table 2.1 Details the composition of the four major placental mammal groupings

Superorder Laurasiatheria	Superorder Euarchontoglires
Order Chiroptera—e.g., bats	Order Rodentia—e.g., mice, rats
Order Perssiodactlya—e.g., horses, rhinos	Order Lagomorpha—e.g., rabbits, hares
Order Eulipotyphla—e.g., hedgehogs, shrews	Order Primates—e.g., Man, monkeys
Order Cetartiodactyla—e.g., whales, deer	Order Dermoptera—e.g., flying lemurs
Order Carnivora—e.g., dogs, lions, seals	Order Scandentia—e.g., tree-shrews
Order Philodota—e.g., pangolins	
Superorder Afrotheria	Order Xenarthra—e.g., armadillo, sloth
Order Afrosoricida—e.g., golden mole, tenrec	
Order Macroscelidea—e.g., elephant shrews	
Order Tubulidentata—e.g., aardvark	
Order Proboscidea—e.g., elephant	
Order Hyracoidea—e.g., hyrax	
Order Sirenia—e.g., manatee	

Analyses of shared retroposon (i.e., mobile DNA elements that originate from RNA molecules) insertion sites (Nishihara et al. 2006) and conserved non-coding elements (McCormack et al. 2012) found support for a sister group relationship between bats and horses, termed Pegasoferae (Nishihara et al. 2006). A recent taxonomically limited, whole genome phylogenetic study that included 10 mammals and 2 bat species (Zhang et al. 2013) also found support for a sister group relationship between bats and horses. However, this finding was contradicted by a more recent phylogenomic study that examined over 2,000 genes in 22 placental mammals including 6 bat species (Tsagkogeorga et al. 2013), the largest investigation of its kind to date. Using coalescent-based methods to accommodate the potential effects of incomplete lineage sorting, Tsagkogeorga et al. (2013) provided strong statistical support for Fereuungulata in which Carnivora is sister to an "ungulate" grouping containing Cetartiodactyla and Perrisodactyla, and bats are sister taxa to this group, a finding similar to other seminal phylogenetic/genomic studies (Figure 2.1) (Murphy et al. 2001b; Zhou et al. 2012).

2.1.2 *Molecular Phylogenetic Relationships Within Chiroptera*

Within the order Chiroptera itself, the past 15 years have seen considerable change and phylogenetic/systematic rearrangement resulting from large molecular studies (for a review see Teeling et al. 2012; for new family descriptions for Cistugidae see Lack et al. 2010; for Rhinonycteridae see Foley et al. 2014). Currently, there are over 1,260 species of bats (Simmons 2005) placed in 21 families. Figure 2.2 depicts these families, their interfamilial relationships, consensus divergence times, and also highlights nodes of controversy that differ between studies. Four superfamilial groups of echolocating bats are typically supported: Rhinolophoidea (Rhinolophidae,

Hipposideridae, Rhinonycteridae, Craseonycteridae, Megadermatidae, Rhinopomatidae); Vespertilionoidea (Vespertilionidae, Molossidae, Miniopteridae, Cistugidae, Natalidae); Emballonuroidea (Nycteridae, Emballonuridae); and Noctilionoidea (Myzopodidae, Mystacinidae, Furipteridae, Thyropteridae, Noctilionidae, Mormoopidae, Phyllostomidae). All non-laryngeal echolocating bats are placed in the family Pteropodidae.

Some discrepancies remain about the phylogenetic relationships between these superfamilies and, at times, their composition (Figure 2.2). Meredith et al. (2011) placed Emballonuroidea and Noctilionoidea as sister taxa, whereas Teeling et al. (2012) placed Emballonuroidea basal within the supordinal group Yangochiroptera, albeit with lower statistical support. The position of the monotypic Myzopoda can also differ between studies: Teeling et al. (2005) found support for a basal position for Myzopoda within the Noctilionoidea; however, Eick et al. (2005) and Meredith et al. (2011) supported a basal position for Myzopoda within the superfamily Vespertilionoidea. Within the superfamily Noctilionoidea, the phylogenetic position of Thyropteridae is still debated (Teeling et al. 2012; Jones et al. 2013), and potentially the Taphozinae should be elevated to familial level status as a sister family to Emballonuridae (Ruedi et al. 2012).

2.1.3 *Yinpterochiroptera and Yangochiroptera*

The most significant phylogenetic rearrangement in bats, especially in relation to the evolution of echolocation, pertains to the position of the non-echolocating family Pteropodidae. Molecular data consistently support a sister group relationship between the non-echolocating Pteropodidae and the echolocating superfamily Rhinolophoidea in the subordinal group Yinpterochiroptera (Teeling et al. 2000; Meredith et al. 2011). This is in stark contrast to previous morphologically based phylogenies, whereby all bats capable of laryngeal echolocation were placed into the monophyletic suborder Microchiroptera, and all non-laryngeal echolocating bats (Pteropodidae) were placed in the suborder Megachiroptera (Simmons and Geisler 1998; Teeling 2009a, b). This non-echolocating phenotype was considered basal within bats and laryngeal echolocation was considered to have evolved once (Teeling 2009a; Jones et al. 2013). In contrast, large molecular and mitochondrial data sets (see Teeling et al. 2012 for a review of previous studies; Meganthan et al. 2012) consistently supported a basal division in bats: Yinpterochiroptera (Pteropodidae + Rhinolophoidea) and Yangochiroptera (all other echolocating lineages) (Figure 2.2).

O'Leary et al. (2013) examined the phylogenetic relationships and divergence times among representative living and fossil mammals using a large supermatrix that combined 4,500 phenomic characters with DNA sequences for segments of 27 nuclear genes [based on the Meredith et al. (2011) molecular data set]. Using this large morphologically based data set and unusual/limited taxonomic sampling in bats (Fossil bats: *Onychonycteris*, *Icaronycteris*; Extant bats: *Pteropus*, *Rhinopoma*,

Pteronotus, *Myotis*, *Nycteris*, *Saccopteryx*), they recovered the classical Microchiroptera/Megachiroptera split (Simmons and Geisler 1998).

Within bats, morphological data have always supported a monophyletic Microchiroptera; however, this result is unsurprising given the known convergence of morphological characters in bats (Teeling et al. 2000, 2005). Bats are particularly prone to morphological convergences as they occupy a narrow ecological space and have constrained bauplans imposed by flight and echolocation. In the case of higher level relationships among placental mammals, there is a long history of discordance between morphological and molecular data (Springer et al. 2008), arguably resulting from rapid morphological mutations and high levels of morphological convergence, causing long-branch attraction and, therefore, inaccurate phylogenetic relationships (Springer et al. 2004, 2013). This problem can potentially be corrected for extant taxa by combining phenomic and genomic data sets, but genomic data are not available for most fossils and relatively few extant taxa have had their genomes sequenced (Springer et al. 2008). For these reasons it is likely that the results obtained by O'Leary et al. (2013) were particularly prone to the problem of long-branch attraction, which contributed to the erroneous support for Microchiroptera (Springer et al. 2013).

Indeed, Tsagkogeorga et al. (2013) is the only study to date to include genome data from echolocating species from both of the proposed suborders Yinpterochiroptera and Yangochiroptera. This study found unequivocal support for these groupings, and thus rejected the traditional subordinal bat clades of Microchiroptera and Megachiroptera (Figure 2.1). It follows that molecular-based phylogenies do not find support for the monophyly for Microchiroptera nor, by inference, a single acquisition of laryngeal echolocation in the ancestor of the echolocating lineages. Rather, molecular phylogenies suggest that laryngeal echolocation must either have evolved once in the ancestor of bats with subsequent loss in the "derived" pteropodids (Figure 2.3a) or have multiple acquisitions in the echolocating lineages (Figure 2.3b) within the two subordinal groups (Figure 2.3) (Jones et al. 2013). Despite the large molecular data sets (e.g., Parker et al. 2013; Tsagkogeorga et al. 2013) and the recovery of key transition bat fossils, such as *Onychonycteris finneyi* (Simmons et al. 2008; Veselka et al. 2010), this question still remains to be answered and represents a grand challenge in biology (see Teeling et al. 2012 for a review).

Potentially, more bat genomes sequenced and analyzed appropriately could uncover loss-of-function mutations in genes required for echolocation in pteropodid bats (Teeling et al. 2012), suggesting that they once had the ability to echolocate but lost it. Conversely, finding multiple genetic bases for echolocation in convergent echolocating bat lineages, but not in the pteropodids, would suggest independent, convergent gain of echolocation. However, before being able to take either of these approaches, the molecular bases for echolocation must be discovered. This is a difficult task, likened to finding the "Holy Grail" in this field (Teeling 2009a, b). However, recent whole genome comparative studies and different targeted gene approaches, focused on mammals with known auditory specializations for echolocation (bats and toothed whales), have finally started to uncover the molecular bases of echolocation.

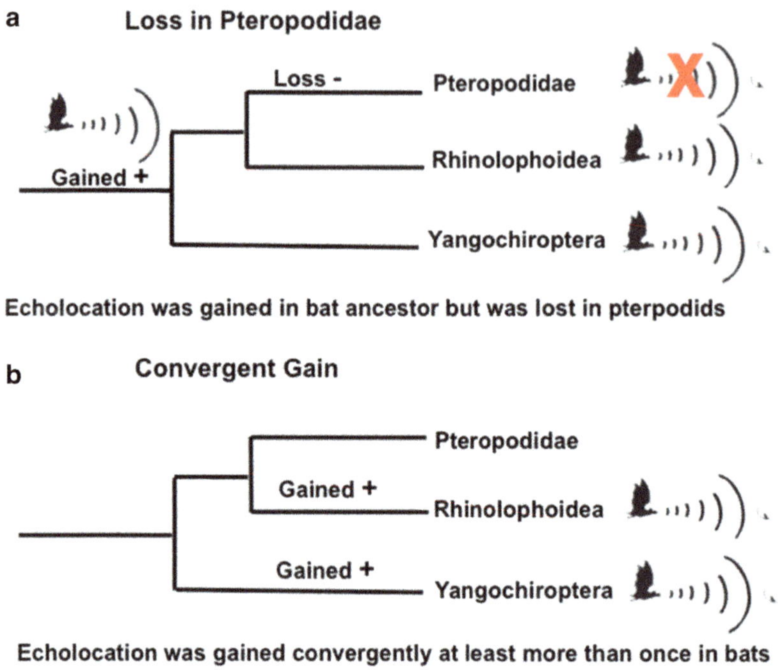

Fig. 2.3 Two alternative scenarios [(**a**) or (**b**)] regarding the gain/loss of echolocation in the pteropodid lineage

2.2 Auditory Specializations for Echolocation

Echolocation can be described as the production of sonar signals followed by comparison of the returning echoes for orientation, obstacle avoidance, and prey detection. It follows that echolocation necessitates a complex interplay of the vocalization, auditory, and neural systems, all of which are likely to have been subject to selection in the evolution of this sensory modality. Given the complexity of echolocation, it is especially remarkable that it has evolved not only in bats—possibly more than once—but also independently in the toothed cetaceans (order Odontoceti), making it a classic text book example of phenotypic convergence.

Attempts to identify the genes that function in echolocation in bats and cetaceans can be informed by studies of their respective key morphological, anatomical, and physiological specializations. Echolocating bats and cetaceans generate, emit, and receive sounds in contrasting ways. In bats, sound is generated in the larynx and, depending on the species, emitted through the mouth or nostrils. Although data on the laryngeal morphology of echolocating bats do not point to any gross adaptations, several features may represent specializations for generating ultrasonic pulses, such as the high degree of ossification of the larynx, calcification of the cricothyroid muscle, and extended vocal folds (Metzner and Schuller 2010; Carter 2014). In contrast, sound generation in odontocetes seems to be initiated in the larynx but is subsequently propagated by specialized nasal plugs and a complex of tissues collectively known as the monkey lips and dorsal bursae (Cranford and Amundin 2003).

To date, most research into the adaptations of echolocating bats and cetaceans has focused on the auditory system (Vater and Kossl 2004; Surlykke et al. 2014). All echolocating mammals possess a broadly standard mammalian inner ear and middle ear. However, while the outer ears of bats consist of large mobile pinnae adapted for receiving and localizing incoming sound, cetaceans have lost their outer ears altogether and, instead, appear to receive and channel sound via fatty tissues in their jaws (Ketten 1997, 2000). The inner ear comprises the cochlea and semicircular canals, constituting the main auditory and vestibular structures, respectively. The cochleae and vestibular systems of echolocating bats are remarkably varied in structure and may compete for space within the petrous bone (Davies et al. 2013a). The relative length of the basilar membrane in the cochlea, and its number of turns, are seen to correlate with echolocation call parameters (Davies et al. 2013b). Comparative morphometric analyses show that horseshoe bats have among the longest relative basilar membranes, probably relating to the fine auditory tuning of their auditory foveae.

Compared to other mammals, the basilar membranes of echolocating bats are thickened and narrow at the basal turn of the cochlea, and the outer hair cells (OHCs) are short and tightly anchored to reinforced, supporting Deiter cells (Dannhof and Bruns 1991; Vater and Kossl 1996, 2004). All of these features occur in odontocetes (Ketten 2000; Vater and Kossl 2004). Similarly, the stereocilia of the hair cells are also unusually short in echolocating bats, possibly improving their sensitivity to incoming sound waves, although no comparative data are available from cetaceans (Vater et al. 1992; Yao et al. 2007). Overall, many of the structural modifications in cochleae of echolocating bats contribute to increased stiffness of the cochlear partition at the basal end, maximizing the transfer of energy from the OHCs to the basilar membrane (Russell 2014). Further apparent specializations for processing echolocation signals are seen in the ascending auditory pathway. For example, compared with other mammals, the brain stems of bats show a greater degree of hypertrophy in several structures, including the superior olivary complex (Grothe and Park 2000), anteroventral cochlear nucleus, and inferior colliculus (Pollak 1992; Covey and Caseday 1995). Parallel expansions of auditory nuclei and auditory regions have also been documented in the dolphin brain (Ridgway 2000).

2.2.1 *The Molecular Basis of Hearing*

Despite considerable knowledge of the phenotypic correlates of echolocation, almost nothing is known about its molecular basis. Yet, because echolocation is so closely linked to audition, the molecular machinery of echolocation is expected to largely overlap with that of hearing.

To date, much of our understanding of the molecular basis of hearing has come from two main approaches: mapping and animal models. In mapping studies, loci are typically identified on the basis that they are linked to informative genetic markers. For example, much of our knowledge of hearing genes has come from identifying microsatellite loci or single-nucleotide polymorphisms (SNPs) that are associated with hearing impairment and then looking for genes near to these markers. By applying these so-called genome-wide association studies (GWAS) to sequence data from humans (Van Laer et al. 2010) and other species (Kluth and Distl 2013), the loci underpinning a large range of deafness conditions have been mapped to chromosomal positions, and, in many cases, the genes themselves have been identified. Once candidate genes have been identified, further insights into their role and importance in hearing have come from work on animal models, in particular mouse models (Leibovici et al. 2008).

As a consequence of these and related techniques, there have been remarkable advances in our understanding of the molecular basis of hearing and hearing loss. Indeed, over 100 genetic syndromes involving hearing loss (SHL) have been described, associated with mutations in one or multiple genes (Steel and Kros 2001; Petit 2006). Similarly, there are now over 100 recognized forms of "nonsyndromic hearing loss" (NSHL) in which hearing impairment is the only symptom present (Shalit and Avraham 2007). NSHL is also characterized by both allelic and locus heterogeneity. By convention, the names of loci associated with nonsyndromic hearing loss are prefixed by DFNX for X-linked deafness, DFNA for autosomal dominant deafness, or DFNB for autosomal recessive deafness. More details of the diversity of hearing genes discovered are available from recent reviews on the subject (Eisen and Ryugo 2007; Dror and Avraham 2010).

2.2.2 *Studying the Molecular Basis of Hearing and Echolocation in Bats*

Unfortunately, the conventional approaches for gene discovery outlined above are largely unsuitable for studying hearing and echolocation genes. In particular, mapping studies rely on the presence of intra-specific phenotypic variation, whereas bat and cetacean conspecifics do not tend to differ markedly in their echolocation call parameters. More fundamentally, mutations in hearing genes in bats and cetaceans are likely to result in lethal phenotypes. Therefore it is unsurprising that there are no reported cases of deafness in bats, although hearing loss has been reported in several stranded echolocating cetaceans (Finneran et al. 2007; Mann et al. 2010; Li et al. 2013). At the present time, conventional gene knockdown and knockout bat and cetacean models are limited, in part due to their slow rates of reproduction and the difficulties of culturing in the laboratory.

2.2.2.1 Candidate Gene Approaches

To date, nearly all attempts to identify genes involved in echolocation have taken a comparative approach, typically examining the molecular evolution of candidate genes that have been identified from humans, mice, and other organisms

(Jones et al. 2013). The rationale of such studies is that if genes show evidence of having undergone positive selection in echolocating taxa, then the gene products are likely to be of particular functional importance (e.g., Kirwan et al. 2013). While finding genes with roles in high-frequency hearing and echolocation might seem like looking for a needle in a haystack, candidate gene approaches have proven surprisingly successful in bats.

The gene encoding the forkhead transcription factor FOXP2 was one of the first putative echolocation genes to be studied. In humans, mapping studies show that mutations in the *FoxP2* gene are associated with deficits in aspects of speech and language, including orofacial coordination (Fisher et al. 1998; Lai et al. 2001), while molecular evolutionary analysis has revealed adaptive amino acid substitutions since the split with chimpanzees (Enard et al. 2002). From work on other non-echolocating species, *FoxP2* has been implicated in vocal learning in songbirds (Haesler et al. 2007) and in both ultrasonic vocalizations (Fujita et al. 2008, 2009) and auditory-motor learning in mice (Kurt et al. 2012). Given that echolocation in bats involves very rapid orofacial (or nasofacial) auditory-motor control, Li et al. (2007) tested whether *FoxP2* has been subject to molecular adaptation in the evolution of echolocation in bats. Gene sequence alignments revealed greater amino acid variation coupled with accelerated and divergent selection pressure in bats compared to other mammals, consistent with a role in echolocation. Further support for the involvement of *FoxP2* in echolocation comes from brain expression data, which show expression in the anterior cingulate cortex (ACC) and suprageniculate nucleus of echolocating bats at much higher levels than in Old World fruit bats (Metzner and Schuller 2010). Building on this result, Chen et al. (2013) have successfully applied lentivirus-based RNA interference (RNAi) to reduce the expression of *FoxP2* in the ACC, paving the way for behavioral studies in the future.

Additional insights into the molecular basis of vocalizations in bats come from experiments on echolocation pulse acoustics. Tressler et al. (2011) used a neurotoxin to target dopamine-producing cells in the basal ganglia and found that high striatal dopamine levels were associated with reductions in echolocation pulse amplitude, duration, and bandwidth. These results support a role for dopamine in the vocal control of echolocation, in line with findings from humans and rats that show an impact of striatal dopamine on the tone of the laryngeal musculature (Feng et al. 2009; Zarzur et al. 2010).

The majority of the studies of putative echolocation molecules to date have focused on hearing genes, in particular those implicated in a reduced sensitivity to high frequencies. Thus far, the best studied of these in bats has been the hearing gene *Prestin* (also known as *SLC26A5*; locus DFNB61); mutations in *Prestin* have been linked to autosomal recessive nonsyndromic hearing loss in humans (Liu et al. 2003). *Prestin* encodes a transmembrane solute carrier protein of the same name that is expressed and distributed in the OHCs (Zheng et al. 2000). In response to changes in membrane potential, *Prestin* undergoes voltage-dependent conformational changes that lead to electromotility of the OHCs; as such, *Prestin* is considered a key component of the cochlear amplifier that underpins the high sensitivity of the mammalian hearing apparatus (Liberman et al. 2002).

Li et al. (2008) undertook phylogenetic reconstructions of bats based on *Prestin* amino acid sequences and found that echolocating yinpterochropteran bats formed a well-supported clade with yangochiropterans to the exclusion of their true non-echolocating sister taxa, the pteropodid Old World fruit bats. The authors attributed this unexpected result to strong sequence convergence that was particularly evident in the cytoplasmic and extracellular domains of the protein, including the C-terminus. Follow-up studies that included cetacean *Prestin* sequences found even more dramatic convergence, this time between odontocetes and echolocating bats, together with evidence of molecular adaptation in multiple ancestral branches of echolocating taxa (Liu et al. 2010a, b). Building on these results, Liu et al. (2014) conducted functional assays and showed that that two parallel amino acid substitutions in the Prestin protein accounted for changes in the voltage-dependent membrane capacitance of cells, which in turn correlated with the frequency of best hearing sensitivity.

A strikingly similar signature of convergence in echolocating bats has been documented in the *Kcnq4* gene (Liu et al. 2011, 2012). In humans *KCNQ4* maps to locus DFNA2A, encodes a voltage-gated potassium channel protein, and is associated with nonsyndromic autosomal dominant deafness (Kubisch et al. 1999; Kharkovets et al. 2000). Liu et al. (2012) found eight parallel amino acid substitutions between the two groups of echolocating bats, while a similar study with slightly fewer taxa found four of these sites (Liu et al. 2011). As with Prestin, most of the parallel changes discovered were distributed in the cytoplasmic C-terminus of the protein. Immunofluorescence data from the mouse indicate that Kcnq4 expression follows both a longitudinal gradient, from the base to apex of the cochlea, as well as a radial gradient, from the IHCs to OHCs, with additional expression in the spiral ganglion neurons (SGNs) and vestibular hair cells (Beisel et al. 2005). The finding that the highest OHCs expression of Kcnq4 occurs at the apex, whereas the highest IHC and SGN expression occurs at the base, casts doubt on earlier speculation that Kcnq4 -associated deafness arises because of disrupted K^+ circulation in the OHCs; instead, this form of deafness might relate to problems of the IHCs and SGNs (Beisel et al. 2005).

Phylogenetic reconstructions based on the amino acid sequences of other hearing genes have also been found to recover erroneous well-supported groupings of laryngeal echolocating species. For example, Davies et al. (2012) reported similar findings from the genes *Tmc1* (locus DFNB7/11) and *Pjvk* (locus DFNB59) and Shen et al. (2012) from *Otof* (locus DFNB9) as well as *Cdh23* (locus DFNB12) and *Pcdh15* (locus DFNB23). The first of these, *Tmc1*, is expressed in both the IHCs and OHCs (Kurima et al. 2002) and encodes a transmembrane protein that functions in hair cell transduction and permeation (Kawashima et al. 2011; Pan et al. 2013). Davies et al. (2012) found particularly strong amino acid convergence at *Tmc1* between two bat species that have independently evolved high duty cycle echolocation with Doppler shift compensation: the horseshoe bat *Rhinolophus ferrumequinum* and the mustached bat *Pteronotus parnellii*. In contrast, comparisons of *Prestin* sequences from the same two taxa have revealed no such parallel changes (Shen et al. 2011).

The genes *Cdh23* and *Pcdh15* (encoding cadhedrin 23 and protocadhedrin 15, respectively) (Siemens et al. 2004; Ahmed et al. 2006) are both distributed in the tip links of the stereocilia and are thought to contribute to hair bundle motility. Shen et al. (2012) found parallel evolution in bats and dolphins, as well as positive selection in several key ancestral branches. The gene *Otof* encodes the protein Otoferlin, which has been shown to act as a calcium sensor mediating neurotransmitter release in cochlear hair cells, although its interactions with other proteins suggest additional functions (Zak et al. 2011). Real-time PCR has shown that in the echolocating common bent-winged bat (*Miniopterus schreibersii*) Otof expression levels are much higher (70-fold) in the auditory cortex compared with the cerebellum, whereas no such pattern was seen in the Old World fruit bat Leschenault's rousette (*Rousettus leschenaultii*) that does not possess laryngeal echolocation. Finally, data from mice suggest that the protein product of *Pjvk* (Pejvakin) probably has a role in the afferent auditory pathway (Delmaghani et al. 2006) rather than in the hair cells, which appear to be unaffected in mutant forms.

2.2.2.2 Genomics Approaches

Despite the success of some recent studies, candidate gene approaches undoubtedly often require considerable luck. New high throughput sequencing technologies offer the means to scale-up comparative approaches to genome scales, allowing thousands of loci to be studied at the same time (Brownstein et al. 2012; Yan et al. 2013). Zhang et al. (2013) tested for molecular adaptation in over 2,400 genes in David's myotis bat (*Myotis davidii*) and found significant positive selection in seven putative echolocation-related genes, including *Prestin*, *FoxP2*, and *Tmc1*, together with *Wnt8a*, *Fos*, *Mmp14* and *Dzip1*. Applying the same approach to the congeneric little brown bat (*Myotis brandtii*), Seim et al. (2013) analyzed 2,600 genes and found positive selection in two additional putative hearing genes, *Rgs7bp* and *Slc45a2*, as well as shared amino acid substitutions with the bottlenose dolphin (*Tursiops truncatus*) in two more hearing genes: *Trpv5*, mutants of which suffer from hair cell death, and *Nox3*, which is expressed in the inner ear and involved in the perception of gravity. Zhou et al. (2013) also investigated gene evolution in echolocating lineages, comparing 74 orthologs of putative hearing or vocalization genes in the Yangtze river dolphin (*Lipotes vexillifer*) and *M. lucifugus*. Of these, accelerated evolution was found in seven (*Prestin*, *Tmc1*, *Dzip1*, *Mmp14*, *Pax2*, *Wnt8a* and *Sparc*), of which parallel evolution was seen in the first three, as well as in 14 other genes, including *Myo15a*, *Otof*, *Notch1* and *Bmp4*.

Building on the findings of candidate gene studies, Parker et al. (2013) developed a bioinformatics pipeline to compare locus-wise support for competing phylogenetic hypotheses at a genome scale. They then used this method to identify all loci along a genome alignment that supported an erroneous grouping of either all unrelated echolocating bats or echolocating bats and the dolphin. This study showed that the strength of support for convergence for the trait echolocation was significantly stronger in hearing genes than in other genes. Moreover, the work identified numerous other genes supporting convergence, including several known hearing and/or deafness genes (e.g., *Slc4a11*, *Coch*, *Itm2b*, *Ercc3*, and *Opa1*). Perhaps more interestingly, the results revealed numerous genes that support "echolocation convergence" but which are poorly known with no known roles in sensory perception, suggesting more investigation is needed. Finally, in the first attempt to identify regulatory sequences underpinning echolocation, Davies et al. (2014) screened ~82,000 mammal-specific conserved non-coding elements (CNEs), and looked for changes in evolutionary rates in those CNEs underlying auditory system development. The authors found clear differences between echolocating and non-echolocating taxa in the substitution rates of four CNEs associated with inner ear development, implying possible roles of these regulatory loci in echolocation.

2.2.2.3 Future Approaches

Future studies of the molecular basis of echolocation now have a rich resource of published genome data sets for bats (Seim et al. 2013; Tsagkogeorga et al. 2013; Zhang et al. 2013), and cetaceans (Gui et al. 2013; Zhou et al. 2013; Yim et al. 2014) with more genomes to be published soon (Genome 10K Community of Scientists 2009). An additional application of high-throughput sequencing that can provide strong verification of a role in organs or structures associated with echolocation is through the collection of expression data. In general, sequencing RNA transcripts (RNA-Seq) offers a cost-effective means of obtaining coding gene sequences, especially for specific tissues, and has already been used in several studies of bats (Shaw et al. 2012; Francischetti et al. 2013; Phillips et al. 2014; Huang et al. 2016). On the other hand, obtaining sufficient yields of non-degraded RNA from the cochleae of bats is technically difficult because of the high degree of mineralization of the cochlea and the small amounts of starting material. Such problems might partially be addressed by the emerging field of single-cell transcriptomics. Obtaining transcriptome data from the ears of cetaceans is arguably even more challenging because RNA degrades rapidly postmortem, thus precluding the use of stranded animals. Currently there is one published study of cochlear transcriptomes of bats, which examined the echolocating Rickett's big-footed bat, *Myotis ricketti* and the Old World greater short-nosed fruit bat, *Cynopterus sphinx* (Dong et al. 2013). Comparisons of expression profiles revealed 987 genes were significantly upregulated in the echolocating species, including 18 known hearing genes. Of these, only *Tmc1* has been studied in bats.

Unfortunately, genome-scale approaches also have problems; for example, they typically require considerable computational resources and present nontrivial analytical challenges. In comparative studies of multiple and often divergent species, aligning and assembling large volumes of short-read sequence data inevitably introduces errors, which if not detected and accounted for will appear as signals of molecular evolution in downstream analyses. Therefore, given that genome resequencing of bats has limited value in light of the absence of sufficient intraspecific natural or pathological phenotypic (i.e., echolocation) variation, perhaps future efforts to discover echolocation genes should focus on cases of very closely related species that have evolved divergent call frequencies or even different systems of echolocation altogether. Indeed members of the mustached bat (*Pteronotus parnellii*) species complex (Clare et al. 2013) are the only examples of the genus to use high duty cycle echolocation, while there are many cases of sister-taxa/clades having undergone dramatic shifts in call frequency (Kingston and Rossiter 2004; Puechmaille et al. 2011). Finally, it is important to recognize that most sequencing approaches, including SNP-based assays, rarely take account of genome architecture, structural rearrangements, or copy number variation (CNV), the latter of which may present the genetic substrate for evolutionary phenotypic innovations (Perry et al. 2008; Paudel et al. 2013).

As studies of molecular evolution continue to add to the growing number of putative echolocation genes, there is a mounting need for functional assays. Mouse models have already been used for studying the impacts of bat gene sequences on limb development (Cretekos et al. 2008) and present enormous potential for studies of hearing. Transient knock down of hearing genes by RNA interference is also feasible (Chen et al. 2013); however, these approaches require access to the tissue of interest, precluding their use for studying cochlea-specific gene function.

The availability of whole genomes and large molecular data sets has enabled a genomic exploration of the consequences of extreme ecological adaptation, i.e., the acquisition of echolocation. Visual inspection of an echolocating bat (e.g., *Rhinolophus*) compared to a non-echolocating bat (e.g., *Pteropus*) reveals obvious morphological differences (small eyes and large nose leaves versus large eyes and no nose leaves) resulting from the acquisition of echolocation. Extreme adaptation typically causes loss of function in another trait. Therefore, these trade-offs in sensory perception should be mirrored within the genome. By using bats as a model of phenotypic plasticity and exploring the genetic bases of their unique and divergent sensory traits, the link between phenotype and genotype can be further elucidated, addressing another grand challenge in biology.

2.3 Are Sensory Trade-Offs Associated with the Evolution of Echolocation?

An assumption that is frequently held in sensory biology is that brains are energetically expensive to maintain, and, therefore, selection acting on the relative allocation of tissue among different regions of the brain is severe. Trade-offs in resource allocation may occur among brain regions that are specialized for specific sensory tasks (Harvey and Krebs 1990). For example, it may not be possible to evolve both sophisticated echolocation and vision; hence, enlargement of brain regions associated with echolocation may occur in tandem with a reduction in the sizes of other brain regions associated with vision. The "mosaic evolution" patterns that emerge independently in functional brain units may be subject to intense selection if brain size is constrained (Cooper et al. 1993), as is likely in bats. Echolocating bat species may need to be small to react quickly to rapidly returning echoes (Barclay and Brigham 1991) or to produce echolocation pulses at high repetition rates

(Jones 1994) using superfast muscles (Elemans et al. 2011; Ratcliffe et al. 2013). Indeed, 70 % of all echolocating bat species are under 20 g and 30 % are under 10 g (Jones 1996). Consequently, their small brains will have limited capacity for neuron populations. Trends in the evolution of brain size appear to be related to foraging strategy. Fast-flying bats may have undergone reductions in brain size over evolutionary time, while the demands of orienting in complex environments may have selected for increased brain size in maneuverable flyers (Safi et al. 2005).

Investigations of sensory trade-offs in animals have largely focused on morphological and anatomical traits. For example, Mexican cavefish (*Astyanax mexicanus*) have lost their eyes because they are of no use in the dark caves they inhabit, though they possess relatively large numbers of taste buds and neuromasts (sensory cells associated with long-distance, tactile-like sensing) and large olfactory bulbs compared with surface-feeding conspecific forms (Gunter and Meyer 2013). Similar trade-offs may be expected in bats if echolocation is more effective than vision for orientation in darkness. Bats that use laryngeal echolocation have relatively enlarged brain regions associated with audition, such as the inferior colliculus and the auditory cortex, while Old World fruit bats (Pteropodidae) have relatively enlarged brain regions associated with vision and olfaction (Dechmann and Safi 2009). Morphological and anatomical traits have a genetic basis, however, and investigating whether sensory trade-offs exist via a molecular evolutionary perspective has great potential.

Loss of function in sensory adaptations can arise through relaxed selection that leads to pseudogenization. Pseudogenes possess DNA sequences similar to related genes that produce functional proteins but have become non-functional from disabling mutations such as premature stop codons or frameshifts. Pseudogenes are the genetic equivalent of vestigial morphological traits, such as the non-functional hind limb bones in cetacean skeletons, and the detection of pseudogenes can provide clues about traits that perhaps were functional in ancestral relatives and became non-functional only more recently. For example, the detection of pseudogenes associated with vision may imply that visual systems have regressed over evolutionary time. Moreover, if pseudogenization is more prevalent in taxa that have evolved sophisticated echolocation, then perhaps sensory trade-offs can be implied as a result of the intense neural demands necessary for complex acoustic imaging. Below we review whether genes associated with olfaction, vision, and taste are more likely to have become pseudogenized in echolocating taxa and discuss whether the detection of sensory trade-offs at the genetic level is feasible in bats.

2.3.1 Olfaction

Olfaction is important in the lives of bats and is used by many species for communication and by some species for finding food (Altringham and Fenton 2003). Tetrapods possess two distinct olfactory systems that operate via different anatomical and neurobiological pathways, but they can overlap in function (see review in Hayden and Teeling 2014). The "main olfactory system" (MOS) is used for the detection of volatile substances and involves olfactory sensory neurons in the nose that transmit information to the main olfactory bulb in the brain, then onwards to the olfactory cortex and other brain regions (Kishida et al. 2007). The "accessory olfactory system" (AOS) detects fluid-based stimuli (including pheromones) by the vomeronasal organ (VNO) located at the base of the nasal cavity. Nerves from the VNO connect to the accessory olfactory bulb from which signals are transmitted to the amygdala and the bed nucleus of the stria terminalis before being transmitted to the hypothalamus (Bhatnagar and Meisami 1998).

In the MOS, olfactory receptors (ORs) are expressed in the cell membranes of sensory neurons in the upper nasal epithelium. The ORs are G protein-coupled receptors and provide information that is translated by the brain into receptor codes representing specific scents (Rinaldi 2007). The OR genes comprise the largest gene family in mammalian genomes (Lindblad-Toh et al. 2005), accounting for 3–6 % of all protein coding genes (Niimura 2012). Animal species that use olfaction extensively typically possess large numbers of functional OR genes, and species that are less dependent on olfaction show high rates of pseudogenization in OR genes. Evidence for sensory trade-offs in other mammals includes high rates of loss of function by pseudogenization of OR genes in primates that evolved trichromatic color vision (Gilad et al. 2004), in the platypus that uses mechanoreception and electrodetection for finding prey, and in echolocating cetaceans (Niimura and Nei 2007; Hayden et al. 2010; Niimura 2012).

Transition to an obligate aquatic environment has resulted in significant modifications and reductions to chemosensory structures within cetaceans, including olfactory bulbs, olfactory nerves, and the cribiform plate, which are lacking postnatally in odontocetes and significantly reduced in mysticetes, compared with terrestrial mammals (McGowen et al. 2014). Indeed, whales have some of the lowest numbers of functional OR genes among mammals (Niimura 2012). However, additional loss of function mutations also may have resulted from the acquisition of echolocation. Odontocetes that echolocate have an extremely high proportion of OR pseudogenes (74–100 %); whereas, Mysticeti, none of which echolocate, have a lower proportion of OR pseudogenes (29–58 %) (McGowen et al. 2008, 2014).

Echolocation in cetaceans appears to have led to reduced investment in olfaction; however, no such trade-off is seen in bats. In fact, there is little evidence that echolocating bats have high rates of pseudogenization of OR genes (10–36 %) compared with other mammals, and the level of pseudogenization is similar in taxa that use laryngeal echolocation compared with the non-echolocating pteropodids. The lesser horseshoe bat, *Rhinolophus hipposideros*, uses echolocation involving Doppler-shift compensation yet shows only 10 % pseudogenization of OR genes, the lowest value among the 11 bat species studied (Hayden et al. 2010). Expanding the taxonomic representation, Hayden et al. (2014) generated and examined 5,517 OR genes from 27 bat species and still found no evidence of a sensory trade-off between echolocation and olfaction. Echolocating bats were not found to possess an OR gene repertoire that is significantly different from that of non-echolocating bats, and the variability in levels of OR pseudogenes could not be attributed to echolocation capabilities. However, there appears to be a trade-off between OR gene families, arguably driven by frugivory (Hayden et al. 2014). In both the Phyllostomidae and Yinpterochiroptera, an increase in the proportion of genes in families OR 1/3/7 and OR 2/13 coupled with a loss of genes in family OR 5/8/9 coincided with frugivory, regardless of echolocation capabilities. This suggests that chemosensory trade-offs are occurring between different OR gene families in bats and are driven by feeding ecology rather than sensory modalities. A second family of receptors in the olfactory epithelium encoded by trace amine-associated receptors (TAARS) deserves further study in bats, especially as these may be associated with the detection of pheromones (Liberles and Buck 2006; Hayden and Teeling 2014).

The loss of the AOS in primates and birds correlates with the acquisition of trichromatic color vision and tetrachromatic color vision, respectively, suggesting sensory trade-offs have taken place (Zhang and Webb 2003). Loss of function in the AOS is widespread in bats, with evidence from anatomical research (Bhatnagar and Meisami 1998) that is consistent with findings of pseudogenization of the *Trpc2* gene (transient receptor potential cation channel, subfamily C, member 2) required for vomeronasal signal transduction (Zhao et al. 2011). Functional VNOs are found only in phyllostomid bats and in some species in the genera *Miniopterus* and *Pteronotus* (Bhatnagar and Meisami 1998). Loss of function in *Trpc2* has occurred several times independently in a number of bat lineages and is not related to whether taxa have dichromatic or monochromatic color vision or whether they echolocate. However, from a wider perspective, perhaps the loss of the AOS in most bat species might be related to the specialized neural demands imposed by both vision and echolocation at night. Extant whales have also lost their VNO and, as expected, show few or no functional VNO-related genes (McGowen et al. 2014); however, these losses have not occurred due to the acquisition of echolocation since the degradation of the VNO is estimated to have occurred before the split of Odontoceti and Mysticeti (McGowen et al. 2014).

2.3.2 Taste

Genes involved with the reception of bitter, sweet, and umami tastes have been sequenced in bats and cetaceans. The extent of pseudogenization of bitter taste (*T2r*) genes in bats does not differ much from those in the human and rat genomes (Zhuo et al. 2009). Zhao et al. (2010a) sequenced a 720 bp portion of the exon of the *Tas1R* gene (associated with the detection of sweetness) from 42 bat species from a wide range of families and found that the gene was pseudogenized only in three species of vampire bats where the detection of sweetness may be redundant. The *Tas1R1* gene associated with the detection of umami (savory or meatlike) taste was pseudogenized, not amplifiable, or absent in 31 bat species studied (Zhao et al. 2012), although it appears intact in all other mammals studied except the giant panda (Zhao et al. 2010b). Hence, the need for detecting umami taste (assumed to be used in detection of amino acids in nutritious foods; Herness and Gilbertson 1999) may be redundant in bats, though pseudogenization is common to echolocating and non-echolocating taxa. Vampire bats are some of the only mammals known so far to lack multiple tastes, and this could be the result of sensory trade-offs associated with a wide range of neural demands from their functional AOS and their use of echolocation and thermoreception. In cetaceans, the tongue is characterized by very few taste buds, and bioinformatics analyses of the bottlenose dolphin genome have revealed many pseudogenized taste receptor genes (Jiang et al. 2012; McGowen et al. 2014). Feng et al. (2014) have confirmed that the genes responsible for tasting sweet, sour, umami, and bitter have all lost their protein-coding function in all cetaceans, whereas the function of the gene responsible for sensing salty tastes has been retained. Consequently, no trade-off can be linked to the acquisition of echolocation.

2.3.3 Vision

Vision is important in the lives of bat species, especially in non-echolocating pteropodids. Even in echolocating bats, vision is more effective than echolocation over long distances, although at the cost of reduced acuity (Boonman et al. 2013). Bat retinae are dominated by rods that confer sensitive monochromatic vision in dim-light conditions. The DNA sequences of the rod opsin (rhodopsin) were intact in 15 bat species, suggesting that the evolutionary advantages of rhodopsin are fundamental to all bat species (Zhao et al. 2009a). Color vision in bats is more complex, with some bat species resembling many other mammal species by being potentially dichromats with intact cone opsin genes sensitive to short (*Sws1*) and medium-to-long wavelengths (*Mws/Lws*) (Zhao et al. 2009b). Sensitivity to short wavelengths allows some bats to see ultraviolet light (Winter et al. 2003), and this could fulfill a range of functions, including the detection of flowers in dimly lit conditions. However, *Sws1* has become a pseudogene in all rhinolophid and hipposiderid bat species studied, as well as in some pteropodids, especially those that roost in caves (Zhao et al. 2009b). The lack of sensitivity to short wavelengths is supported by immunohistochemical evidence showing that the primary visual cortex does not respond to UV light in a pteropodid that roosts in caves (*Rousettus leschenaultii*) or in *Hipposideros armiger* (Xuan et al. 2012).

Perhaps the strongest evidence for sensory trade-offs comes from studies investigating genes associated with vision and audition in rhinolophid and hipposiderid bats. These bats use high-duty echolocation that is arguably the most specialized form of biosonar used in nature. Echolocation calls are constant frequency (CF), allowing the detection of fluttering targets in clutter and even allowing the bats to classify different types of fluttering targets and make adaptive decisions in prey selection (Koselj et al. 2011). Rhinolophid and hipposiderid bats also use frequency-modulated tails at the ends of their calls for target ranging, and they adjust the frequency of the CF component from call to call to compensate for Doppler shifts in echoes resulting from their flight speed (Trappe and Schnitzler 1982).

Other genes associated with visual perception have also become pseudogenes in rhinolophid and hipposiderid bats. Gja10 is a gap junction protein expressed in retinal horizontal cells in mammals and is important in horizontal cell coupling. Irbp (encoded by the *Rbp3* gene) is the interphotoreceptor retinoid-binding protein critical for normal functioning of the visual cycle in most mammals. *Gja10* was pseudogenized in all ten rhinolophid and hipposiderid bat species studied but was intact in 14 of 19 species using low duty cycle echolocation and in six pteropodid species (Shen et al. 2013). *Rb3p* has also become a pseudogene in all six rhinolophid and hipposiderid bats investigated and in *Pteronotus parnellii*, which also uses high duty cycle echolocation with Doppler-shift compensation. The gene was nevertheless intact in three pteropodid species and in six of eight bat species using low duty cycle echolocation (Shen et al. 2013).

Although the loss of traits that are no longer necessary is well-documented in animals at both the anatomical and genetic levels (e.g., Carroll 2006), whether such "regressive evolution" (Jeffery 2009) occurs as a coevolutionary event associated with concomitant enhancement of other sensory modalities is more difficult to demonstrate. Perhaps the brains of bats are not as energetically demanding as those of primates, and sensory trade-offs may not be important during evolution of their brains (Dechmann and Safi 2009). Given their specialized echolocation abilities, the pseudogenization of *Rbp3*, *Gja10*, and *Sws1* in rhinolophid and hipposiderid bats is strongly suggestive of a sensory trade-off arising from an increased investment in echolocation at the expense of vision. Indeed, these taxa also show accelerated evolution of the hearing gene *Prestin* (Li et al. 2008) that may enhance their sensitivity to high frequencies (Liu et al. 2010a, b). Identification of such trade-offs is problematic because most bats are reliant on a wide range of senses in their lives, and so trade-offs need to be considered among a wide range of sensory modalities rather than between echolocation and vision alone.

2.4 Summary

Genetic and genomic data sets have provided great insights into our understanding of the position of bats among mammals, as well as the main sub-ordinal groupings of bats. Today there is overwhelming agreement, based on molecular data, that echolocating bats are paraphyletic, indicating that echolocation has either been lost in the Pteropodidae or has evolved multiple times in bats. In contrast, phylogenetic signals obtained from some morphological data sets have persisted to support the monophyly of echolocating bats. This latter arrangement is likely to arise from strong phenotypic convergence for echolocation that pervades multiple morphological characters, which appears to have confounded bat taxonomists for hundreds of years. Bat phylogenetics represents a "hot-bed" of future research to untangle which characters are homoplastic and which are not, ultimately advancing our understanding of the constraints of morphological evolution.

Additional problems probably stem from the limitations of current analytical approaches and knowledge regarding morphological evolution. For example, typically morphological characters are considered to be independent from each other, whereas they might well be shaped by the same sets of underlying genes. Furthermore, uniting both morphological and molecular characters in a phylogenetic framework can lead to conflicting signals and violations of parameters; however, these analytical problems must be overcome to gain phylogenetic information from all the available data. As genome sequencing becomes cheaper and more accessible and more lineages of bats are included, it is inevitable that outstanding questions in bat sub-ordinal systematics will be resolved over time. However, analyses of genomic data also are fraught with analytical difficulties, and thus studying bats will enable the advancement of these methods required for all modern fields of biology. Arguably the greatest advances in the field of echolocation genomics among mammals will be in the genetic dissection of phenotypes, which, for bats and cetaceans, will include greater elucidation of the molecular basis of echolocation via functional assays, expression studies, and more complete locus-wide and taxonomic-wide surveys of molecular evolution. This is an exciting time for the study of bat evolution and mammalian echolocation. Indeed, within the next decade the study of some of nature's most highly specialized mammalian species, the bats and the whales, will ultimately illuminate links between genotype and phenotype never uncovered before.

Acknowledgements Stephen Rossiter and Emma Teeling are grateful to the European Research Council for current funding. Gareth Jones thanks the Biotechnology and Biological Sciences Research Council (BBSRC) and the Darwin Initiative (Defra); Emma Teeling thanks the Irish Research Council and Science Foundation Ireland; and Stephen Rossiter thanks the Royal Society, BBSRC, and the Natural Environment Research Council for funding work that contributed to this review. All authors thank Kalina Davies, Nicole Foley, Hugo Parker, Stefan Paul, and Sebastien Puechmaille for useful discussions, and Arthur Popper, Brock Fenton, and Conor Whelan for editorial suggestions and comments on original drafts.

References

Ahmed, Z. M., Goodyear, R., Riazuddin, S., Lagziel, A., Legan, P. K., Behra, M., Burgess, S. M., Lilley, K. S., Wilcox, E. R., Riazuddin, S., Griffith, A. J., Frolenkov, G. I., Belyantseva, I. A., Richardson, G. P., & Friedman, T. B. (2006). The tip-link antigen, a protein associated with the transduction complex of sensory hair cells, is protocadherin-15. *Journal of Neuroscience*, 26(26), 7022–7034.

Altringham, J. D., & Fenton, M. B. (2003). Sensory ecology and communication in the Chiroptera. In T.H. Kunz & M.B. Fenton (Eds.), *Bat ecology* (pp. 90–127). Chicago, IL: University of Chicago Press.

Barclay, R. M. R., & Brigham, R.M. (1991). Prey detection, dietary niche breadth, and body size in bats: Why are aerial insectivorous bats so small? *American Naturalist*, 137, 693–703.

Beisel, K. W., Rocha-Sanchez, S. M., Morris, K. A., Nie, L. P., Feng, F., Kachar, B., Yamoah, E. N., & Fritzsch, B. (2005). Differential expression of KCNQ4 in inner hair cells and sensory neurons is the basis of progressive high-frequency hearing loss. *Journal of Neuroscience*, 25(40), 9285–9293.

Bhatnagar, K. P., & Meisami, E. (1998). Vomeronasal organ in bats and primates: Extremes of structural variability and its phylogenetic implications. *Microscopy Research and Technique*, 43, 465–475.

Boonman, A., Bar-On, Y., Cvikel, N., & Yovel, Y. (2013). It's not black or white – on the range of vision and echolocation in echolocating bats. *Frontiers in Physiology*, 4, 248, doi: 10.3389/fphys.2013.00248

Brownstein, Z., Bhonker, Y., & Avraham, K. B. (2012). High-throughput sequencing to decipher the genetic heterogeneity of deafness. *Genome Biology*, 13(5), doi: 10.1186/gb-2012-13-5-245

Carroll, S. B. (2006). *The making of the fittest. DNA and the ultimate forensic record of evolution*. London: Quercus.

Carter, R. T. (2014). Ontogeny of the larynx and flight ability in Jamaican fruit bats (Phyllostomidae) with considerations for the evolution of echolocation. *The Anatomical Record*, doi: 10.1002/ar.22934

Chen, Q., Zhu, T., Jones, G., Zhang, J., & Sun, Y. (2013). First knockdown gene expression in bat (*Hipposideros armiger*) brain mediated by lentivirus. *Molecular Biotechnology*, 54(2), 564–571.

Clare, E. L., Adams, A. M., Maya-Simões, A. Z., Eger, J. L., Hebert, P. D. N, & Fenton, M. B. (2013). Diversification and reproductive isolation: Cryptic species in the only New World high-duty cycle bat, *Pteronotus parnellii*. *BMC Evolutionary Biology*, 13, 26.

Cooper, H. M., Herbin, M., & Nevo, E. (1993). Ocular regression conceals adaptive progression of the visual system in a blind subterranean mammal. *Nature*, 361(6408), 156–159.

Covey, E. & Caseday, J.H. (1995). *The lower brainstem auditory pathways*. Heidelberg, New York: Springer.

Cranford, T., & Amundin, M. (2003). Biosonar pulse production in odontocetes: The state of our knowledge. In J. A. Thomas, C. F. Moss, & M. Vater (Eds.), *Echolocation in bats and dolphins* (pp. 27–35). Chicago: The University of Chicago Press.

Cretekos, C. J., Wang, Y., Green, E. D., Martin, J. F., Rasweiler, J. J., Behringer, R. R., & Progra, N. C. S. (2008). Regulatory divergence modifies limb length between mammals. *Genes & Development*, 22(2), 141–151.

Dannhof, B. J., & Bruns, V. (1991). The organ of Corti in the bat *Hipposideros bicolor*. *Hearing Research*, 53(2), 253–268.

Davies, K. T. J., Cotton, J. A., Kirwan, J. D., Teeling, E. C., & Rossiter, S. J. (2012). Parallel signatures of sequence evolution among hearing genes in echolocating mammals: An emerging model of genetic convergence. *Heredity*, 108(5), 480–489.

Davies, K. T. J., Bates, P. J. J., Maryanto, I., Cotton, J. A., & Rossiter, S. J. (2013a). The evolution of bat vestibular systems in the face of potential antagonistic selection pressures for flight and echolocation. *PloS ONE*, 8(4), e61998, doi: 10.1371/journal.pone.0061998

Davies, K. T. J., Maryanto, I., & Rossiter, S. J. (2013b). Evolutionary origins of ultrasonic hearing and laryngeal echolocation in bats inferred from morphological analyses of the inner ear. *Frontiers in Zoology*, 10, doi: 10.1186/1742-9994-10-2

Davies, K. T. J., Tsagkogeorga, G., & Rossiter, S. J. (2014). Divergent evolutionary rates in vertebrate and mammalian specific conserved non-coding elements (CNEs) in echolocating mammals. *BMC Evolutionary Biology*, 14, 261.

Dechmann, D. K. N., & Safi, K. (2009). Comparative studies of brain evolution: A critical insight from the Chiroptera. *Biological Reviews*, 84(1), 161–172.

Delmaghani, S., del Castillo, F. J., Michel, V., Leibovici, M., Aghaie, A., Ron, U., Van Laer, L., Ben-Tal, N., Van Camp, G., Weil, D., Langa, F., Lathrop, M., Avan, P., & Petit, C. (2006). Mutations in the gene encoding pejvakin, a newly identified protein of the afferent auditory pathway, cause DFNB59 auditory neuropathy. *Nature Genetics*, 38(7), 770–778.

Dong, D., Lei, M., Liu, Y., & Zhang, S. (2013). Comparative inner ear transcriptome analysis between the Rickett's big-footed bats (*Myotis ricketti*) and the greater short-nosed fruit bats (*Cynopterus sphinx*). *BMC Genomic*, 14, doi: 10.1186/1471-2164-14-916

Dror, A. A., & Avraham, K. B. (2010). Hearing impairment: A panoply of genes and functions. *Neuron*, 68(2), 293–308.

Eick, G. N., Jacobs, D. S., & Matthee, C. A. (2005). A nuclear DNA phylogenetic perspective on the evolution of echolocation and historical biogeography of extant bats (Chiroptera). *Molecular Biology and Evolution*, 22(9), 1869–1886.

Eisen, M. D., & Ryugo, D. K. (2007). Hearing molecules: Contributions from genetic deafness. *Cellular and Molecular Life Sciences*, 64(5), 566–580. doi: 10.1007/s00018-007-6417-3

Elemans, C.P.H., Mead, A.F., Jakobsen, L., & Ratcliffe, J.M. (2011). Superfast muscles set maximum call rate in echolocating bats. *Science*, 333(6051), 1885–1888.

Enard, W., Przeworski, M., Fisher, S. E., Lai, C. S. L., Wiebe, V., Kitano, T., Monaco, A. P., & Paabo, S. (2002). Molecular evolution of FOXP2, a gene involved in speech and language. *Nature*, 418(6900), 869–872.

Feng, P., Zheng, J., Rossiter, S. J., Wang, D. & Zhao, H. (2014). Massive losses of taste receptor genes in toothed and baleen whales. *Genome Biology and Evolution*, 6(6),1254–1265. doi: 10.1093/gbe/evu095

Feng, X., Henriquez, V. M., Walters, J. R., & Ludlow, C. L. (2009). Effects of dopamine D-1 and D-2 receptor antagonists on laryngeal neurophysiology in the rat. *Journal of Neurophysiology*, 102(2), 1193–1205. doi: 10.1152/jn.00121.2009

Finneran, J. J., London, H. R., & Houser, D. S. (2007). Modulation rate transfer functions in bottlenose dolphins (*Tursiops truncatus*) with normal hearing and high-frequency hearing loss. *Journal of Comparative Physiology A: Neuroethology, Sensory, Neural, and Behavioral Physiology*, 193(8), 835–843.

Fisher, S. E., Vargha-Khadem, F., Watkins, K. E., Monaco, A. P., & Pembrey, M. E. (1998). Localisation of a gene implicated in a severe speech and language disorder. *Nature Genetics*, 18(2), 168–170. doi: 10.1038/ng0298-168

Foley, N. M., Thong, V. D., Soisook, P., Goodman, S. M., Armstrong, K., Jacobs, D., Puechmaille S. J., & Teeling E. C. (2014). How and why overcome the impediments to resolution: Lessons from rhinolophid and hipposiderid bats. *Molecular Biology and Evolution*, 32(2), 313–33. doi: 10.1093/molbev/msu329

Francischetti, I. M. B., Assumpcao, T. C. F., Ma, D., Li, Y., Vicente, E. C., Uieda, W., & Ribeiro, J. M. C. (2013). The "Vampirome": Transcriptome and proteome analysis of the principal and accessory submaxillary glands of the vampire bat *Desmodus rotundus*, a vector of human rabies. *Journal of Proteomics*, 82, 288–319.

Fujita, E., Tanabe, Y., Shiota, A., Ueda, M., Suwa, K., Momoi, M. Y., & Momoi, T. (2008). Ultrasonic vocalization impairment of Foxp2 (R552H) knockin mice related to speech-language disorder and abnormality of Purkinje cells. *Proceedings of the National Academy of Sciences of the USA*, 105(8), 3117–3122.

Fujita, E., Tanabe, Y., Fujiwara, Y., Momoi, M., & Momoi, T. (2009). Ultrasonic vocalization of the knock-in mice with mutated Foxp2 related to speech-language disorder and normal Foxp2 expressed in Purkinje cells. *Neuroscience Research*, 65, S241-S241.

Genome 10 K Community of Scientists. (2009). Genome10K: A proposal to obtain whole-genome sequence for 10,000 vertebrate species. *Journal of Heredity*, 100(6), 659–674.

Gilad, Y., Wiebe, V., Przeworski, M., Lancet, D., & Pääbo, S. (2004). Loss of olfactory receptor genes coincides with the acquisition of full trichromatic vision in primates. *PLoS Biology*, 2(1), 120–125.

Grothe, B., & Park, T. J. (2000). Structure and function of the bat superior olivary complex. *Microscopy Research and Technique*, 51(4), 382–402.

Gui, D., Jia, K., Xia, J., Yang, L., Chen, J., Wu, Y., & Yi, M. (2013). De novo assembly of the indo-pacific humpback dolphin leucocyte transcriptome to identify putative genes involved in the aquatic adaptation and immune response. *PLoS ONE*, 8(8), doi: 10.1371/journal.pone.0072417

Gunter, H., & Meyer, A. (2013). Trade-offs in cavefish sensory capacity. *BMC Biology*, 11, 5, doi: 10.1186/1741-7007-11-5

Haesler, S., Rochefort, C., Georgi, B., Licznerski, P., Osten, P., & Scharff, C. (2007). Incomplete and inaccurate vocal imitation after knockdown of FoxP2 in songbird basal ganglia nucleus area X. *PLoS Biology*, 5(12), 2885–2897. doi: 10.1371/journal.pbio.0050321

Harvey, P. H., & Krebs, J.R. (1990). Comparing brains. *Science*, 249(4965), 140–146.

Hayden, S., & Teeling, E. C. (2014). The molecular biology of olfaction in vertebrates. *Anatomical Records* 297(11), 2216–2226.

Hayden, S., Bekaert, M., Crider, T. A., Mariani, S., Murphy, W. J., & Teeling, E. C. (2010). Ecological adaptation determines functional mammalian olfactory subgenomes. *Genome Research*, 20(1), 1–9. doi: 10.1101/gr.099416.109

Hayden, S., Davalos, L., Bekaert, M., Goodbla, A., Murphy, W. J., &. Teeling E. C. (2014). A cluster of olfactory receptor genes linked to frugivory in bats. *Molecular Biology and Evolution*, 31(4):917–27. doi: 10.1093/molbev/msu043

Herness, M. S., & Gilbertson, T.A. (1999). Cellular mechanisms of taste transduction. *Annual Review of Physiology*, 61, 407–419.

Huang, Z., Gallot, A., Lao, N. T., Puechmaille, S. J., Foley, N. M., Jebb, D., Bekaert, M., & Teeling, E. C. (2016). A non-lethal sampling method to obtain, generate and assemble whole-blood transcriptomes from small, wild mammals. *Molecular Ecology Resources,* 16(1), 150–162.

Jeffery, W. R. (2009). Regressive evolution in *Astyanax* cavefish. *Annual Review of Genetic*, 43, 25–47.

Jiang, P., Josue, J., Li, X., Glaser, D., Li, W., Branda, J. G., Margolskeea, R. F., Reed, D. R., & Beauchamp, G. K. (2012). Major taste loss in carnivorous mammals. *Proceedings of the National Academy of Sciences of the USA*, 109(13), 4956–4961.

Jones, G. (1994). Scaling of wingbeat frequency and echolocation pulse emission rates in bats: Why are aerial insectivorous bats so small? *Functional Ecology*, 8, 450–457.

Jones, G. (1996). Does echolocation constrain the evolution of body size in bats? *Symposia of the Zoological Society of London*, 69, 111–128.

Jones, G., & Teeling, E.C. (2006). The evolution of echolocation in bats. *Trends in Ecology and Evolution*, 21(3), 149–156.

Jones, G., Teeling, E. C., & Rossiter, S. J. (2013). From the ultrasonic to the infrared: Molecular evolution and the sensory biology of bats. *Frontiers in Physiology*, 4, 117–117. doi: 10.3389/fphys.2013.00117

Kawashima, Y., Geleoc, G. S. G., Kurima, K., Labay, V., Lelli, A., Asai, Y., Makishima, T., Wu, D. K., Della Santina, C. C., Holt, J. R., & Griffith, A. J. (2011). Mechanotransduction in mouse inner ear hair cells requires transmembrane channel-like genes. *Journal of Clinical Investigation*, 121(12), 4796–4809.

Ketten, D. R. (1997). Structure and function in whale ears. *Bioacoustics*, 8(1–2), 103–135.

Ketten, D. R. (2000). Cetacean ears. In W. W. L. Au, A. N. Popper, & R. R. Fay (Eds.), *Hearing by whales and dolphins* (pp. 43–108). New York: Springer.

Kharkovets, T., Hardelin, J. P., Safieddine, S., Schweizer, M., El-Amraoui, A., Petit, C., & Jentsch, T. J. (2000). KCNQ4, a K^+ channel mutated in a form of dominant deafness, is expressed in the inner ear and the central auditory pathway. *Proceedings of the National Academy of Sciences of the USA*, 97(8), 4333–4338.

Kingston, T., & Rossiter, S. J. (2004) Harmonic-hopping in Wallace's bats. *Nature*, 429(6992), 654–657.

Kirwan, J. D., Bekaert, M., Commins, J. M., Davies, K. T. J., Rossiter, S. J., & Teeling, E. C. (2013). A phylomedicine approach to understanding the evolution of auditory sensory perception and disease in mammals. *Evolutionary Applications*, 6(3), 412–422.

Kishida, T., Kubota, S., Shirayama, Y., & Fukami, H. (2007). The olfactory receptor gene repertoires in secondary-adapted marine vertebrates: Evidence for reduction of the functional proportions in cetaceans. *Biology Letters*, 3(4), 428–430.

Kluth, S., & Distl, O. (2013). Congenital sensorineural deafness in dalmatian dogs associated with quantitative trait loci. *PLoS ONE*, 8(12), doi: 10.1371/journal.pone.0080642

Koselj, K., Schnitzler, H.-U., & Siemers, B.M. (2011). Horseshoe bats make adaptive prey-selection decisions, informed by echo cues. *Proceedings of the Biological Society of London*, 278B, 3034–3041.

Kubisch, C., Schroeder, B. C., Friedrich, T., Lutjohann, B., El-Amraoui, A., Marlin, S., Petit, C., & Jentsch, T. J. (1999). KCNQ4, a novel potassium channel expressed in sensory outer hair cells, is mutated in dominant deafness. *Cell*, 96(3), 437–446. doi: 10.1016/s0092-8674(00)80556-5

Kunz, T. H., &. Fenton, M. B. (2003). *Bat ecology*. Chicago: The University of Chicago Press.

Kurima, K., Peters, L. M., Yang, Y. D., Riazuddin, S., Ahmed, Z. M., Naz, S., Arnaud, D., Drury, S., Mo, J. H., Makishima, T., Ghosh, M., Menon, P. S. N., Deshmukh, D., Oddoux, C., Ostrer, H., Khan, S., Deininger, P. L., Hampton, L. L., Sullivan, S. L., Battey, J. F., Keats, B. J. B., Wilcox, E. R., Friedman, T. B., & Griffith, A. J. (2002). Dominant and recessive deafness caused by mutations of a novel gene, TMC1, required for cochlear hair-cell function. *Nature Genetics*, 30(3), 277–284.

Kurt, S., Fisher, S. E., & Ehret, G. (2012). Foxp2 mutations impair auditory-motor association learning. *PLoS ONE*, doi: 10.1371/journal.pone.0033130

Lack, J. B., Roehrs, Z. P., Stanley, C. E., Ruedi, M., & Van Den Bussche, R.A. (2010). Molecular phylogenetics of *Myotis* indicate familial-level divergence for the genus *Cistugo* (Chiroptera). *Journal of Mammalogy*, 91(4), 976–992.

Lai, C. S. L., Fisher, S. E., Hurst, J. A., Vargha-Khadem, F., & Monaco, A. P. (2001). A forkhead-domain gene is mutated in a severe speech and language disorder. *Nature*, 413(6855), 519–523.

Leibovici, M., Safieddine, S., & Petit, C. (2008). Mouse models of human hereditary deafness. *Mouse Models of Developmental Genetic Disease*, 84, 385–429.

Li, G., Wang, J., Rossiter, S. J., Jones, G., & Zhang, S. (2007). Accelerated FoxP2 evolution in echolocating bats. *PLoS ONE*, doi: 10.1371/journal.pone.0000900

Li, G., Wang, J., Rossiter, S. J., Jones, G., Cotton, J. A., & Zhang, S. (2008). The hearing gene *Prestin* reunites echolocating bats. *Proceedings of the National Academy of Sciences of the USA*, 105(37), 13959–13964.

Li, S., Wang, D., Wang, K., Hoffmann-Kuhnt, M., Fernando, N., Taylor, E. A., Lin, W., Chen, J., & Ng, T. (2013). Possible age-related hearing loss (presbycusis) and corresponding change in echolocation parameters in a stranded Indo-Pacific humpback dolphin. *Journal of Experimental Biology*, 216(22), 4144–4153.

Liberles, S.D., & Buck, L.B. (2006). A second class of chemosensory receptors in the olfactory epithelium. *Nature*, 442(7103), 645–650.

Liberman, M. C., Gao, J. G., He, D. Z. Z., Wu, X. D., Jia, S. P., & Zuo, J. (2002). Prestin is required for electromotility of the outer hair cell and for the cochlear amplifier. *Nature*, 419(6904), 300–304.

Lindblad-Toh, K., Wade, C. M., Mikkelsen, T. S., Karlson, E. K., Jaffe, D. B., Kamal, M., Clamp, M., Chang, J. L., Kulbokas, E.J. III., Zody, M. C., et al. (2005). Genome sequence, comparative analysis and haplotype structure of the domestic dog. *Nature*, 438(7069), 803–819.

Liu, X. Z., Ouyang, X. M., Xia, X. J., Zheng, J., Pandya, A., Li, F., Du, L. L., Welch, K. O., Petit, C., Smith, R. J. H., Webb, B. T., Yan, D., Arnos, K. S., Corey, D., Dallos, P., Nance, W. E., & Chen, Z. Y. (2003). Prestin, a cochlear motor protein, is defective in non-syndromic hearing loss. *Human Molecular Genetics*, 12(10), 1155–1162.

Liu, Y., Cotton, J. A., Shen, B., Han, X., Rossiter, S. J., & Zhang, S. (2010a). Convergent sequence evolution between echolocating bats and dolphins. *Current Biology*, 20(2), R53-R54.

Liu, Y., Rossiter, S. J., Han, X., Cotton, J. A., & Zhang, S. (2010b). Cetaceans on a molecular fast track to ultrasonic hearing. *Current Biology*, 20, 1834–1839.

Liu, Y., Han, N., Franchini, L. F., Xu, H., Pisciottano, F., Elgoyhen, A. B., Rajan, K. E., & Zhang, S. (2012). The voltage-gated potassium channel subfamily KQT Member 4 (KCNQ4) displays parallel evolution in echolocating bats. *Molecular Biology and Evolution*, 29(5), 1441–1450.

Liu, Z., Li, S., Wang, W., Xu, D., Murphy, R. W., & Shi, P. (2011). Parallel evolution of KCNQ4 in echolocating bats. *PLoS ONE*, doi: 10.1371/journal.pone.0026618

Liu, Z., Qi, F., Zhou, Z., Ren, H, & Shi, P. (2014). Parallel sites implicate functional convergence of the hearing gene *prestin* among echolocating mammals. *Molecular Biology and Evolution*, doi: 10.1093/molbev/msu194

Madsen, O., Scally, M., Douady, C. J., Kao, D. J., DeBry, R. W., Adkins, R., Amrine, H. M., Stanhope, M. J., de Jong, W. W., & Springer, M. S. (2001) Parallel adaptive radiations in two major clades of placental mammals. *Nature*, 409(6820), 610–614.

Mann, D., Hill-Cook, M., Manire, C., Greenhow, D., Montie, E., Powell, J., Wells, R., Bauer, G., Cunningham-Smith, P., Lingenfelser, R., DiGiovanni, R., Jr., Stone, A., Brodsky, M., Stevens, R., Kieffer, G., & Hoetjes, P. (2010). Hearing loss in stranded odontocete dolphins and whales. *PLoS ONE*, doi: 10.1371/journal.pone.0013824

McCormack, J. E., Faircloth, B. C., Crawford, N. G., Gowaty, P. A., Brumfield, R. T., & Glenn, T. C. (2012). Ultraconserved elements are novel phylogenomic markers that resolve placental mammal phylogeny when combined with species-tree analysis. *Genome Research*, 22(4), 746–754.

McGowen, M. R., Clark, C., & Gatesy, J. (2008) The vestigial olfactory receptor subgenome of odontocete whales: Phylogenetic congruence between gene-tree reconciliation and supermatrix methods. *Systematic Biology*, 57(4), 574–590.

McGowen, M. R, Gatesy, J., & Wildman D. E. (2014). Molecular evolution tracks macroevolutionary transitions in Cetacea. *Trends in Ecology and Evolution*, doi.org/10.1016/j.tree.2014.04.001

Meganthan, P.R., Pagan, H. J. T., McCulloch E. S., Stevens R. D., & Ray, D. A. (2012). Complete mitochondrial genome sequences of three bats species and whole genome mitochondrial analyses reveal patterns of codon bias and lend support to a basal split in Chiroptera. *Gene*, 492(1), 121–129.

Meredith, R. W., Janečka, J. E., Gatesy, J., Ryder, O. A., Fisher, C. A., Teeling, E. C., Goodbla, A., Eizirik, E., Simão, T. L. L., Stadler, T., Rabosky, D.L., Honeycutt, R. L., Flynn, J. J., Ingram, C. M., Steiner, C., Williams, T. L., Robinson, T.J., Burk-Herrick, A., Westerman, M., Ayoub, N. A., Springer, M. S., & Murphy W. J. (2011). Impacts of the Cretaceous terrestrial revolution and KPg extinction on mammal diversification. *Science*, 334(6055), 521–524.

Metzner, W., & Schuller, G. (2010). Vocal control in echolocating bats. *Handbook of Behavioral Neuroscience*, 19, 403–415.

Miller-Butterworth, C. M., Murphy, W. J., O'Brien, S. J., Jacobs, D. S., Springer, M. S., & Teeling, E. C. (2007). A family matter: Conclusive resolution of the taxonomic position of the long-fingered bats, *Miniopterus*. *Molecular Biology and Evolution*, 24(7), 1553–1561.

Morgan, C. C., Foster, P. G, Webb, A. E., Pisani, D., McInerney, J. O., & O'Connell, M. J. (2013). Heterogeneous models place the root of the placental mammal phylogeny. *Molecular Biology Evolution*, 30(9), 2145–2156.

Murphy, W. J., Eizirik, E., Johnson, W. E., Zhang, Y. P., Ryder, O. A., & O'Brien, S. J. (2001a). Molecular phylogenetics and the origins of placental mammals. *Nature*, 409(6820), 614–618.

Murphy, W. J., Eizirik, E., O'Brien, S. J., Madsen, O., Scally, M., Douady, C. J., Teeling, E., Ryder, O. A., Stanhope, M. J., de Jong, W. W., & Springer, M. S. (2001b). Resolution of the early placental mammal radiation using Bayesian phylogenetics. *Science*, 294(5550), 2348–2351.

Niimura, Y. (2012). Olfactory receptor multigene family in vertebrates: From the viewpoint of evolutionary genomics. *Current Genomics*, 13(2), 103–114.

Niimura, Y., & Nei, M. (2007). Extensive gains and losses of olfactory receptor genes in mammalian evolution. *PLoS ONE*, 2(8), e708.

Nishihara, H., Hasegawa, M., & Okada, N. (2006). Pegasoferae, an unexpected mammalian clade revealed by tracking ancient retroposon insertions. *Proceedings of the National Academy of Sciences of the USA*, 103(26), 9929–9934.

O'Leary, M.A., Bloch, J. I., Flynn, J. J., Gaudin, T. J., Giallombardo, A., Giannini, N. P., Goldberg, S. L., Kraatz, B. P., Luo, Z. X., Meng, J., Ni, X., Novacek, M. J., Perini, F. A., Randall, Z. S., Rougier, G. W., Sargis, E. J., Silcox, M. T., Simmons, N. B., Spaulding, M., Velzaco, P. M., Weksler, M., Wible, J. R., & Cirranello, A. L. (2013). The placental mammal ancestor and the post-K-Pg radiation of placentals. *Science*, 339(6120), 662–667.

Pan, B., Geleoc, G. S., Asai, Y., Horwitz, G. C., Kurima, K., Ishikawa, K., Kawashima, Y., Griffith, A. J., & Holt, J. R. (2013). TMC1 and TMC2 are components of the mechanotransduction channel in hair cells of the mammalian inner ear. *Neuron*, 79(3), 504–515.

Parker, J., Tsagkogeorga, G., Cotton, J. A., Liu, Y., Provero, P., Stupka, E., & Rossiter, S. J. (2013). Genome-wide signatures of convergent evolution in echolocating mammals. *Nature*, 502(7470), 228–231.

Paudel, Y., Madsen, O., Megens, H.-J., Frantz, L. A. F., Bosse, M., Bastiaansen, J. W. M., Crooijmans, R. P. M. A., & Groenen, M. A. M. (2013). Evolutionary dynamics of copy number variation in pig genomes in the context of adaptation and domestication. *BMC Genomics*, doi: 10.1186/1471-2164-14-449

Perry, G. H., Yang, F., Marques-Bonet, T., Murphy, C., Fitzgerald, T., Lee, A. S., Hyland, C., Stone, A. C., Hurles, M. E., Tyler-Smith, C., Eichler, E. E., Carter, N. P., Lee, C., & Redon, R.

(2008). Copy number variation and evolution in humans and chimpanzees. *Genome Research*, 18(11), 1698–1710.
Petit, C. (2006). From deafness genes to hearing mechanisms: harmony and counterpoint. *Trends in Molecular Medicine*, 12(2), 57–64.
Phillips, C. J., Phillips, C. D., Goecks, J., Lessa, E. P., Sotero-Caio, C. G., Tandler, B., Gannon, M. R., & Baker, R. J. (2014). Dietary and flight energetic adaptations in a salivary gland transcriptome of an insectivorous bat. *PloS ONE*, doi: 10.1371/journal.pone.0083512
Pollak, G. D. (1992). Adaptation of basic structures and mechanisms in the cochlea and central auditory pathway of the mustache bat. In D. B. Webster, R. R. Fay, & A. N. Popper (Eds.), *The evolutionary biology of hearing* (pp. 751–778). New York: Springer.
Puechmaille, S. J., Ar Gouilh, M., Piyapan, P., Yokubol, M., Mie, K. M., Bates, P. J., Satasook, C., Nwe, T., Bu, S. S. H, Mackie, I. J., Petit, E. J, & Teeling, E. C. (2011). The evolution of sensory divergence in the context of limited gene flow in the bumblebee bat. *Nature Communications*, 2, 573. doi: 10.1038/ncomms1582
Ratcliffe, J.M., Elemans, C.P.H., Jakobsen, L., & Surlykke, A. (2013). How the bat got its buzz. *Biology Letters*, 9(2), 20121031.
Ridgway, S. (2000). The auditory central nervous system of dolphins. In W. W. L. Au, A. N. Popper & R. R. Fay (Eds.), *Hearing by whales and dolphins* (pp. 273–293). New York: Springer.
Rinaldi, A. (2007). The scent of life. The exquisite complexity of the sense of smell in animals and humans. *EMBO Reports*, 8, 629–633.
Romiguier, J., Ranwez, V., Delsuc, F., Galtier, N., & Douzery, E. J. P. (2013). Less is more in mammalian phylogenomics: AT-rich genes minimize tree conflicts and unravel the root of placental mammals. *Molecular Biology and Evolution*, 30(9), 2134–2144.
Ruedi, M., Friedli-Weyeneth, N., Teeling, E. C., Puechmaille, S. J., & Goodman, S. M. (2012). Biogeography of Old World emballonurine bats (Chiroptera: Emballonuridae) inferred with mitochondrial and nuclear DNA. *Molecular Phylogenetics and Evolution*, 64(1), 204–211.
Russell, I. (2014). Roles for prestin in harnessing the basilar membrane to the organ of Corti. In C. Köppl, G. A. Manley, A. N. Popper & R. R. Fay (Eds.), *Insights from comparative hearing research* (pp. 37–67). New York: Springer.
Safi, K., Seid, M. A. & Dechmann, D. K. N. (2005). Bigger is not always better: When brains get smaller. *Biology Letters*, 1(3), 283–286.
Seim, I., Fang, X., Xiong, Z., Lobanov, A. V., Huang, Z., Ma, S., Feng, Y., Turanov, A. A., Zhu, Y., Lenz, T. L., Gerashchenko, M. V., Fan, D., Yim, S. H., Yao, X., Jordan, D., Xiong, Y., Ma, Y., Lyapunov, A. N., Chen, G., Kulakova, O. I., Sun, Y., Lee, S.-G., Bronson, R. T., Moskalev, A. A., Sunyaev, S. R., Zhang, G., Krogh, A., Wang, J., & Gladyshev, V. N. (2013). Genome analysis reveals insights into physiology and longevity of the Brandt's bat *Myotis brandtii*. *Nature Communications*, doi: 10.1038/ncomms3212
Shalit, E. & Avraham, K. B. (2007). Genetics of hearing loss. In J. Schacht, A. N. Popper & R. R. Fay (Eds.), *Trauma, protection and treatment* (pp. 9–47). New York: Springer.
Shaw, T. I., Srivastava, A., Chou, W.-C., Liu, L., Hawkinson, A., Glenn, T. C., Adams, R., & Schountz, T. (2012). Transcriptome sequencing and annotation for the Jamaican fruit bat (*Artibeus jamaicensis*). *PLoS ONE*, doi: 10.1371/journal.pone.0048472
Shen, B., Avila-Flores, R., Liu, Y., Rossiter, S. J., & Zhang, S. (2011). Prestin shows divergent evolution between constant frequency echolocating bats. *Journal of Molecular Evolution*, 73(3–4), 109–115.
Shen, B., Fang, T., Dai, M., Jones, G., & Zhang, S. (2013). Independent losses of visual perception genes *Gja10* and *Rbp3* in echolocating bats (Order: Chiroptera). *PLoS ONE*, 8, e68867.
Shen, Y. Y., Liang, L., Li, G. S., Murphy, R. W., & Zhang, Y. P. (2012). Parallel evolution of auditory genes for echolocation in bats and toothed whales. *PLoS Genetics*, 8, e1002788. doi:10.1371/journal.pgen.1002788
Siemens, J., Lillo, C., Dumont, R. A., Reynolds, A., Williams, D. S., Gillespie, P. G., & Muller, U. (2004). Cadherin 23 is a component of the tip link in hair-cell stereocilia. *Nature*, 428(6986), 950–955.

Simmons, N. B. (2005). Order Chiroptera. In D. E. Wilson and D. M Reeder (Eds.), *Mammal species of the World: A taxonomic and geographic reference*. Vol. 1 (pp. 312–529). Washington, DC: Johns Hopkins University Press.

Simmons, N. B., & Geisler, J. H. (1998). Phylogenetic relationships of *Icaronycteris*, *Archaeonycteris*, *Hassianycteris*, and *Palaeochiropteryx* to extant bat lineages, with comments on the evolution of echolocation and foraging strategies in Microchiroptera. *Bulletin of the American Museum of Natural History*, 235, 1–182.

Simmons, N. B., Seymour, K.L., Habersetzer, J., and Gunnell, G.F. (2008) Primitive early Eocene bat from Wyoming and the evolution of flight and echolocation. *Nature*, 451 (7180), 818–821.

Song, S., Liu, L., Edwards, S. V., & Wu, S. (2012). Resolving conflict in eutherian mammal phylogeny using phylogenomics and the multispecies coalescent model. *Proceedings of the National Academy of Sciences of the USA*, 109(37), 14942–14947.

Springer, M. S., Murphy, W. J., Eizirik, E., & O'Brien, S. J. (2003). Placental mammal diversification and the Cretaceous-Tertiary boundary. *Proceedings of the National Academy of Sciences of the USA*, 100(3), 1056–1061.

Springer, M. S., Stanhope, M. J., Madsen, O., & de Jong, W. W. (2004). Molecules consolidate the placental mammal tree. *Trends in Ecology and Evolution*, 19(8), 430–438.

Springer, M. S., Meredith, R. W., Eizirik, E., Teeling, E., & Murphy, W. J. (2008) Morphology and placental mammal phylogeny. *Systematic Biology*, 57(3), 499–503.

Springer, M. S., Meredith, R. W., Teeling, E. C., & Murphy, W. J. (2013). Technical comment on "The placental mammal ancestor and the post-K-Pg radiation of placentals". *Science*, 341(6146), 613.

Steel, K. P., & Kros, C. J. (2001). A genetic approach to understanding auditory function. *Nature Genetics*, 27(2), 143–149.

Surlykke, A., Nachtigall, P. E., Fay, R. R., & Popper, A. N. (2014). *Biosonar.* New York: Springer.

Teeling, E. C. (2009a). Hear, hear: The convergent evolution of echolocation in bats? *Trends in Ecology and Evolution*, 24(7), 351–354.

Teeling, E. C. (2009b). Chiroptera. In B. Hedges & S. Kumar (Eds.) *The time tree of life* (pp. 499–503). Oxford, UK: Oxford University Press.

Teeling, E. C., Scally, M., Kao, D. J., Romangnoli, M. L., Springer, M. S., & Stanhope, M. J. (2000). Molecular evidence regarding the origin of echolocation and flight in bats. *Nature*, 403(6766), 188–192.

Teeling, E. C., Springer, M. S., Madsen, O., Bates, P., O'Brien, S. J., & Murphy, W. J. (2005). A molecular phylogeny for bats illuminates biogeography and the fossil record. *Science*, 307(5709), 580–584.

Teeling, E.C., Dool, S., & Springer, M.S. (2012). Phylogenies, fossils and functional genes: The evolution of echolocation in bats. In G. F. Gunnell & N.B. Simmons (Eds.), *Evolutionary history of bats: Fossils, molecules and morphology* (pp. 1–21). Cambridge: Cambridge University Press.

Trappe, M., and Schnitzler, H.-U. (1982). Doppler shift compensation in insect-catching horseshoe bats. *Naturwissenschaften*, 69, 193–194.

Tressler, J., Schwartz, C., Wellman, P., Hughes, S., & Smotherman, M. (2011). Regulation of bat echolocation pulse acoustics by striatal dopamine. *Journal of Experimental Biology*, 214(19), 3238–3247.

Tsagkogeorga, G., Parker, J., Stupka, E., Cotton, J. A., & Rossiter, S. J. (2013). Phylogenomic analyses elucidate the evolutionary relationships of bats. *Current Biology*, 23(22), 2262–2267.

Van Laer, L., Huyghe, J. R., Hannula, S., Van Eyken, E., Stephan, D. A., Maki-Torkko, E., Aikio, P., Fransen, E., Lysholm-Bernacchi, A., Sorri, M., Huentelman, M. J., & Van Camp, G. (2010). A genome-wide association study for age-related hearing impairment in the Saami. *European Journal of Human Genetics*, 18(6), 685–693.

Vater, M., & Kossl, M. (1996). Further studies on the mechanics of the cochlear partition in the mustached bat. 1. Ultrastructural observations on the tectorial membrane and its attachments. *Hearing Research*, 94(1–2), 63–77.

Vater, M., & Kossl, M. (2004). Introduction: The ears of whales and bats. In J. A. Thomas, C. F. Moss, & M. Vater (Eds.), *Echolocation in bats and dolphins* (pp. 89–98). Chicago and London: The University of Chicago Press.

Vater, M., Lenoir, M., & Pujol, R. (1992). Ultrastructure of the horseshoe bats organ of Corti. 2. Transmission electron-microscopy. *Journal of Comparative Neurology*, 318(4), 380–391.

Veselka, N., McErlain, D. D., Holdsworth, D. W., Eger, J. L., Chem, R. K., Mason, M. J., Brain, K. L., Faure, P. A., & Fenton, M. B. (2010). A bony connection signals laryngeal echolocation in bats. *Nature*, 463(7283), 939–942.

Winter, Y., Lopez, J., & von Helversen, O. (2003). Ultraviolet vision in a bat. *Nature*, 425(6958), 612–614.

Xuan, F., Hu, K., Zhu, T., Racey, P., Wang, X., Zhang, S., & Sun, Y. (2012). Immunohistochemical evidence of cone-based ultraviolet vision in divergent bat species and implications for its evolution. *Comparative Biochemistry and Physiology*, 161B, 398–403.

Yan, D., Tekin, M., Blanton, S. H., & Liu, X. Z. (2013). Next-generation sequencing in genetic hearing loss. *Genetic Testing and Molecular Biomarkers*, 17(8), 581–587.

Yao, Q., Zeng, J., Zheng, Y., Latham, J., Liang, B., Jiang, L., & Zhang, S. (2007). Characteristics of echolocating bats' auditory stereocilia length, compared with other mammals. *Science in China Series C: Life Sciences*, 50(4), 492–496.

Yim, H.-S., Cho, Y. S., Guang, X., Kang, S. G., Jeong, J.-Y., Cha, S.-S., Oh, H.-M., Lee, J.-H., Yang, E. C., Kwon, K. K., Kim, Y. J., Kim, T. W., Kim, W., Jeon, J. H., Kim, S.-J., Choi, D. H., Jho, S., Kim, H.-M., Ko, J., Kim, H., Shin, Y.-A., Jung, H.-J., Zheng, Y., Wang, Z., Chen, Y., Chen, M., Jiang, A., Li, E., Zhang, S., Hou, H., Kim, T. H., Yu, L., Liu, S., Ahn, K., Cooper, J., Park, S.-G., Hong, C. P., Jin, W., Kim, H.-S., Park, C., Lee, K., Chun, S., Morin, P. A., O'Brien, S. J., Lee, H., Kimura, J., Moon, D. Y., Manica, A., Edwards, J., Kim, B. C., Kim, S., Wang, J., Bhak, J., Lee, H. S., & Lee, J.-H. (2014). Minke whale genome and aquatic adaptation in cetaceans. *Nature Genetics*, 46(1), 88–92. doi: 10.1038/ng.2835

Zak, M., Pfister, M., & Blin, N. (2011). The otoferlin interactome in neurosensory hair cells: Significance for synaptic vesicle release and trans-Golgi network (Review). *International Journal of Molecular Medicine*, 28(3), 311–314.

Zarzur, A. P., Duarte, I. S., H. Goncalves, G. D. N., & Ueda Russo Martins, M. A. (2010). Laryngeal electromyography and acoustic voice analysis in Parkinson's disease: a comparative study. *Brazilian Journal of Otorhinolaryngology*, 76(1), 40–43.

Zhang, G., Cowled, C., Shi, Z., Huang, Z., Bishop-Lilly, K. A., Fang, X., Wynne, J. W., Xiong, Z., Baker, M. L., Zhao, W., Tachedjian, M., Zhu, Y., Zhou, P., Jiang, X., Ng, J., Yang, L., Wu, L., Xiao, J., Feng, Y., Chen, Y., Sun, X., Zhang, Y., Marsh, G. A., Crameri, G., Broder, C. C., Frey, K. G., Wang, L.-F., & Wang, J. (2013). Comparative analysis of bat genomes provides insight into the evolution of flight and immunity. *Science*, 339(6118), 456–460.

Zhang, J., & Webb, D.M. (2003). Evolutionary deterioration of the vomeronasal pheromone transduction pathway in catarrhine primates. *Proceedings of the National Academy of Sciences of the USA*, 100(14), 8337–8341.

Zhao, H., Rossiter, S. J., Teeling, E. C., Li, C., Cotton, J. A., & Zhang, S. (2009a). The evolution of color vision in nocturnal mammals. *Proceedings of the National Academy of Sciences of the USA*, 106(22), 8980–8985.

Zhao, H., Ru, B., Teeling, E. C., Faulkes, C. G., Zhang, S., & Rossiter, S. J. (2009b). Rhodopsin molecular evolution in mammals inhabiting low light environments. *PLoS ONE*, 4(12), e8326. doi: 10.1371/journal.pone.0008326

Zhao, H., Zhou, Y., Pinto, C.M., Charles-Dominique, P., Galindo-González, J., Zhang, S., & Zhang, J. (2010a). Evolution of the sweet taste receptor gene *Tas1r2* in bats. *Molecular Biology and Evolution*, 27(11), 2642–2650.

Zhao, H., Yang, J.R., Xu, H., & Zhang, J. (2010b). Pseudogenization of the umami taste receptor gene *Tas1r1* in the giant panda coincided with its dietary switch to bamboo. *Molecular Biology and Evolution*, 27(12), 2669–2673.

Zhao, H., Xu, D., Zhang, S., & Zhang, J. (2011). Widespread losses of vomeronasal signal transduction in bats. *Molecular Biology and Evolution*, 28(1), 7–12.

Zhao, H., Xu, D., Zhang, S., & Zhang, J. (2012). Genomic and genetic evidence for the loss of umami taste in bats. *Genome Biology and Evolution*, 4(1), 73–79.

Zheng, J., Shen, W. X., He, D. Z. Z., Kevin, B. L., Madison, L. D., & Dallos, P. (2000). Prestin is the motor protein of cochlear outer hair cells. *Nature*, 405(6783), 149–155.

Zhou, X., Xu, S., Xu, J., Chen, B., Zhou, K., & Yang, G. (2012). Phylogenomic analysis resolves the interordinal relationships and rapid diversification of the laurasiatherian mammals. *Systematic Biology*, 61(1), 150–164.

Zhou, X., Sun, F., Xu, S., Fan, G., Zhu, K., Liu, X., Chen, Y., Shi, C., Yang, Y., Huang, Z., Chen, J., Hou, H., Guo, X., Chen, W., Chen, Y., Wang, X., Lv, T., Yang, D., Zhou, J., Huang, B., Wang, Z., Zhao, W., Tian, R., Xiong, Z., Xu, J., Liang, X., Chen, B., Liu, W., Wang, J., Pan, S., Fang, X., Li, M., Wei, F., Xu, X., Zhou, K., Wang, J., & Yang, G. (2013). Baiji genomes reveal low genetic variability and new insights into secondary aquatic adaptations. *Nature Communications*, 4, doi: 10.1038/ncomms3708

Zhuo, Y., Dong, D., Zhang, S., & Zhao, H. (2009). Positive selection drives the evolution of bat bitter taste receptor genes. *Biochemical Genetics*, 47(3–4), 207–215.

Chapter 3
Ultrasound Production, Emission, and Reception

Walter Metzner and Rolf Müller

Keywords Auditory-vocal feedback • Communication • Dynamic changes • Labial echolocation • Laryngeal echolocation • Larynx • Nose leafs • Pinnae • Static complexity • Sensory feedback control • Vocal control

3.1 Introduction

Common to all bats is their ability to actively fly and their mostly nocturnal lifestyle. All bats use acoustic signals for passive listening to ambient sounds and for intra- and inter-species communication. In addition, most bats have evolved an active sonar system that uses ultrasonic calls for acoustically exploring the environment. Echolocation behavior uses sonar pulses produced by the larynx in the majority of bats that used to form the suborder Microchiroptera (Gunnell and Simmons 2005); however, the taxonomic status of "microchiropteran bats" is still in dispute (Teeling, Jones, and Rossiter, Chapter 2). The category will be used here to simply refer to all bats using laryngeal echolocation. Laryngeal echolocation is absent in the family of Old World fruit bats, Pteropodidae, which rely on their keen eyesight and sense of smell for orientation and foraging. One genus within this group, *Rousettus*, however, "reinvented" echolocation by producing sonar pulses using tongue clicks (Schuller and Moss 2004; for a general overview see Thomas et al. 2004).

W. Metzner (✉)
Department of Integrative Biology and Physiology, University of California, Los Angeles, CA 90095, USA
e-mail: metzner@ucla.edu

R. Müller
Department of Mechanical Engineering, Virginia Tech, Blacksburg, VA 24061, USA

Shandong University—Virginia Tech International Laboratory, Shandong University, Jinan 250100, China
e-mail: rolf.mueller@vt.edu

M.B. Fenton et al. (eds.), *Bat Bioacoustics*, Springer Handbook of Auditory Research 54, DOI 10.1007/978-1-4939-3527-7_3

The distinct tasks of communication on one side and acoustical imaging of the environment on the other demand a rather different spectrotemporal composition of signals. Information transmitted in communication calls is generally encoded with complex time-varying modulations of amplitude and frequency in calls that are long enough to transmit sufficient spectrotemporal information about the intended "meaning" of these calls (Figure 3.1) (see overview in Kanwal and Ehret 2006). Acoustical signals used for imaging obstacles that a bat may encounter in its flight path or for imaging its prey, however, require a completely different spectrotemporal design. Echolocation pulses (see Figure 3.1) are generally shorter to optimize the resolution of target distance and maximize target range accuracy. They contain higher sound energies to yield sufficiently strong echoes even from small, distant objects. Their spectrotemporal characteristics, therefore, are rather simple and stereotypical with a well-defined time course. This aids in the detection and discrimination of target-induced modulations in the echo by being easily recognizable as deviations from the call template. In addition, echolocation calls are usually very intense and can reach above 130 dB SPL in most species (measured at 10 cm distance; e.g., Hartley 1992), although some bats, so-called "whispering bats," especially phyllostomatids, produce much fainter sounds (70–80 dB SPL), especially when hunting close to vegetation and having to avoid being heard by their insect prey. However, more recent work, using non-invasive multi-microphone array recordings in the natural environment of "whispering bats" revealed considerably higher sound levels than those obtained in a laboratory setting (e.g., Brinkløv et al. 2009, 2011).

Figure 3.1a illustrates some extreme examples of different types of microchiropteran sonar pulses. Echolocation calls can range from very short, broadband signals that are usually composed of several, and often overlapping, harmonic downward frequency-modulated (FM) components (Figure 3.1a, Rickett's big-footed bat or Chinese fishing bat, *Myotis ricketti*) to long-lasting, constant-frequency (CF) calls with an extraordinarily small bandwidth (Figure 3.1a, greater horseshoe bat, *Rhinolophus ferrumequinum*). Whereas the former yield high temporal resolution and range accuracy, the latter produce long-ranging echolocation calls, especially for low-frequency signals (Figure 3.1a, Chinese noctule, *Nyctalus plancyi*). With high-frequency, narrowband pulses, they encode small frequency variations induced by moving targets ("flutter detection," e.g., *R. ferrumequinum*) (Schnitzler and Denzinger 2011). In contrast, communication signals are much longer in duration, contain more complex spectral and amplitude modulations, and are usually also emitted at lower frequencies than the sonar pulses of the same species (Figure 3.1b) (Gillam and Fenton, Chapter 5).

This chapter first considers how sounds are produced by the bat larynx by outlining its characteristic morphological features and discussing general sound production mechanisms, including non-linear features that play a key role in enabling echolocating bats to switch between echolocation and communication sounds. Subsequently, the neural basis for sound production in echolocating bats is discussed, including the innervation pattern of the larynx and how that

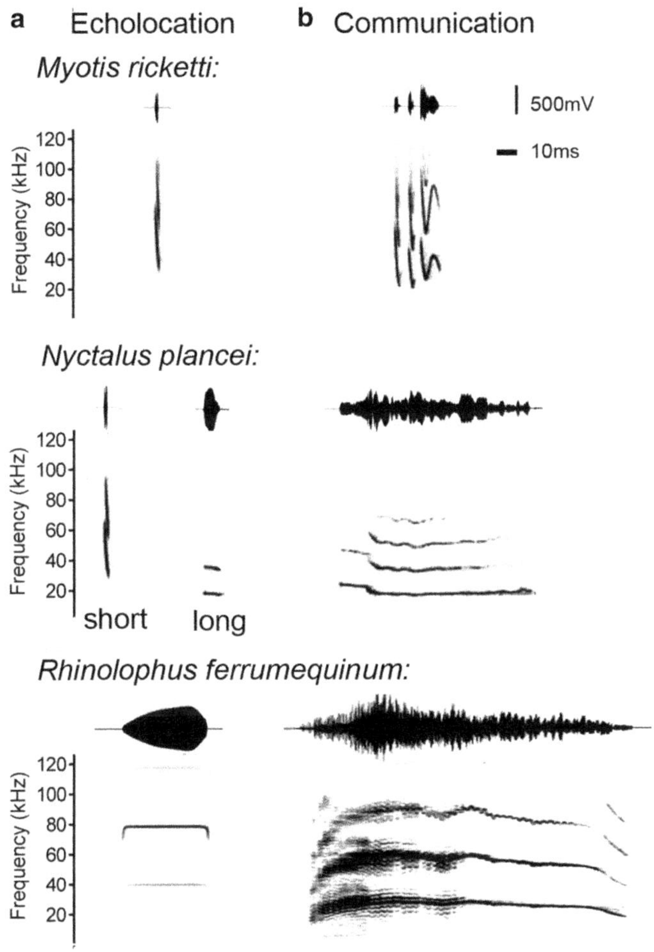

Fig. 3.1 Waveforms and sonagrams of different types of echolocation pulses (**a**) and communication calls (**b**)

affects the spectral and temporal features of sounds, followed by a summary of the connection patterns and mechanisms within the brain stem as well as higher order brain structures. The chapter then explores the role of various forms of sensory feedback in sound production, such as auditory and somatosensory stimulation, and highlights how sound production is intertwined with other motor patterns, such as flying. Finally, this chapter explores the static and dynamic complexity in face characteristics, i.e., the nose leafs and pinnae of echolocating bats, and discusses what role dynamic changes may play in echolocation.

3.2 Sound Production by the Bat Larynx

3.2.1 Morphology of the Bat Larynx

Echolocation and communication signals in microchiropteran bats are generated in the larynx during expiration like vocalizations in most other mammals. During expiration, air pressure builds up beneath the closed glottis, and when the glottis opens, sound is produced. Bat echolocation calls can be very intense. Accordingly, the subglottic pressures are more than five times that measured in humans (Neuweiler 2000). The air-filled space above the larynx (throat, mouth, nostrils, and nose leaf) acts as an acoustic filter that affects the spectral composition of the emitted sound as well as its directional characteristics.

It appears that the design principle of the bat larynx is very similar to that in other non-echolocating mammals (e.g., review by Metzner and Schuller 2010). Nevertheless, sonar pulses in bats show several distinct features when compared with vocalizations in other mammals, such as very high call frequencies, short durations, and high repetition rates, as well as high intensities.

Figure 3.2 provides a simplified sketch of the major components of the larynx as exemplified in horseshoe bats. The larynx contains a large, unpaired thyroid cartilage, which in humans forms the "Adam's apple"; a large cartilage ring, the cricoid, forming the inferior wall of the larynx; and a pair of arytenoid cartilages, located at the anterio-dorsal portion of the cricoid. The vocal folds (and vocal membranes) stretch between the ventral part of each arytenoid and the inner portion of the thyroid. The opening between the vocal folds, the glottis, allows air to flow by them and cause their oscillations, thus producing the calls.

Numerous internal larynx muscles connect the different cartilages, most importantly the cricothyroid muscle, which spans between the cricoid and thyroid. The arytenoid cartilages in bats are relatively rigid (Denny 1976; Suthers et al. 1988), and the cricothyroid muscle is virtually the only larynx muscle in bats that affects the tension of the vocal fold and thus controls call frequencies. Other internal larynx muscles, most of them attached to the arytenoids, control the opening (abduction) and closing (adduction) of the glottis (Novick and Griffin 1961; Suthers and Fattu 1973, 1982).

Other potentially important adaptations of the bat larynx include its large relative size (see Frey and Gebler 2010); calcified or ossified laryngeal cartilages; a hypertrophied, superfast cricothyroid muscle (see below); the configuration of the laryngeal membranes; and, in some species, the presence of air-filled tracheal cavities and their filter characteristics (Schuller and Moss 2004; Metzner and Schuller 2010).

Production of high-intensity sonar pulses necessitates that the lips of the vocal cords, as well as the cricothyroid membranous wall, be much thicker relative to those in other mammals of similar body size to enable them to withstand and maintain these high subglottic pressures even during the emission of several calls during one expiratory cycle (Neuweiler 2000). In addition, a relatively high degree of calcification or ossification occurs during early development in many bats

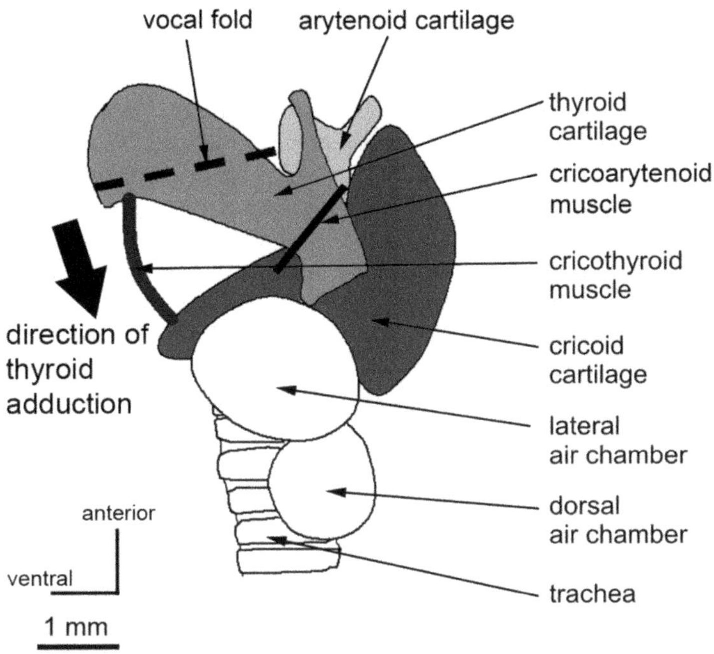

Fig. 3.2 Side view of a schematized bat larynx. Instead of using direct action of a vocalis muscle to change vocal fold tension, echolocating bats tilt the thyroid cartilage (see *arrow*) through action of the cricothyroid muscle in a visor-like fashion. This makes the vocal folds more taut, resulting in higher frequencies

(Carter and Adams 2014) that may reinforce the larynx and enable bats to tolerate the high tension of the vocal folds needed to produce the high-intensity, ultrasonic pulses. Ossification of laryngeal cartilages occurs to a greater degree in species in the Rhinolophoidea and Emballonuroidea, whereas some other bats that emit rather low-intensity sonar pulses, such as phyllostomids (Novick 1963), exhibit a lower degree of calcification (Carter and Adams 2014).

Another morphological adaptation for generating high-frequency calls is that the cricothyroid muscle, a very fast twitching muscle, is enlarged in most bats with laryngeal echolocation (Suthers 2004). As an echolocating bat closes in on its flying prey insect, it usually emits a "terminal buzz" by increasing its emission rate above 100 calls/s or, in some species, even 200 calls/s while reducing call duration below 1 ms (Schnitzler and Kalko 2001; Ratcliffe 2009). Only recently it was discovered that superfast muscles in the cricothyroid enable the bats to produce such high call rates (Elemans et al. 2011; Ratcliffe et al. 2013). Superfast muscles have previously been unknown in mammals. Histological studies of larynx muscles in bats revealed some modifications, including ultrastructural features that are characteristic for high-speed excitation-contraction coupling seen in other fast twitching muscles, such as increased sarcoplasmic reticulum (Revel 1962) and increased mitochondrial density (Reger 1978).

Yet another morphological specialization in many bat larynges (and in some primates) is the existence of vocal membranes or "vocal lips," which are membranous extensions of the vocal fold and positioned along its rims (Brown and Cannito 1995; Mergell et al. 1999). The vocal lips are only a few microns thick and contain no muscles or fibers. These vocal membranes are believed to allow for the generation of the high frequencies typical for bat echolocation signals, especially since echolocating bats lack the vocalis muscle that is responsible for controlling sound frequencies in many other mammals. It is unclear if laryngeal airflow directly induces vibrations of the vocal membranes, and/or if they are biomechanically coupled to the oscillations of the underlying vocal folds. In any case, this mechanism enables bats to emit high frequency calls at elevated sound pressure levels of more than 100 dB SPL with only moderately high subglottic pressures (Fattu and Suthers 1981), and some calls can reach above 140 dB SPL (Surlykke and Kalko 2008).

Finally, the trachea of the horseshoe bat exhibits a peculiar morphological specialization not found in other bats (review by Metzner and Schuller 2010). It is enlarged below the larynx by three tracheal pouches (two lateral and one medial) that are connected by small openings with the trachea. These tracheal enlargements were initially believed to possess resonant properties that aid in the creation of the unusually long-lasting, constant-frequency portion of horseshoe bat echolocation calls. It appears, however, that the tracheal chambers suppress the fundamental frequency component in the trachea and prevent it from being reflected from the lungs back toward the cochlea (Hartley and Suthers 1988; Suthers et al. 1988). This could prevent cochlear stimulation along multiple indirect pathways (via tissue conduction), which would severely smear clear reception of sound. As a result of this vocal tract filtering, the dominant component in horseshoe bat echolocation pulses is usually the second harmonic. Under certain circumstances, however, it appears that horseshoe bats can shift the dominant sound energy to the fundamental frequency (Neuweiler et al. 1987). This would enable them to increase the detection distance for targets, such as prey, because of the reduced attenuation for lower frequencies (Hiryu, Mora, and Riquimaroux, Chapter 9).

An important non-laryngeal adaptation that enables microchiropteran bats to produce the very intense echolocation pulses is found in the arrangement of muscular and tendinous elements of the abdominal wall, suggesting that they are analogous to the diaphragm (Lancaster et al. 1995). Most skeletal muscles deliver a linear force to a bone that is translated into rotation about a joint, and muscles are typically arranged in antagonistic pairs to effect opposite motions. The dome-shaped diaphragm differs in that its function does not require the direct movement of bones: contractions flatten the dome and thus result in a reduction of thoracic pressure. Since contractions of the abdominal wall of bats do not produce skeletal translation, then flank muscle action is both analogous with and antagonistic to the diaphragm and provides the pressure necessary to power intense biosonar vocalizations. The thin flank muscles observed in some phyllostomid bats may correlate to the faint biosonar vocalizations used by these bats (Lancaster et al. 1995).

3.2.2 Sound Production Mechanisms

Instead of using direct action of a vocalis muscle to change vocal fold tension, and thus call frequency (as in many other mammals), bats with laryngeal echolocation use an indirect mechanism to change the tension of the vocal folds. They tilt the thyroid cartilage downward toward the cricoid in a visor-like fashion by contracting the cricothyroid muscle (see Figure 3.2; the arrow indicates the direction of adduction of the thyroid toward the cricoid). This makes the vocal folds more taunt, resulting in higher frequencies. Conversely, relaxation of the cricothyroid results in lowering of call frequencies (review by Metzner and Schuller 2010; Elemans et al. 2011). The airflow through the larynx during expiration induces self-sustained oscillations of the vocal folds and vocal membranes, which in turn generate fluctuations in the ultrasonic pressure. The vocal fold tension determines the fundamental frequency of the oscillations and thus of the emitted calls.

Whereas in bats the contraction of the cricothyroid muscle determines the call's frequency, other intrinsic larynx muscles, such as the posterior cricoarytenoid and the lateral cricoarytenoid, control opening (abduction) and closing (adduction) of the glottis, respectively (Novick and Griffin 1961; Suthers and Fattu 1973, 1982). Precise timing of the inner laryngeal muscles determines the opening and closing of the glottis, yielding the beginning and end of the call as well as its frequency. This coordinated activity of the different muscle sets could then yield the various types of echolocation calls emitted by bats (see Figure 3.1a).

This simple model of a pulse-gating mechanism for sound production is illustrated in Figure 3.3 (Pye 1967; Metzner 2008). By varying the timing of glottal opening and/or closing relative to the cricothyroid contraction/relaxation cycle, bats would be able to produce all known types of echolocation calls. For example, in bats that produce FM calls, the vocal folds are tensed prior to emission of the sonar pulse by contracting the cricothyroid muscle. This contraction exerts a simultaneous transverse force, which closes the glottis. As the cricothyroid muscle relaxes, the glottis opens and simultaneously relaxes the vocal folds. This allows air to flow past the vocal folds and results in the emission of a downward frequency-modulated call (Carter and Adams 2014). Correspondingly, the FM-CF-FM call type that is characteristic for horseshoe bats (Figure 3.1a) is a result of glottal opening while the muscle is contracting and then remaining open during the contracted state until finally closing during muscle relaxation (Figure 3.3).

The intrinsic larynx muscles are innervated by two branches of the tenth cranial nerve (vagus nerve): the superior laryngeal nerve (SLN) innervates the cricothyroid muscle, whereas the inferior or recurrent laryngeal nerve (RLN) innervates the other intrinsic laryngeal muscles, particularly the thyroarytenoid and cricoarytenoid muscles (Schuller and Suga 1976; Schweizer et al. 1981). No communicating fibers were found between the SLN and the RLN in bats (Bowden and Scheuer 1961).

Although transection of the SLN has little effect on normal speech in humans, in bats it results in a marked reduction of pulse frequency and loss of frequency modulation, as well as reduced sound intensity (Novick and Griffin 1961; Suthers and Fattu 1982). Transection of the RLN has little if any effect on call frequency, but lowers mean peak intensity of the sonar pulses (Suthers and Fattu 1982).

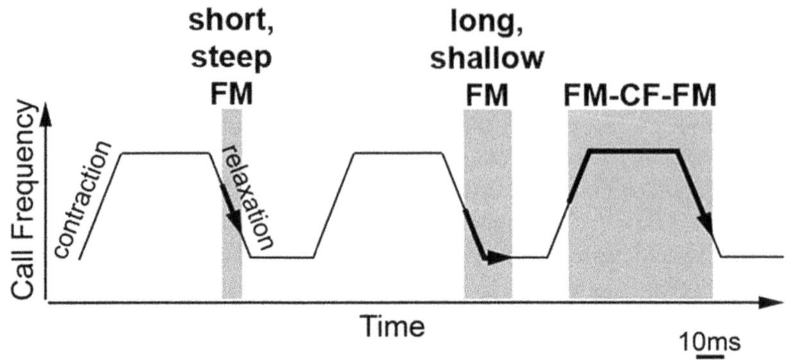

Fig. 3.3 Model of sound production by the larynx (modified after Pye 1980 and Metzner 2008). The timing and the duration of glottis openings and closings (*gray areas*) with contraction and relaxation of the cricothyroid muscle (indicated by the trapezoidal contraction and relaxation pattern) controls call frequency and creates calls with different spectrotemporal patterns (*thick arrows*). Compare with Fig. 3.1a for echolocation call types produced by bats

3.2.3 Non-Linear Phenomena in Sound Production: Echolocation Versus Communication Calls ("Yodeling")

Mammalian vocalizations exhibit large variations in their spectrotemporal features, although it is still largely unknown which features result from intrinsic biomechanical properties of the larynx and which features are under direct neuromuscular control. Echolocating bats produce a wide variety of different sonar pulses and also have a rich repertoire of communication calls (Figure 3.1). In horseshoe bats, for example, a long CF-component characterizes their echolocation pulses. When the bat is perched (at rest), it emits the CF-portion of its echolocation calls at its so called "resting frequency," RF_2, which normally represents the second harmonic of the calls (Schnitzler 1968). The pulses usually terminate with a brief and rapid drop in frequency, and they sometimes also contain an initial brief frequency rise, each extending 10–15 kHz below RF_2. During flight, horseshoe bats change their CF-frequency by up to 5 kHz below and 1–2 kHz above RF_2 (Schnitzler 1968; Metzner et al. 2002) to compensate for Doppler-shifts in the returning echo signal. This maintains the echo frequency virtually constant (Doppler-shift compensation, DSC) (Schnitzler 1968) and maintains echoes within the frequency range that the bats hear best (auditory fovea) (Schuller and Pollak 1979).

Communication signals of horseshoe bats, on the other hand, are spectrotemporally more complex. Their second harmonic frequencies are more than 15 kHz below the RF_2 of echolocation pulses (Ma et al. 2006; Kobayasi et al. 2012). Interestingly, neurophysiological recordings of the spike activity in various motor structures demonstrate that their activity virtually ceases at call frequencies only 4–6 kHz below RF_2 (Kobayasi et al. 2012). While this frequency range is sufficient to account for call frequencies produced during DSC, it poses a challenge for explaining how such extremely low discharge patterns can encode for the low frequencies and large spectral modulations seen in communication calls.

Recent work in an isolated larynx preparation from horseshoe bats, which lacked any neuronal innervation, addressed the role of non-linear and linear phenomena in generating this large range of frequencies in larynx phonations (Kobayasi et al. 2012). Indeed, mere changes in laryngeal air flow yield several non-linear effects on sound production, most notably, sudden jumps of up to 20 kHz (for the second harmonics) between the two frequency bands that are used for either echolocation or communication in the bats' natural vocalizations. These jumps resemble changes in "registers," as in yodeling. It is therefore tempting to speculate that a behavior similar to yodeling enables horseshoe bats to switch between echolocation and communication. Borrowing terminology used for the human voice, horseshoe bats would then emit communication sounds, for example, in "modal" or "chest voice" and echolocation pulses in "falsetto." However, it is important to note that in many cases echolocation signals may also have a communicative function (Gillam and Fenton, Chapter 5).

Other non-linear phenomena included rapid transitions between periodic ("tonal") and aperiodic ("chaotic") signals, which occurred within less than 200 μs or only 2–3 cycles of the tonal signal, as well as sinusoidal frequency modulations. These sound features are all prominent characteristics of communication calls in this species. The non-linear effects occurred in isolated larynx preparations indicating that direct neuromuscular control by larynx muscles was not necessary. Instead, the production of these sound attributes largely resulted from intrinsic biomechanical properties of the larynx.

On the other hand, simulated contractions of the main larynx muscle in bats, the cricothyroid, produced highly linear frequency changes covering less than 10 kHz (for the second harmonics). Thus, they were limited to the range of frequency modulations observed in either communication calls or during echolocation (DSC). These linear frequency changes, therefore, correspond to the effects normally produced by direct neuromuscular control of the cricothyroid contraction.

Hence, only by combining non-linear and linear properties can this larynx produce sounds covering the entire frequency range of natural calls. This may give behavioral meaning to yodeling-like vocal behavior and reshape our thinking about how the brain controls the multitude of spectral vocal features in bats and, perhaps, mammals in general (see below).

Non-linear phenomena in sound production, such as non-harmonically related call components or chaotic features in calls, are also known in other mammals, for example, in dogs, monkeys, and humans (Fitch et al. 2002), as well as in birds and frogs (Fee et al. 1998; Suthers et al. 2006). These phenomena yield such call features as subharmonics and biphonation. In cases where such non-linear vocal features have been investigated, they were most often related to coupled oscillations of vocal membranes adjacent to the vocal folds. Modeling approaches (Mergell et al. 1999) corroborated that vocal membranes can indeed account for non-linear laryngeal properties and cause many of the complex spectrotemporal features seen in a wide range of mammalian vocalizations. The presence of pronounced "vocal lips" in bats, however, represents a prime example for how purely biomechanical adaptations in the bat larynx can generate the particularly rich variation of call features observed in bat echolocation and communication signals (e.g., Kanwal et al. 1994; Ma et al. 2006).

Laryngeal vocalizations are transmitted through the supralaryngeal tract and are emitted either through the mouth, as in other mammals, or through the nostrils. Approximately one-third of all echolocating bats emit their calls through the nose, such as rhinolophids, hipposiderids, and megadermatids, and perhaps also the stenodermatines of the New World phyllostomids. Transmission through the mouth allows for spectral as well as directional shaping, which is quite variable and depends on the opening constellation of the mouth (Mogensen and Møhl 1979; Ghose et al. 2007). Call emission through the nostrils and its resulting filtering properties are spectrally and directionally less variable due to the rather fixed structure of the transmitting ducts. However, they dramatically alter the frequency composition of the emitted calls, and protrusions ("nose leafs") surrounding the nostrils in these bats act as acoustic beam-forming baffles that influence the shape of the emitted biosonar beam (see below; Zhuang and Müller 2006, 2007).

Resonating cavities of the supraglottic air passages are also involved in sound production by introducing harmonic components of the fundamental frequencies generated in the larynx. Especially in bats that emit echolocation calls through the nostrils, such as horseshoe and leaf-nosed bats, these resonance cavities can shift the dominant energy component away from the fundamental frequency component to higher integer multiples of it, such as the second harmonic in horseshoe bats.

3.3 Neural Control of Sound Production in Bats

3.3.1 Neural Control of the Bat Larynx

Non-linear spectral features of call emission are likely determined by the biomechanical properties of the larynx (see above). Laryngeal nerve-cut experiments (Suthers and Fattu 1982; Durant 1988) and the discovery of superfast properties of the cricothyroid (Elemans et al. 2011; Ratcliffe et al. 2013) indicate that the calls a bat emits, especially the temporal features, are under active neuromuscular control and thus are governed by neural control from the brain. Numerous neurophysiological and behavioral investigations revealed that the spectrotemporal fine control in longer calls (>10 ms) as well as call-by-call changes in the frequency, duration, and call rate are also under neuronal control. Such changes in call design, which are usually caused by changing echolocation demands, most likely are achieved by temporal delay through neuronal feedback loops rather than by the differences in the response time of the various internal larynx muscles (Schuller 1977; Metzner and Schuller 2010).

Intriguingly, each call appears to be individually controlled by the brain, even during the extremely fast call emissions that occur during terminal buzzes, which all aerial hawking bats produce during the final stages of their attack on prey and in which call rates can exceed 200 calls/s (Schnitzler and Kalko 2001; Ratcliffe et al. 2013). Consequently, muscle performance places an upper limit on the rate at which bats produce calls. Superfast cricothyroid muscles in bats allow, but also limit, this extraordinary laryngeal motor performance (Elemans et al. 2011; Ratcliffe et al. 2013).

Similar to other mammals (Berke and Long 2010), the brain stem exerts laryngeal motor control through the external branch of the superior laryngeal nerve (SLN) and via the inferior or recurrent laryngeal nerve (RLN). The SLN exclusively innervates the cricothyroid muscle, which controls the tension of the vocal fold and thus the fundamental frequency of the calls (Schuller and Suga 1976). The RLN provides the motor input to all other laryngeal muscles and is primarily responsible for the temporal fine structuring of the calls (Rübsamen and Schuller 1981).

The vocalization behavior of echolocating bats is embedded into a complex mesh with other motor acts such as active flight and highly directional movements of the head and pinnae (Ghose and Moss 2003, 2006; Moss and Surlykke 2010). Although the functional organization and anatomical interconnection of motor subsystems (such as those for locomotion or vocalization) with respiratory control are not unique for bats, they are nevertheless more apparent and pronounced and, therefore, more accessible to investigation than in other mammals (Smotherman et al. 2006; see below).

3.3.2 Brain Stem Circuits and Mechanisms

As in all mammals (Berke and Long 2010), the SLN and RLN originate from the nucleus ambiguus (NA), a rostro-caudally elongated structure in the medulla oblongata (e.g., *Rhinolophus*: Rübsamen and Schweizer 1986). Similar to other mammals (Kitamura et al. 1993), laryngeal motor neurons appear to be myotopically organized in bats (*Rhinolophus* spp.: Schweizer et al. 1981; *Pteronotus parnellii*: Kobler 1983).[1] Somata that give rise to the SLN are located more rostrally and those of the RLN are distributed over the central and more caudal portions of NA. This myotopic organization of the bat NA represents a functional topography of motor neurons in a rostral-to-caudal direction, controlling spectral and temporal parameters of echolocation calls, respectively (Rübsamen and Schuller 1981; Schuller and Rübsamen 1981).

The NA is part of a final common network of motor pathways in which coordination of all subsystems for vocalization is taken over by an area in the reticular formation that receives major inputs from the periaqueductal gray (PAG), but also from

[1] Note that the taxonomic identity of "*Pteronotus parnellii*" has recently been challenged by the recognition of several cryptic species (Clare et al. 2013).

other areas (Hage 2010b; Metzner and Schuller 2010). So far, however, no study has addressed the cytoarchitecture, anatomical connectivity, or function of a nucleus retroambiguus in bats.

The interaction between the PAG and other brainstem areas involved in vocal control is multifaceted in bats and does not follow a strictly hierarchical, almost exclusive, dominance of the PAG as the controlling input to the final common motor pathway, as proposed for other mammals (e.g., Jürgens 2002, 2009). Electrical or pharmacological stimulation in the PAG elicits vocalizations in several bat species (Suga et al. 1973; Fenzl and Schuller 2007), as it does in other mammals (e.g., Jürgens 2009; Gruber-Dujardin 2010). In bats, PAG stimulation yields both echolocation calls and communication calls, although at slightly different locations (Suga et al. 1973; Fenzl and Schuller 2002). In contrast, in several quite distinct areas of the brainstem below the level of the PAG, electrical and pharmacological stimulation yielded only echolocation calls (Schuller and Radtke-Schuller 1990; Schuller and Moss 2004). Since stimulation of these areas did not distort the calls, these regions most likely do not contain the central pattern generator that controls the distinct spectrotemporal structure of echolocation pulses (Hage 2010a). Stimulation of some other areas within the reticular formation and caudal brainstem of bats, however, did yield such distorted calls (Schuller and Radtke-Schuller 1990; Behrend and Schuller 2000). Similar to sites in the brainstem of squirrel monkeys (Hage and Jürgens 2006), stimulation of these brain areas in bats not only triggered call emission but also interfered with functional processes organizing the spectrotemporal pattern of the calls. This may be an indication that these sites within the reticular formation contain the pattern generator for those vocalizations (Hage 2010a).

The complex interaction between the bat PAG and other vocal motor structures within the brainstem is further illustrated by the fact that vocalizations caused by stimulation in the paralemniscal area and in the nucleus of the brachium of the inferior colliculus (the only regions tested so far) did not depend on a functioning PAG (Fenzl and Schuller 2002, 2005). The premotor vocal control of echolocation pulses and communication calls must therefore be differently organized. The descending vocalization pathway as described for mammals other than bats (Holstege 1989; Hage 2010a) does not adequately represent the condition for the control of bat echolocation calls. Instead, it appears that a network of various brainstem regions interacting in parallel with the proposed pathway from PAG to the ventral reticular formation (including NA and nucleus retroambiguus) portrays the condition in bats more appropriately.

Various neurophysiological and pharmacological studies indicate that other brainstem structures may also play a central role in mediating vocalizations in bats. Drug injections into the parabrachial area and the immediately adjacent tegmentum, for example, demonstrated that this area is essential for DSC in horseshoe bats (Smotherman et al. 2003).

In summary, the network that controls vocalizations in bats appears to be rather complex. The midbrain network can function independently from higher order structures of vocalization control, such as the cingulate cortex, and lesions at the level of the midbrain dramatically affect sound production in various mammals

(Jürgens 2009; Metzner and Schuller 2010). Earlier, often neglected, studies of *R. ferrumequinum* suggested that after bilateral ablation within the auditory midbrain, call frequencies emitted at rest and during DSC became less stable, DSC behavior was eliminated, or the DSC response was inverted. Instead of decreasing its vocalization frequency in response to increasing echo frequencies, the bat's vocalization frequency increased, on average 1 kHz above RF (Movchan 1984; Konstantinov et al. 1988). Similarly, DSC behavior in horseshoe bats could be reversibly suppressed by electrically stimulating an area ventral to the inferior colliculus (Schuller 1986), and DSC could be diminished by lesions of the nucleus of the central acoustic tract (Kobler et al. 1987; Behrend and Schuller 2000). In addition to relaying auditory information relevant for DSC to areas outside the classical auditory pathway, the central acoustic tract probably represents a sensory pathway that affects other behaviors as well (Behrend and Schuller 2000).

Further neurophysiological and pharmacological investigations will aid in elucidating the functional significance of the complicated interconnections of brain areas involved in triggering or controlling bat echolocation calls. This will clarify their integration within the descending vocal control pathway and also determine their role as recipients of auditory information from nuclei of the ascending auditory system.

3.3.3 Higher Order Brain Structures Involved in Vocal Control

3.3.3.1 Anterior Cingulate Cortex (ACC)

Although midbrain structures play a central and complex role in the generation and patterning of mammalian vocalizations, they are also under the influence of higher-order structures. Most notably, the anterior cingulate cortex (ACC) affects vocalizations in all mammals investigated and appears to control the "readiness to vocalize" (Gruber-Dujardin 2010; Hage 2010b). Electrical and chemical stimulation of the primate ACC very reliably yielded vocalizations. In the mustache bat, which performs DSC, the ACC exhibited a topographic representation of echolocation call frequencies (Gooler and O'Neill 1987). Chemical lesions of the bat's ACC with muscimol, however, did not abolish DSC and had no significant effect on call emission in general (Gaioni et al. 1988).

The ACC also played a major role in the echolocation behavior of echolocating bats when investigating the expression of FoxP2, a transcription factor implicated in the development and neural control of vocalization (Li et al. 2007), in different brain areas of bats using different echolocation systems (Sun et al., unpublished data). The study compared three bat species that have well-developed echolocation with two species of Old World fruit bats that lack laryngeal echolocation. The first group comprised greater horseshoe bats and great roundleaf bats (*Hipposideros armiger*), two species that emit CF pulses and exhibit DSC behavior, and *Myotis ricketti*, a species that produces short frequency-modulated sonar pulses. Species of

Old World fruit bats studied were Leschenault's rousette, *Rousettus leschenaulti*, which uses a sonar system based on tongue-clicks (see below), and the greater short-nosed fruit bat, *Cynopterus sphinx*, which lacks any echolocation. The ACC exhibited FoxP2 expression only in bats with laryngeal echolocation and not in Old World fruit bats. A recent study demonstrated knockdown of FoxP2 expression both in vitro and *in vivo* in *Hipposideros*, thus achieving effective silencing of FoxP2 (Chen et al. 2013). Such knockdown of FoxP2 expression in the ACC of *Hipposideros* had marked consequences for echolocation call control (Sun et al., unpublished data). Whereas it had no effect on the bat's resting frequency (i.e., the call frequency emitted when not performing DSC), it did significantly alter the DSC behavior by reducing the amount of frequency compensation and the call activity. This suggests that FoxP2 expression in the ACC of a bat with a highly sophisticated echolocation behavior, such as *Hipposideros*, is involved in controlling call parameters during more complex echolocation tasks (DSC), whereas it has no significant effect on the control of more basic vocalization parameters, such as the call frequency emitted when not performing DSC (i.e., RF).

3.3.3.2 Basal Ganglia

Several lines of evidence from studies of birdsong and human speech (e.g., Doupe and Kuhl 1999; Jarvis 2004) have suggested that a suite of forebrain nuclei, known collectively as the basal ganglia, may play an important role in vocal plasticity among vertebrates. The basal ganglia represent a feedback loop that provides the motor cortex with information needed to generate the proper motor commands that are ultimately sent to the phonatory motor neurons. The motor cortex, together with its feedback loops, is involved in the production of learned vocal patterns (Doupe and Kuhl 1999; Jarvis 2004). Therefore, the basal ganglia appear to have the potential to contribute to some general aspects of mammalian vocal plasticity. Nevertheless, these circuits are usually considered dependable, and they are ignored in most models of non-human mammalian vocal production (e.g., Jürgens 2002).

Recent work with Brazilian free-tailed bats (*Tadarida brasiliensis*) using c-fos immunohistochemistry revealed that key sites in the basal ganglia and thalamus are active during vocalizing (Schwartz and Smotherman 2011). This suggests a striato-thalamic feedback loop that may be involved in the control of echolocation pulse production. Additionally, the study presents evidence of a motivation pathway. These data provide novel indirect evidence that the basal ganglia are active during bat echolocation, but their precise functions remain unknown (Schwartz and Smotherman 2011). The basal ganglia are characterized by their dependence on dopamine levels. Pharmacological manipulations of brain dopamine levels in Mexican free-tailed bats corroborated the results obtained from the c-fos study (Tressler and Smotherman 2011). The study showed that basal ganglia circuits have the capacity to influence the spectrotemporal characteristics of echolocation pulses by shortening their duration, narrowing their bandwidth and reducing their amplitude. It also eliminated the bat's ability to make compensatory changes in pulse amplitude in response to background noise, i.e., the Lombard response (Tressler and Smotherman 2011; see below). Therefore, contrary to previous belief, the basal ganglia may represent an important mechanism for vocal control in bats.

3.4 Sensory Feedback for the Control of Echolocation Calls

3.4.1 Auditory Feedback

Hearing one's own voice is critical for the maintenance of stable vocalizations in humans and songbirds, even in adulthood after human speech and bird song have already been established (Doupe and Kuhl 1999; Boughman and Moss 2003). While in most other mammals, such as non-human primates, rodents, or cats, such auditory feedback appears to play a somewhat minor role (Janik and Slater 1997; Boughman and Moss 2003), auditory feedback is quintessential in echolocating bats. They continuously fine tune the spectrotemporal features of their sonar pulses in order to optimize the detection of the corresponding echo signals. The DSC behavior in bats whose echolocation calls are dominated by CF-portions, such as the horseshoe (*Rhinolophus* spp.) and the mustache bat (*Pteronotus parnellii*; but see Clare et al. 2013), exhibits this audio-vocal interaction particularly well. To optimize signal detection, they compensate for changes in the echo frequency and amplitude by constantly adjusting these parameters (and also temporal features) in their echolocation calls (Hiryu, Mora, and Riquimaroux, Chapter 9).

In addition to producing echolocation pulses, bats generate a large repertoire of social calls to communicate with one another (Gillam and Fenton, Chapter 5). At least some of these communication signals are also dependent on auditory feedback, as effects on vocal learning suggest (Boughman and Moss 2003). Not only do bats modify the fine structure of their vocalizations based on social experience, they can even acquire new vocalizations through vocal imitation (Knörnschild et al. 2010).

Echolocation requires very fast reactions involving adjustments within less than a few tens of milliseconds and, therefore, cannot tolerate long neural delays. To avoid lengthy feedback loops, audio-vocal feedback may take place at rather low levels in the brain. Several multimodal structures have been identified within the brain stem of bats and other mammals that may serve as candidates for audio-vocal integration. Many of these neurons are situated in areas that do not belong to the ascending auditory pathway but are areas that respond to acoustic stimuli (review by Metzner and Schuller 2010).

In bats, so far only neurons in the paralemniscal area and the pretectal area were tested simultaneously during active vocalization and acoustical stimulation. In the paralemniscal area, neuronal activity before and during vocalization, as well as acoustically induced activity, can occur in the same neuron (Metzner 1989, 1993). This indicates that audio-vocal interfacing can occur directly at the level of single neurons. Neurons that are active before and during self-produced vocalization and respond to acoustic stimulation are not restricted to bats, but also have been recorded in the brainstem of squirrel monkeys (Hage et al. 2006). It appears, however, that this area cannot be homologized with the more dorsally located paralemniscal area in the bat's brain (Hannig and Jürgens 2005).

In several mammals, neuroanatomical studies of connectivity in brainstem nuclei that are involved in vocalization have focused mostly on the final common vocal motor pathway (using stimulation or inactivation). Studies with the objective of determining the connectivity that mediates audio-vocal feedback in mammals are sparse and limited mostly to bats and monkeys (e.g., Metzner 1996; Hannig and Jürgens 2005).

The audio-vocal interconnections are complex. For example, neurophysiological and anatomical investigations suggested that the paralemniscal tegmentum may be involved in auditory feedback control of call frequencies (Metzner 1989, 1993, 1996). Yet electrolytic and pharmacological lesions did not support a direct involvement (Schuller et al. 1997; Pillat and Schuller 1998). Similarly, injections of excitatory or inhibitory drugs into the paralemniscal tegmentum yielded no changes in DSC or RF (Metzner, personal observation). Therefore, although there are reliable indications that this tegmental region is functionally associated with some aspects of vocalization and echolocation, its normal functioning does not appear to be critical to DSC or the control over RF.

Although behavioral evidence suggests that natural adjustments in call pitch typically occur in concert with a definable set of related changes in temporal calling patterns, there is no evidence that one is functionally dependent upon the other. Yet pharmacological work addressing the role of the parabrachial area in bat echolocation (Smotherman and Metzner 2005) indicates that, at least at the level of this brain area, some overlap exists between the neural circuits underlying the control of at least two autonomous parameters of vocalization in a mammal: call frequency and call temporal patterns. Subtle changes in the temporal pattern of call emissions did occur coincident with changes in the control of call pitch. Pharmacological manipulations of synaptic transmission in the parabrachial area caused significant changes in the way horseshoe bats adjusted their call durations and interpulse-intervals during DSC. This suggests that call duration and interpulse-intervals are coupled parameters within the broader neuronal network that controls echolocation, and that the control of call temporal patterns and call frequency shares some similarities in their mutual dependence on echo frequency and delay. Therefore, the parabrachial area may play a role in the integration of similar combinations of auditory cues that represent mutual triggers for the control of different vocal parameters during echolocation.

The amplitude of vocalizations is also significantly affected by auditory input. This becomes particularly evident in the Lombard effect, an involuntary rise in call amplitude in response to masking ambient noise. It represents one of the most efficient mechanisms to optimize signal-to-noise ratio, and it occurs in birds and mammals, including humans and bats. It is often associated with changes in the spectrotemporal composition of calls. Only a few recent investigations in horseshoe bats have tackled the questions about how the adaptive changes in call amplitude relate to associated vocal changes such as frequency shifts, how the underlying mechanisms are linked, and if auditory feedback from the changing vocal output is needed (Hage et al. 2013, 2014). These studies showed that the upward shifts in amplitude and CF-frequency of horseshoe bat echolocation calls were controlled independently, depending on which frequency band within the bat's hearing range was masked. Both amplitude and CF-frequency rises were extremely fast and occurred in the first call uttered after noise onset, suggesting that no auditory feedback from the bat's own voice was required to trigger these changes (Hage et al. 2013). Other spectrotemporal features of horseshoe bat calls were affected as well. For example, the FM-portions of the echolocation calls became larger when calls were masked with noise, and for some noise conditions, the bandwidth of the CF-portion also increased (Hage et al. 2014). These different effects that masking with noise had on CF- and FM-components of the bats' calls indicated that not only the Lombard effect itself but also its associated acoustic parameters are controlled by different neuronal mechanisms or circuits.

It appears that the proper production of both echolocation and communication calls depends on some form of auditory feedback (Boughman and Moss 2003; Knörnschild et al. 2010). Recently, an attempt was made to determine if and how auditory feedback control differs in echolocation and communication. Liu et al. (2013) analyzed echolocation and communication in greater horseshoe bats, whose echolocation pulses are dominated by a constant frequency component (Figure 3.1a) that matches the frequency range they hear best. To maintain echoes within this auditory fovea (Schuller and Pollak 1979), horseshoe bats constantly adjust their echolocation call frequency depending on the frequency of the returning echo signal (DSC behavior). DSC represents one of the most precise forms of sensory-motor feedback known (Grinnell 1989). Liu et al. (2013) examined the variability of echolocation pulses emitted at rest (resting frequencies, RFs) and one type of communication signal that resembles an echolocation pulse but is much shorter (short constant frequency communication calls, SCFs) and is produced only during social interactions. While RFs varied from day to day, corroborating earlier studies in other constant frequency bats (Huffman and Henson 1993a, b; Hiryu et al. 2006), SCF frequencies remained unchanged. In addition, RFs overlapped for some bats, whereas SCF frequencies were always distinctly different (Liu et al. 2013). This indicates that auditory feedback during echolocation changed with varying RFs but remained constant, or may have been absent, during emission of SCF calls for communication. This fundamentally different feedback mechanism for echolocation and communication may have enabled these bats to use SCF calls for individual recognition, whereas they adjusted RF calls to accommodate the daily shifts of their auditory fovea.

3.4.1.1 Vocal-Respiratory Coupling and Somatosensory Feedback

Mammalian vocalizations require the precise coordination of separate laryngeal and respiratory motor pathways. Somewhere within the mammalian midbrain, the descending vocal motor pathway is presumed to divide into separate laryngeal and respiratory components (Jürgens 2002). Precisely how and where in the brain the vocal motor pathways interact with the respiratory rhythm centers is unclear. However, it is known that the lateral, and especially the ventrolateral, region of the parabrachial area is centrally involved in the coordination of respiratory reflexes in mammals in general, and that it is also considered an integral component of the mammalian vocal motor pathway (e.g., reviews by Smotherman 2007; Smotherman et al. 2010). Thus, just as appears to be the case in the control of birdsong, several lines of evidence suggest that the mammalian parabrachial area may be a likely site for vocal-respiratory interactions.

Pharmacological manipulations of the synaptic activity in the parabrachial area in spontaneously vocalizing horseshoe bats caused a very specific change in vocal motor patterns that could be attributed to an equally specific deficiency in the respiratory control circuitry (Smotherman et al. 2006, 2010). Depending on the pharmacological agent applied to the parabrachial area, both calls and inter-call intervals became shorter, and the bats no longer emitted two calls per breath in response to frequency-shifted playback or the bats increased both call durations and inter-call intervals. These changes may be accounted for by a change in the underlying mechanisms coordinating breathing patterns with calling. During DSC, the expiratory phase must be lengthened to accommodate the additional calls associated with this behavior (Smotherman and Metzner 2005), and it follows that this is achieved by postponing the next inspiratory onset. Following drug injections, however, inspirations were either no longer postponed sufficiently to allow for more than one syllable per breath or the drugs extended expiratory durations, presumably by postponing inspiratory onset. Hence, vocal-respiratory rhythm entrainment may be achieved at least in part via mechanisms similar to other forms of locomotor-respiratory coupling, namely, the influence of somatosensory feedback on respiratory phase switching in the parabrachial area.

These experiments (Smotherman et al. 2006), of course, do not determine whether or not the respiratory phase switching observed here was being modulated by either descending vocal motor commands to the parabrachial area or by somatosensory feedback arising from the larynx. Indeed, both are likely to contribute under normal circumstances. For example, the respiratory component of the vocal motor pathway must also include activation of the lateral abdominal muscles to generate the high subglottic pressures needed to vocalize (Jürgens and Schriever 1991; Lancaster and Speakman 2001), and these likely derive from central vocal motor commands. However, the evidence from drug injections into the parabrachial area in horseshoe bats demonstrates that vocal-respiratory coupling can be disrupted by pharmacological manipulations of the parabrachial area, which has also been shown to be centrally involved in other forms of locomotor-respiratory coordination that rely upon somatosensory feedback control.

3.4.1.2 Vocalization and Flying

Bat echolocating is intertwined with numerous other motor actions, most notably active flight and highly directional movements of the head (Ghose and Moss 2003; Moss and Surlykke 2010) and pinnae (Schneider and Möhres 1960; Pye and Roberts 1970; Ghose and Moss 2003). The emission of echolocation pulses is locked to distinct phases of the respiratory cycle which itself is tightly linked temporally to wing stroke cycles during active flight (e.g., Suthers et al. 1972; Wong and Waters 2001). During search flight, wingbeat, respiration, and echolocation are synchronized in a 1:1 relationship. Given the high intensity levels of echolocation calls in most microchiropteran bats, the energetic demands of producing such intense sounds are high. For instance, the cost of echolocation in resting common pipistrelles, *Pipistrellus pipistrellus*, was 9.5 times the basal metabolic rate and comparable with the high cost of flight itself (Speakman et al. 1989). Surprisingly, however, the cost of echolocation during flight seems to be not much higher than the cost of flight alone (Speakman and Racey 1991). Echolocation during flight is energetically cheaper because the efficiently integrated locomotor–respiratory system facilitates echolocation as a by-product of flight. Echolocation during active flight adds little or no energy costs. Considering that laryngeal echolocation evolved in relatively few animal species, the low costs of echolocation for flying bats may have been a significant factor favoring its evolution in this group (Speakman and Racey 1991).

3.5 Static and Dynamic Diversity in Sound-Diffracting Structures

Sound production in the larynx and sound propagation in the vocal tract happen in confined internal spaces. As a consequence, the effects of these stages in the bats' biosonar system are limited to the time-frequency domain and cannot directly affect the encoding of spatial information. Biosonar is turned into a spatial sense only at the interfaces between the animal and its environment. On the emission side, this interface is provided by the mouth or, in the case of bats with nasal emission, the nose leaf. On the reception side, the interface is provided by the external ear (pinna).

Very much like the emitters and receivers of technical sonar and radar, the biosonar's periphery gives the system a direction-dependent response, which can be used to accomplish sensory tasks such as determining the direction of a sound source or imposing a weighting on signals impinging from different directions. The most common example for the latter would be the suppression of an unwanted signal from a certain direction.

Biosonar differs from technical sonar/radar in the way it produces the direction-dependent system responses. Engineered systems consist of arrays with a large number of transducer elements that are distributed over a large aperture area. Each individual element typically has a simple device characteristic that does not need to be direction dependent. The direction dependence of the overall system response is the result of combining the signals from the individual elements (Johnson and Dugeon 1993). Increasing the number of elements and the area of the array allows for a stronger dependence of the response on direction.

As flying mammals, bats are limited in size as well as in the number of transducers that they can employ for ultrasound emission and reception. Hence, the biosonar systems of bats had to find different ways to encode sensory information with a small number of transducers. Probably as a result of this, the external interfaces of the bat sonar system have a much higher complexity than those of technical systems. In bats, this complexity comes in two forms: static complexity and dynamic complexity. Static complexity refers to the geometry of the diffracting surfaces.

Whereas technical antennas and transducer baffles usually have simple geometries such as symmetrical horns, the diffracting baffles found in bats have more complicated overall geometries to which further complexities are often added in the form of local shape features such as ridges or cavities. Many of these features have dimensions that are not negligible when compared to the wavelengths used by the bats and, hence, could have an effect on the encoding of sensory information.

In some bat groups, such as rhinolophids and hipposiderids, the interfaces between the biosonar system and the external world have an additional temporal dimension of complexity. In these bat species, the surfaces that diffract the outgoing and incoming ultrasonic wave packets can be actively deformed by the animals. As is the case for static complexity, dynamic complexity can be found on the emission side as well as on the reception side.

3.5.1 Static Complexity

The pinnae of bats exhibit a large diversity not only involving size but also the orientation and overall shape of the pinnae as well as the presence of local shape features such as ridges and grooves. In bat species with nasal emission, the same is true for the nose leafs.

Some of the diversity seen in the nose leaf and pinna shapes could have a functional significance related to biosonar. Other possible explanations for the diversity would be social signaling, e.g., in the context of sexual selection, or random variations in traits that are not subject to selective pressures. The hypothesis that the ear or nose leaf shapes have a social signaling function is not readily reconciled with the nocturnal life style and comparatively poor visual performance in the respective bat groups (Eklöf 2003). The scarcity of sexual dimorphisms in the nose leafs as well as in the pinna shapes is an argument against sexual selection as a general driving force behind the diversity in these structures. Hence, if non-random patterns can be found in the variability of these shapes, it is highly likely that they are related to biosonar function.

Variability in pinna shape in bats (Ma and Müller 2011) as well as in their acoustic properties (Motamedi and Müller 2014) have been studied using principal component analysis (PCA; Jolliffe 2002). For the pinnae, this was accomplished using a geometrical transformation of the pinna surfaces into cylindrical coordinates. In spherical coordinates, the pinna surface can be described by a radius that is a function of height and direction. After discretization, the values of this two-dimensional function can be represented as a point in a high-dimensional vector space. PCA can then be used in this vector space to obtain a set of orthogonal axes that are ordered by the amount of the variability that is captured along each axis. The principal component axes can be viewed as *eigenears* when transformed back into the coordinate space of the original pinna geometries.

Applying this transformation approach to a set of 100 pinnae from at least 59 different bat species (Ma and Müller 2011) has yielded patterns that can be readily understood. When all pinnae are averaged in cylindrical coordinates, the resulting average pinna has the shape of an obliquely truncated horn. This geometry had been proposed previously as a general model for the vertebrate pinna (Fletcher and Thwaites 1988). The average pinna can be seen as the origin of the coordinate system about which the natural variability occurs along the axes defined by the eigenears. Like the average ear, the first eigenear was found to resemble a cone, hence adding (or subtracting) the surface radii of the first eigenear to the radii of the average ear will result in a widening (or narrowing) of the resulting pinna shape. This means that the largest single portion of the variability in the bat pinnae of the analyzed sample can be explained by different pinna opening angles, e.g., large opening angles versus small opening angles.

The second eigenear was divided laterally into two regions for which the radii had different signs. This means that when the second eigenear is added to the average ear, it results in a breaking of symmetry because the surface radii of the pinna are increased on one side and decreased on the other. The pinna shapes that result from movement along the axis defined by the second eigenear are hence asymmetric by either leaning to the right or leaning to the left. Like the second eigenear, the third eigenear is divided into two spatial regions defined by opposite signs of the radii, but in this case the division is between the base and the tip of the pinna. Hence, adding the third eigenear to the average ear results in pinnae that are either wide at the base and narrow at the tip, or vice versa. Taken together, the first three eigenears accounted for about half the variability in the sample, the first seven eigenears accounted for two-thirds of the variability, and three-quarters of the variability was explained by the first 12 eigenears (Figure 3.4).

The functional relevance of the pinnae shape to the biosonar system is described by the acoustic beam patterns. Beam patterns are complete (linear) characterizations of the behavior of a pinna in the acoustic far field. The beam pattern represents the device characteristics by scalar gain value as a function of direction and frequency. Since beam patterns are functions of three independent variables (two for direction, e.g., azimuth and elevation, and one for frequency) instead of two for the pinna surfaces, it is advantageous to compress them prior to analyzing the natural variability. This has been done using a "power spectrum" based on real spherical harmonics (Figure 3.5) (Motamedi and Müller 2014).

This method uses a set of real spherical harmonics that contains all degrees up to a fixed maximum, as well as all orders for each of the degrees included. The spherical harmonics are used as basis functions to decompose the beam pattern function, and the coefficients obtained in this decomposition for all orders of the same degree are summed to produce the "power spectrum." Using a power spectrum truncated at only 15 values, i.e., the lowest 15° of the spherical harmonics, it was possible to distinguish functional as well as phylogentic beam pattern groups. Specifically, it was possible to distinguish between emission and reception beam patterns, which indicates the existence of a functional asymmetry between these two interfaces of the biosonar system. Since the transfer function of the entire biosonar system is the product of the emission and reception transfer function, it is difficult to see a reason for this functional specialization from a purely linear-systems point of view (Figure 3.6).

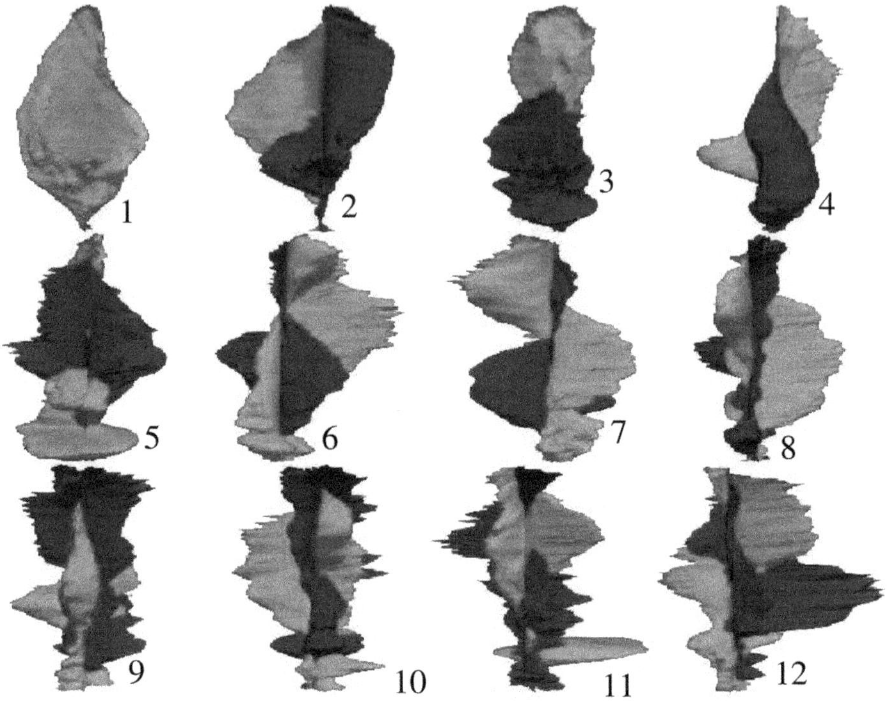

Fig. 3.4 Renderings of the eigenear shapes associated with the eight largest eigenvalues. Different shades of gray (*light* and *dark*) denote surface regions with opposing signs of the radius values

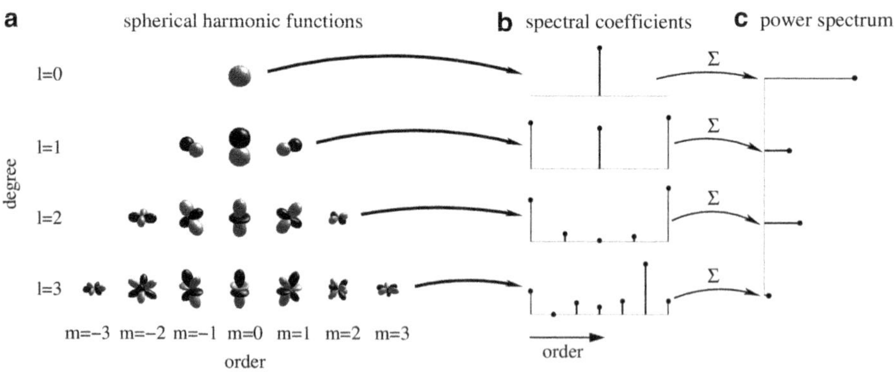

Fig. 3.5 Beam pattern representation based on real spherical harmonics. (**a**) Real spherical harmonics; (**b**) spectral coefficients obtained by using the harmonics as basis functions; (**c**) power spectrum (sum of coefficients across all orders belonging to a given degree)

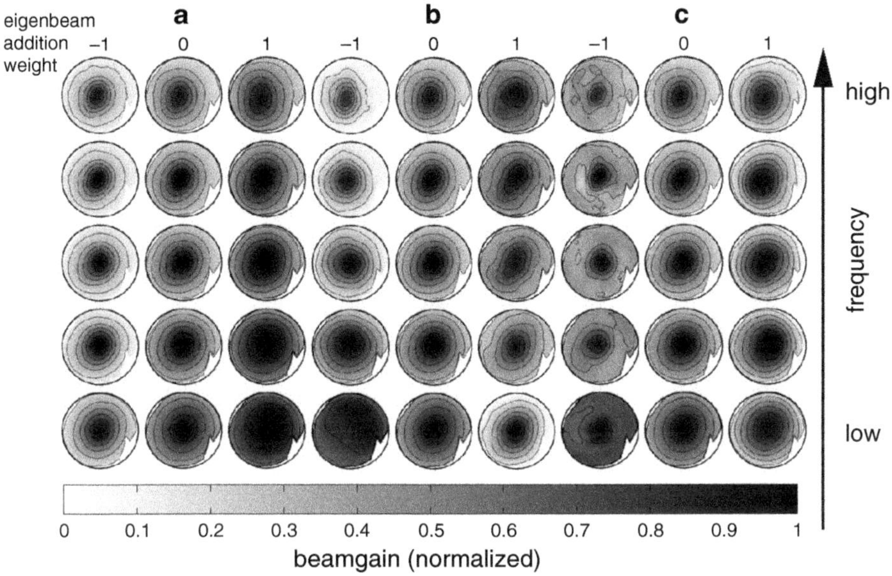

Fig. 3.6 Effect of the first three eigenbeams: (**a**) effect of subtracting/adding the first eigenbeam from/to the mean beam pattern; (**b**) effect of subtracting/adding the second eigenbeam from/to the mean beam pattern; and (**c**) effect of subtracting/ adding the third eigenbeam from/to the mean beam pattern

As in pinna shapes, a PCA of the full beam pattern functions has yielded trends that could be readily interpreted (Caspers and Müller 2015). The average beam pattern obtained for a data set of 267 beam patterns from 98 different species had a very simple structure with a single main lobe. This main lobe captured the basic physical relationship between aperture width and frequency: the higher the frequency, the narrower the beam. It was found that the first three principal components (*eigenbeams*) were all related to beam width but in different ways (Figure 3.6). The first eigenbeam had its largest gain values in a ring-shaped region, i.e., in the surround of the average beam pattern's maximum. Adding the first eigenbeam to the average beam pattern with a positive gain results in a beam widening that is symmetric in all directions and uniform across frequency. Adding the first eigenbeam with a negative sign inverts these changes and hence results in a uniform beam narrowing. The second eigenbeam differed from the first in that its effect depended on frequency. For each sign of the eigenbeam, the effect on the beam pattern at low frequencies was opposite to that at high frequencies. For example, if the sign of the second eigenbeam was chosen to widen the beams at lower frequencies, the beams at higher frequencies were narrowed. Finally, the third eigenbeam could be seen as the complement of the first in that its largest gain values occurred at the center of the average beam pattern.

It is noteworthy that the studies on the eigenears (Ma and Müller 2011) and eigenbeams (Caspers and Müller 2015) produced different results with respect to the portion of the variability that was explained by the first few eigen components. The captured portion of the variance was much faster to converge toward 100 % for the eigenears than for the eigenbeams. For example, capturing 90 % of the variance required only the first 32 of the eigenears but the first 73 of the eigenbeams. This difference could be the result of differences in the composition, size, and diversity of the analyzed sample (100 pinna shapes from 59 species versus 267 beam patterns from 98 species). Alternatively, it could reflect a fundamental difference in pinna or nose leaf shapes and the beam patterns that they produce. A hypothesis for such a fundamental difference could be that relatively small local shape features of the pinna can have a profound impact on the beam pattern. If this hypothesis is correct, it could be that because of their comparatively small size, the local features represent only a small portion of the variability in the shapes, whereas their acoustic effects are much more prominent in the beam patterns.

3.5.2 Dynamic Complexity

Research over the last few years has shown that, at least in some bat groups, the shapes of the nose leafs and pinna are not static but can be actively deformed by the animals to produce acoustic effects (Gao et al. 2011; Feng et al. 2012; Müller et al. 2012, 2013; Meymand et al. 2013; Pannala et al. 2013).

Rhinolophids and hipposiderids are closely related bat families with about 90 species in each (Simmons 2005) in which dynamic changes to the nose leaf and pinna shapes appear to be particularly prevalent. Species from both families have been reported to orient and hunt in confined natural spaces (e.g., Bell and Fenton 1984; Neuweiler et al. 1987; Bontadina et al. 2002). Hence, these bats must obtain all the sensory information they need from echoes that are often composed of contributions from a large number of scatterers. How the animals are able to accomplish this in general remains an open question. An exception is the target classification problem or, more precisely, the task of deciding whether an echo originates from a prey or any other target (e.g., reflecting facets in the vegetation). For this task, it has been shown that the bats can rely on non-linear signal transformations due to Doppler shifts that are specific to the fast moving wings (Emde and Schnitzler 1990; Kober and Schnitzler 1990). The ability to detect these nonlinear signatures already requires a dynamic control mechanism (Schnitzler 1968; Simmons, 1974; Metzner 1989), which could be an indication that dynamic principles are part of many components of the horseshoe bat biosonar system.

On the emission side, horseshoe bats have elaborate nose leafs that consist of three main parts: the anterior leaf, a cone-shaped baffle that surrounds the nostrils to the front and to the sides; the sella, a central peg situated behind/above the nostrils; and the lancet, located behind the sella. The anterior leaf has been shown to carry out non-rigid deformation in tight synchrony with the ultrasonic emission (Feng et al. 2012). In biosonar sequences where deformations of the anterior leaf occur, the walls of the leaf perform an inward twitching motion with each pulse. The onset of this motion coincides with the beginning of the pulse, and a fast return to the initial position occurs during or immediately after the end of the pulse. The magnitude of the change in nose leaf diameter was found to frequently exceed one-quarter of a wavelength (one-eighth of a wavelength on either side). Motions of the anterior leaf did not accompany all recorded biosonar sequences, and there was no evidence that the signals in sequences with anterior-leaf motion were different from those without, e.g., in terms of amplitude (Feng et al. 2012). Hence, it is unlikely that the anterior-leaf motions are a by-product of sound production, e.g., due to the passage of air through the nostrils. Instead, the motions appear to be an active behavior over which the animals have control and a behavior that they can switch on or off depending on the behavioral context.

In addition to the anterior leaf, horseshoe bats can also set the lancet in motion while they emit trains of biosonar pulses (He et al. 2015). Motions of the lancet take the form of rotations where the entire lancet is rotated in the direction toward the sella.

The reception side of the biosonar system in horseshoe bats appears to be no less dynamic than the emission side. Early anatomical work (Schneider and Möhres 1960) had shown that horseshoe bat pinnae have a unique muscular actuation apparatus that is characterized by a larger number of muscles than in other mammals. Furthermore, several of these pinna muscles have both of their insertion points on the pinna proper, and contracting them has to lead to a deformation of the pinna shape. Pinna deformations in greater horseshoe bats were observed to be fast, transitioning from one extreme shape configuration to the other in as little as 100 ms (Gao et al. 2011). The changes in pinna shape were characterized by a bending of the pinna tip, as well as a lateral rotation of the pinna walls. The pinna tip moved by about one-quarter of the total pinna height, which corresponded to about one wavelength (at the pulse frequency with the largest power). The changes in pinna shape were found to cause qualitative changes in the acoustic properties of the pinnae, i.e., in the beam patterns. For the shape configuration with an upright pinna tip, the beam patterns were dominated by a single main lobe that showed little dependence on frequency (over the range of frequencies occupied by the second, i.e., strongest, harmonic of the biosonar pulses). At the other extreme configuration (pinna tip fully bent), the beam pattern showed strong side lobes.

The characterization of a deforming, diffracting baffle is a function of four independent variables: a scalar gain as a function of direction, which is two independent variables, azimuth and elevation; time in the deformation cycle; and time lag in the device response. Beam patterns are frequency-domain descriptions that are obtained by virtue of an integration over time, in this case over the time-lag dimension of the four-dimensional device characterization. Since time integration tends to smooth out temporal changes, beam patterns are not the best way to capture temporal dynamics. In particular, beam patterns do not preserve differences in directional gain that may exist for different time lags as long as the overall frequency content of the responses is the same. It has been shown for a biomimetic pinna prototype (Meymand et al. 2013) that time-time domain device characterizations reveal much larger time dependences than were visible in the respective beam patterns. This could be taken as an indication that time-variant signal features are much more prevalent in the periphery of the bats' biosonar system, and they could play an important role in the encoding of sensory information.

In baffle structures with dynamic shapes, static shape features, such as grooves or ridges, could have effects that depend on the shape configuration of the overall baffle. Evidence for such interactions between static and dynamic shape features have been found in biomimetic microphone baffle shapes that had the geometry of an obliquely truncated conical horn augmented with simplified local shape features of the horseshoe bat pinna (Pannala et al. 2013). In particular, the local shape features that were investigated had very little impact on the acoustic device characteristics when the biomimetic pinna was in an upright configuration but resulted in strong beam pattern modifications for the bent configuration.

Finally, shape complexity on the emission side appears not to be limited to bat species with nasal emission. Several bat species that emit their biosonar pulses through the mouth have conspicuous lip shapes that can be augmented by local shape features such as prominent appendages. A good example for the latter is the ghost-faced (*Mormoops megalophylla*) and the Antillean ghost-faced bats (*Mormoops blainvillii*). At present, no acoustic analysis is available for any of these features, so it remains to be seen if the overall shape of the mouth and local shape features can have an impact.

3.6 Evidence for a Functional Role for Dynamic Complexity

It is well established across mammalian species, including bats and humans (Blauert 1997), that the static complexity of the pinna is an important feature for spatial hearing because it facilitates monaural direction finding through direction-dependent transfer functions. However, it is far less clear whether there is a functional role for the dynamic complexity of pinna motion and what this role could be (Gao and Müller 2014; Vanderelst et al. 2014). At present, there is no direct evidence to prove or disprove such a functional role. Experiments in which pinna motions were blocked by cutting the pinna innervation (Schneider and Möhres 1960; Mogdans et al. 1988) demonstrated a drop in biosonar performance, but since these surgeries eliminated changes in pinna shape along with the large overall rotation of the pinnae, the relative role of dynamic shape changes versus overall orientation of the pinna was not clear. However, there is circumstantial evidence from different sources that suggests a functional role, which is summarized below.

3.6.1 Dedicated Specializations

Ear muscles of horseshoe bats are specialized to produce non-rigid changes of the pinna geometry (Schneider and Möhres 1960). These specializations suggest an evolutionary advantage associated with non-rigid changes in pinna shape. It may be hypothesized that this advantage is not related to biosonar function, e.g., if it provided an efficient protection of the ears. However, this hypothesis is contradicted by the tight correlation between the deformations and biosonar operation (Schneider and Möhres 1960).

3.6.2 Prevalence Within the Biosonar System

Dynamic non-rigid shape changes are found on the emission as well as on the reception side of the biosonar system. If these effects were random features without functional relevance, they would have to have arisen two times independently and in two very different contexts.

3.6.3 Prevalence Across Species

Pinna and nose leaf motions are found in rhinolophids and hipposiderids. Since these two families are closely related (Simmons 2005), it is conceivable that the nose leaf and ear deformations are a common phylogenetic heritage that was preserved without an associated evolutionary advantage. However, pronounced differences in the shape of the nose leaf have evolved between these two families, and the same is true for the pinnae, although to a lesser extent. Despite those differences in baffle shape, the dynamic dimension of the shapes is present in both families. This could be taken as an indication that the dynamics have been preserved as an essential feature of these biosonar systems.

The most intriguing piece of evolutionary evidence for a functional role of the biosonar dynamics is the observation that mustached bats have pinna deformations that appear to be quite similar to those seen in horseshoe bats. Mustached bats and horseshoe bats share a biosonar system that utilizes narrowband signal components to detect the Doppler shifts caused by the moving wings of prey among the clutter echoes reflected from the environment. Since horseshoe and mustached bats are phylogenetically distant taxa, they are believed to have evolved their Doppler-sensitive biosonar systems independently. Hence, the presence of an entire suite of essentially identical adaptations for the detection of Doppler shifts in the biosonar system of these unrelated bat groups can be seen as a case of convergent evolution driven by a highly advantageous solution to a sonar-sensing problem. The presence of non-rigid ear deformations in both bat groups suggests that these dynamic features are an integral part of a sophisticated biosonar solution that has been "discovered" independently by these two bat groups.

3.7 Lingual Echolocation in *Rousettus*

Only one genus of Old World fruit bats, *Rousettus*, is known to have evolved a sonar system by using tongue clicking (Möhres 1953; Kulzer 1958). All Old World fruit bats have excellent night vision and a keen sense of smell, which they rely on for orientation and foraging (e.g., Neuweiler 1962; Raghuram et al. 2009). What distinguishes *Rousettus* from other Old World fruit bats is that members of this genus do not roost in the open (e.g., in trees), where they can rely on their eyesight, but instead, they roost in dark places, such as caves, where vision is insufficient for orientation. Bats of the genus *Rousettus* have evolved an active orientation system that enables them to find their way around in their dark roosts using tongue click-based (lingual) echolocation, though they still rely on olfaction as well as eyesight when the light levels are high enough (e.g., Möhres and Kulzer 1956; Raghuram et al. 2009). Labial echolocation has been described so far in several species of *Rousettus*, such as Egyptian fruit bats, *R. aegyptiacus* (von Herbert 1985; Holland and Waters 2005), Geoffroy's rousette, *R. amplexicaudatus* (Roberts 1975) and *R. leschenaulti* (Raghuram et al. 2007). Clicks produced by *Rousettus* species are very short, with durations far under a few 100 μs (Holland et al. 2004; Yovel et al. 2011a), which is much shorter than initial observations determined (up to several ms long) (Möhres and Kulzer 1956; von Herbert 1985). Their sound intensities reach up to 85–90 dB SPL (measured at 1 m distance, which is equivalent to >105 dB SPL at 10 cm distance; compared with >130 dB SPL in microchiropteran bats). The peak frequencies are 30–40 kHz, and their bandwidths cover around 60 kHz, expanding from approximately 10–70 kHz (Holland et al. 2004; Yovel et al. 2011a, b).

The most data are available for *R. aegyptiacus*; therefore, the following will focus on this species. As in microchiropteran bats, the frequency range of *Rousettus*' echolocation sounds matches the bat's hearing range (Grinnell and Hagiwara 1972), which should enable these bats to detect the full range of spectral information contained in the returning echo signal. Nevertheless, lingual echolocation has historically been considered to be rudimentary, providing only crude biosonar information and low performance. This was mainly for two reasons. First, clicks that have no spectrotemporal fine structure (no frequency modulation) and that were originally believed to be several ms long (Möhres and Kulzer 1956; von Herbert 1985) were considered inappropriate for "sophisticated" echolocation and would result in a much lower ranging accuracy of *Rousettus* echolocation compared to the frequency-modulated sonar signals of most microchiropterans (Simmons and Stein 1980). Second, it was believed that tongue click-based echolocation lacked the flexibility to change signal parameters, such as duration or frequency modulation.

Recent studies, however, have provided novel evidence that may overturn this traditional view of labial echolocation in *Rousettus* (Yovel et al. 2010, 2011a, b). While flying, *Rousettus* emits its tongue clicks usually in pairs with intervals of 20–30 ms between successive clicks (Kulzer 1958; Yovel et al. 2010). The intervals between double clicks are around 100 ms (Yovel et al. 2010). As pointed out above, more recent re-measurements of click durations revealed that the clicks are actually about one order of magnitude shorter than originally believed (Holland et al. 2004; Yovel et al. 2011a). In addition, the echolocation system in *Rousettus* proved to exhibit much more adaptive variability than previously assumed.

Egyptian fruit bats modulated the intensity of their biosonar emissions, as well as the spatial region they sampled, in a task-dependent manner (Yovel et al. 2010, 2011a). While producing double clicks, *Rousettus* pointed its sonar beam for each of the individual clicks in an alternating manner toward the left or right. In addition, they selectively changed the angle between the beam axes of sequentially emitted clicks, and the emission intensity depended on the distance to the target as well as the level of environmental complexity. In so doing, they effectively adjusted the spatial sector sampled by a pair of clicks, i.e., the "field of view" (Yovel et al. 2010). *Rousettus* also did not center the sonar beam on the target but instead pointed it off axis, accurately directing the maximum slope (edge) of the beam onto the target. Information-theoretic calculations showed that using the maximum slope is optimal for localizing the target, at the cost of detection. This trade-off between detection (optimized at stimulus peak) and localization (optimized at maximum slope) appears to be fundamental to spatial orientation and tracking (Yovel et al. 2011a). Finally, *Rousettus* also adjusts the interval between successive single or double clicks, as well as between the two clicks within a double click (Raghuram et al. 2007; Yovel et al. 2011b).

A combination of high-speed video analysis of the sound production, denervation studies of the hypoglossal nerve, local anesthesia of various muscles (including the tongue), and surgical manipulations of the mouth cavity (Kulzer 1958) revealed the basic mechanism of click production in *Rousettus*. Click production involves lifting of the tongue from the bottom of the mouth cavity followed by a short ejection of the tongue tip 1–2 mm through a small opening of the lips on each side of the mouth. The entire process takes 180–230 ms. The double clicks appear the be produced when the sides of the tongue are lifted up from the bottom of the mouth cavity; the roof of the mouth cavity is not involved. This early study did not allow determining which mechanism generates the two successive clicks that give rise to double clicks. Recent analysis of the behavioral significance of double-click emission (Yovel et al. 2010, 2011a), however, suggests that it may involve alternating activation of the left and the right sides of the tongue.

It had been suggested that the clicks had the characteristics of "shock-excited resonators" (Pye 1967), i.e., the spectral composition of the clicks would reflect resonance properties of the mouth cavity. Subsequent investigations measuring the spectra of tongue clicks produced by bats flying in a helium-oxygen mixture did indeed reveal that the concentrations of energy at certain frequencies in the broadband transient tongue clicks of *Rousettus* are brought about by buccal cavity resonances (Roberts 1975). Interestingly, the basic spectral composition of laryngeal echolocation calls produced by microchiropteran bats remained unaltered when produced in a helium-oxygen mixture (Roberts 1973), again highlighting the fascinating convergent evolution of laryngeal and labial echolocation.

References

Behrend, O., & Schuller, G. (2000). The central acoustic tract and audio-vocal coupling in the horseshoe bat, *Rhinolophus rouxi. European Journal of Neuroscience*, 12, 4268–4280.

Bell, G. P., & Fenton, M. B. (1984). The use of Doppler-shifted echoes as a flutter detection and clutter rejection system: The echolocation and feeding behavior of *Hipposideros ruber* (Chiroptera: Hipposideridae). *Behavioral Ecology and Sociobiology*, 15, 109–114.

Berke, G. S., & Long, J. L. (2010). Functions of the larynx and production of sounds. In S. M. Brudzynski (Ed.), *Handbook of mammalian vocalization* (pp. 419–426). Oxford: Academic Press.

Blauert, J. (1997). *Spatial hearing. The psycho physics of human sound localization*. Cambridge, MA: MIT Press.

Bontadina, F., Schofield, H., & Naef-Daenzer, B. (2002). Radio-tracking reveals that lesser horseshoe bats (*Rhinolophus hipposideros*) forage in woodland. *Journal of Zoology*, 258, 281–290.

Boughman, J. & Moss, C. (2003). Social sounds: Vocal learning and development of mammal and bird calls. In A. Simmons, A. N. Popper, & R. R. Fay (Eds.), *Acoustic communication* (pp. 138–213). Berlin: Springer.

Bowden, R. E. M., & Scheuer, J. L. (1961). Comparative studies of the nerve supply of the larynx in eutherian mammals. *Proceedings of the Zoological Society London*, 136, 325–330.

Brinkløv, S., Kalko, E.K.V., & Surlykke, A (2009). Intense echolocation calls from two 'whispering' bats, *Artibeus jamaicensis* and *Macrophyllum macrophyllum* (Phyllostomidae). *Journal of Experimental Biology*, 212, 11–20.

Brinkløv S., Jakobsen, L., & Ratcliffe, J.M. (2011). Echolocation call intensity and directionality in flying short-tailed fruit bats, *Carollia perspicillata* (Phyllostomidae). *Journal of the Acoustical Society of America,* 129, 427–435.

Brown, C. H., & Cannito, M. P. (1995). Modes of vocal variation in Sykes's monkey, *Cercopithecus albogularis'* squeals. *Journal of Comparative Psychology*, 109, 398–415.

Carter, R. T., & Adams, R. A. (2014). Ontogeny of the larynx and flight ability in Jamaican fruit bats (Phyllostomidae) with considerations for the evolution of echolocation. *Anatomical Record*, 297, 1270–1277.

Caspers, P., & Müller, R. (2015). Eigenbeam analysis of the diversity in bat biosonar beam patterns. *Journal of the Acoustical Society of America*, 137, 1081–1087.

Chen, Q., Zhu, T., Jones, G., Zhang, J., & Sun, Y. (2013). First knockdown gene expression in bat (*Hipposideros armiger*) brain mediated by lentivirus. *Molecular Biotechnology*, 54, 564–571.

Clare, E.L., Adams, A.M., Maya-Simoes, A.Z., Eger, J.L. Hebert, P.D.N., & Fenton, M.B. (2013). Diversification and reproductive isolation, cryptic species in the only New World high-duty cycle bat, *Pteronotus parnellii. BMC Evolutionary Biology*, 13, 26.

Denny, S.P. (1976). The bat larynx. In R. Hinchcliffe & D. F. Harrison (Eds.), *Scientific foundations of otolaryngology* (pp. 346–370). London: Heinemann Med Books Ltd.

Doupe, A. J., & Kuhl, P. K. (1999). Birdsong and human speech: Common themes and mechanisms. *Annual Reviews of Neuroscience*, 22, 567–631.

Durant, G. E. (1988). Laryngeal control of the duration and frequency of emitted sonar pulses in the echolocating bat, *Eptesicus fuscus*. PhD dissertation. Indiana University, Bloomington, IN.

Eklöf, J. (2003). *Vision in echolocating bats*. PhD dissertation. University of Gothenburg, Sweden.

Elemans, C. P. H., Mead, A. F., Jakobsen, L., & Ratcliffe, J. M. (2011). Superfast muscles set maximum call rate in echolocating bats. *Science*, 333, 1885–1888.

Emde, G. V. D., & Schnitzler, H.-U. (1990). Classification of insects by echolocating greater horseshoe bats. *Journal of Comparative Physiology A: Neuroethology, Sensory, Neural, and Behavioral Physiology*, 167, 423–430.

Fattu, J. M., & Suthers, R. A. (1981). Subglottic pressure and the control of phonation by the echolocating bat, *Eptesicus. Journal of Comparative Physiology A: Neuroethology, Sensory, Neural, and Behavioral Physiology*, 143, 465–475.

Fee, M. S., Shraiman, B., Pesaran B., & Mitra, P. P. (1998). The role of nonlinear dynamics of the syrinx in the vocalizations of a songbird. *Nature*, 395, 67–71.
Feng, L., Gao, L., Lu, H., & Müller, R. (2012). Noseleaf dynamics during pulse emission in horseshoe bats. *PLoS ONE*, 7, e34685.
Fenzl, T., & Schuller, G. (2002). Periaqueductal gray and the region of the paralemniscal area have different functions in the control of vocalization in the neotropical bat, *Phyllostomus discolor. European Journal of Neuroscience*, 16, 1974–1986.
Fenzl, T., & Schuller, G. (2005). Echolocation calls and communication calls are controlled differentially in the brainstem of the bat *Phyllostomus discolor. BMC Biology*, doi:10.1186/1741-7007-3-17.
Fenzl, T., & Schuller, G. (2007). Dissimilarities in the vocal control over communication and echolocation calls in bats. *Behavioral Brain Research*, 182, 173–179.
Fitch, W.T., Neubauer, J., & Herzel, H. (2002). Calls out of chaos: The adaptive significance of nonlinear phenomena in mammalian vocal production. *Animal Behavior*, 63: 407–418.
Fletcher, N. H., & Thwaites, S. (1988). Obliquely truncated simple horns, idealized models for vertebrate pinnae. *Acustica*, 65, 194–204.
Frey, R., & Gebler, A. (2010). Mechanisms and evolution of roaring-like vocalizations in mammals. In S. M. Brudzynski (Ed.), *Handbook of mammalian vocalization* (pp. 439–450). New York: Academic Press.
Gaioni, S. J., Suga, N., & Riquimaroux, H. (1988) Effects of bilateral ablation of the auditory cortex and/or cingulate cortex on the biosonar behavior of the mustached bat. *Abstract, Annual Meeting, Society for Neuroscience, Toronto, Canada.*
Gao, L., & Müller, R. (2014). Gao and Müller reply. *Physical Review Letters*, 112, 109401.
Gao, L., Balakrishnan, S., He, W., Yan, Z., & Müller, R. (2011). Ear deformations give bats a physical mechanism for fast adaptation of ultrasonic beam patterns. *Physical Review Letters*, 107, 214301.
Ghose, K., & Moss, C. F. (2003). The sonar beam pattern of a flying bat as it tracks tethered insects. *Journal of the Acoustical Society of America*, 114, 1120–1131.
Ghose, K., & Moss, C. F. (2006). Steering by hearing: A bat's acoustic gaze is linked to its flight motor output by a delayed, adaptive linear law. *Journal of Neuroscience*, 26, 1704–1710.
Ghose, K., Moss, C. F., & Horiuchi, T. K. (2007). Flying big brown bats emit a beam with two lobes in the vertical plane. *Journal of the Acoustical Society of America*, 122, 3717–3724.
Gooler, D. M., & O'Neill, W. E. (1987). Topographic representation of vocal frequency demonstrated by microstimulation of anterior cingulate cortex in the echolocating bat, *Pteronotus parnelli parnelli. Journal of Comparative Physiology A: Neuroethology, Sensory, Neural, and Behavioral Physiology*, 161, 283–294.
Grinnell, A. D. (1989). Sensory-motor control, listening to the voice within. *Nature*, 341, 488–489.
Grinnell, A. D., & Hagiwara, S. (1972). Studies of auditory neurophysiology in non-echolocating bats and adaptations for echolocation in one genus, *Rousettus. Zeitschrift für Vergleichende Physiologie*, 76, 82–96.
Gruber-Dujardin, E. (2010). Role of periaqueductal gray in expressing vocalization. In S. M. Brudzynski (Ed.), *Handbook of mammalian vocalization* (pp. 313–327). New York: Academic Press.
Gunnell G. F., & Simmons, N. B. (2005). Fossil evidence and the origin of bats. *Journal of Mammalian Evolution*, 12, 209–246.
Hage, S. R. (2010a). Localization of the central pattern generator for vocalization. In S. M. Brudzynski (Ed.), *Handbook of mammalian vocalization* (pp. 329–338). New York: Academic Press.
Hage, S. R. (2010b). Neuronal networks involved in the generation of vocalization. In S. M. Brudzynski (Ed.), *Handbook of mammalian vocalization* (pp. 339–349). New York: Academic Press.
Hage, S. R., & Jürgens, U. (2006). Localization of a vocal pattern generator in the pontine brainstem of the squirrel monkey. *European Journal of Neuroscience*, 23, 840–844.

Hage, S. R., Jürgens, U., & Ehret, G. (2006). Audio-vocal interaction in the pontine brainstem during self-initiated vocalization in the squirrel monkey. *European Journal of Neuroscience*, 23, 3297–3307.

Hage, S. R., Jiang, T., Berquist, S., Feng, J., & Metzner, W. (2013). Ambient noise induces independent shifts in call frequency and amplitude within the Lombard effect in echolocating bats. *Proceedings of the National Academy of Sciences of the USA*, 110, 4063–4068.

Hage, S. R., Jiang, T., Berquist, S., Feng, J., & Metzner, W. (2014). Ambient noise causes independent changes in distinct spectrotemporal features of echolocation calls in horseshoe bats. *Journl of Experimental Biology*, doi:10.1242/jeb.102855.

Hannig, S., & Jürgens, U. (2005). Projections of the ventrolateral pontine vocalization area in the squirrel monkey. *Experimental Brain Research*, 169, 92–105.

Hartley, D. J. (1992). Stabilization of perceived echo amplitudes in echolocating bats. II. The acoustic behavior of the big brown bat, *Eptesicus fuscus*, when tracking moving prey. *Journal of the Acoustical Society of America*, 91, 1133–1149.

Hartley, D. J., & Suthers, R. A. (1988). The acoustics of the vocal tract in the horseshoe bat, *Rhinolophus hildebrandti. Journal of the Acoustical Society of America*, 84, 1201–1213.

He, W., Gupta, A., Pedersen, S., Simmons, J. & Müller, R. (2015). Lancet dynamics in greater horseshoe bats, *Rhinolophus ferrumequinum. PLoS ONE*, 10(4), e0121700.

Hiryu, S., Kastsura, K., Nagato, T., Yamazaki, H., Lin, L. K., Watanabe, Y., & Riquimaroux, H. (2006). Intra-individual variation in the vocalized frequency of the Taiwanese leaf-nosed bat, *Hipposideros terasensis*, influenced by conspecific colony members. *Journal of Comparative Physiology A: Neuroethology, Sensory, Neural, and Behavioral Physiology*, 192, 807–815.

Holland, R. A., & Waters, D. A. (2005). Echolocation signals and pinnae movement in the fruitbat *Rousettus aegyptiacus. Acta Chiropterologica*, 7, 83–90.

Holland, R. A., Waters, D. A., & Rayner, J. M. V. (2004). Echolocation signal structure in the megachiropteran bat *Rousettus aegyptiacus. Journal of Experimental Biology*, 207, 4361–4369.

Holstege, G. (1989). Anatomical study of the final common pathway for vocalization in the cat. *Journal of Comparative Neurology*, 284, 242–252.

Huffman, R. F., & Henson, O. W., Jr. (1993a). Labile cochlear tuning in the mustached bat. I. Concomitant shifts in biosonar emission frequency. *Journal of Comparative Physiology A: Neuroethology, Sensory, Neural, and Behavioral Physiology*, 171, 725–734.

Huffman, R. F., & Henson, O. W., Jr. (1993b). Labile cochlear tuning in the mustached bat. II. Concomitant shifts in neural tuning. *Journal of Comparative Physiology A: Neuroethology, Sensory, Neural, and Behavioral Physiology*, 171, 735–748.

Janik, V. M., & Slater, P. J. B. (1997). Vocal learning in mammals. In P. J. B. Slater, J. S. Rosenblatt, C. T. Snowdon & M. Milinski (Eds.), *Advances in the Study of Behavior*. Vol. 26 (pp. 59–99). San Diego and London: Academic Press.

Jarvis, E. D. (2004). Learned birdsong and the neurobiology of human language. *Annals of the New York Academy of Sciences*, 1016, 749–777.

Johnson, D. H., & Dugeon, D. E. (1993). *Array signal processing, concepts and techniques*. New Jersey: Prentice Hall PTR.

Jolliffe, T. (2002). *Principal component analysis*. New York: Springer.

Jürgens, U. (2002). Neural pathways underlying vocal control. *Neuroscience & Biobehavioral Reviews*, 26, 235–258.

Jürgens, U. (2009). The neural control of vocalization in mammals: A review. *Journal of Voice*, 23, 1–10.

Jürgens, U., & Schriever, S. (1991). Respiratory muscle activity during vocalization in the squirrel monkey. *Folia Primatologica*, 56, 121–132.

Kanwal, J. S., & Ehret, G. (2006). *Behaviour and neurodynamics for auditory communication*. Cambridge, MA: Cambridge University Press.

Kanwal, J. S., Matsumura, S., Ohlemiller, K., & Suga, N. (1994). Analysis of acoustic elements and syntax in communication sounds emitted by mustached bats. *Journal of the Acoustical Society of America*, 96, 1229–1254.

Kitamura, S., Nagase, Y., Chen, K., & Shigenaga, Y. (1993). Nucleus ambiguus of the rabbit: Cytoarchitectural subdivision and myotopical and neurotopic representations. *Anatomical Records*, 237, 109–123.

Knörnschild, M., Nagy, M., Metz, M., Mayer, F., & von Helversen, O. (2010). Complex vocal imitation during ontogeny in a bat. *Biology Letters*, 6, 156–159.

Kobayasi, K., Hage, S., Berquist, S., Feng, J., Zhang, S., & Metzner, W. (2012). Behavioural and neurobiological implications of linear and non-linear features in larynx phonations of horseshoe bats. *Nature Communications*, 3, 1184

Kober, R., & Schnitzler, H.-U. (1990). Information in sonar echoes of fluttering insects available for echolocating bats. *Journal of the Acoustical Society of America*, 87, 882–896.

Kobler, J. B. (1983). The nucleus ambiguus of the bat, *Pteronotus parnellii*: Peripheral targets and central inputs. PhD dissertation. University of North Carolina, Chapel Hill, NC.

Kobler, J. B., Isbey, S. F., & Casseday, J. H. (1987). Auditory pathways to the frontal cortex of the mustache bat, *Pteronotus parnellii. Science*, 236, 824–826.

Konstantinov, A. I., Makarov, A. K, Movchan, E. V. Sokolov, B. V., & Goriniskii, I. A. (1988). *Sensory system for echolocation in horseshoe bats* (in Russian). Moscow: Russian Academy of Sciences.

Kulzer, E. (1958). Untersuchungen über die Biologie von Flughunden der Gattung Rousettus. *Zeitschrift für Morphologie und Ökologie der Tiere*, 47, 374–402.

Lancaster, W. C., & Speakman, J. R. (2001). Variations in respiratory muscle activity during echolocation when stationary in three species of bat (Microchiroptera, Vespertilionidae). *Journal of Experimental Biology*, 204, 4185–4197.

Lancaster, W. C., Henson, O. W., Jr., & Keating, A. W. (1995). Respiratory muscle activity in relation to vocalization in flying bats. *Journal of Experimental Biology*, 198, 175–191.

Li, G., Wang, J., Rossiter, S.J., Jones, G., & Zhang, S. (2007). Accelerated FoxP2 evolution in echolocating bats. *PLoS ONE*, 2, e900.

Liu, Y., Feng, J., & Metzner, W. (2013). Different auditory feedback control for echolocation and communication in horseshoe bats. *PLoS ONE*, 8(4), e62710.

Ma, J. & Müller, R. (2011). A method for characterizing the biodiversity in bat pinnae as a basis for engineering analysis. *Bioinspired Biomimetics*, 6, 026008.

Ma, J., Kobayasi, K., Zhang, S., & Metzner, W. (2006). Vocal communication in adult greater horseshoe bats, *Rhinolophus ferrumequinum. Journal of Comparative Physiology A: Neuroethology, Sensory, Neural, and Behavioral Physiology*, 192, 535–550.

Mergell, P., Fitch, W. T. & Herzel, H. (1999). Modeling the role of nonhuman vocal membranes in phonation. *Journal of the Acoustical Society of America,* 105, 2020–2028.

Metzner, W. (1989). A possible neuronal basis for Doppler-shift compensation in echo-locating horseshoe bats. *Nature*, 341, 529–532.

Metzner, W. (1993). An audio-vocal interface in echolocating horseshoe bats. *Journal of Neuroscience*, 13, 1899–1915.

Metzner, W. (1996). Anatomical basis for audio-vocal integration in echolocating horseshoe bats. *Journal of Comparative Neurology*, 368, 252–269.

Metzner, W. (2008). Bat bioacoustics. In D. Havelock, S. Kuwano, & M. Vorländer (Eds.), *Handbook of signal processing in acoustics* (pp. 1835–1849). New York: Springer.

Metzner, W., & Schuller, G. (2010). Vocal control in echolocating bats. In S. M. Brudzynski (Ed.), *Handbook of mammalian vocalization* (pp. 403–416). New York: Academic Press.

Metzner, W., Zhang, S. Y., & Smotherman, M. S. (2002). Doppler-shift compensation behavior in horseshoe bats revisited: Auditory feedback controls both a decrease and an increase in call frequency. *Journal of Experimental Biology*, 205, 1607–1616.

Meymand, S. Z., Pannala, M., & Müller, R. (2013). Characterization of the timevariant behavior of a biomimetic beamforming baffle. *Journal of the Acoustical Society of America*, 133, 1141–1150.

Möhres, F. P. (1953). Über die Ultraschallorientierung der Hufeisennasen (Chiroptera-Rhinolophinae). *Zeitschrift für Vergleichende Physiologie*, 34, 547–588.

Möhres, F. P. & Kulzer, E. (1956). Über die Orientierung der Flughunde (Chiroptera-Pteropodidae). *Zeitschrift für Vergleichende Physiologie*, 38, 1–29.
Mogdans, J., Ostwald, J. & Schnitzler, H.-U. (1988). The role of pinna movement for the localization of vertical and horizontal wire obstacles in the greater horseshoe bat, *Rhinolophus ferrumequinum. Journal of the Acoustical Society of America*, 84, 1676–1679.
Mogensen, F., & Møhl, B. (1979). Sound radiation patterns in the frequency domain of cries from a Vespertilionid bat. *Journal of Comparative Physiology A: Neuroethology, Sensory, Neural, and Behavioral Physiology*, 134, 165–171.
Moss, C. F., & Surlykke, A. (2010). Probing the natural scene by echolocation in bats. *Frontiers in Behavioral Neuroscience*, 4, 33.
Motamedi, M., & Müller, R. (2014). Characterization of the biodiversity in bat biosonar beam patterns with spherical harmonics power spectra. *Journal of the Acoustical Society of America*, 135, 3613. http://dx.doi.org/10.1121/1.4874595.
Movchan, E. V. (1984). Role of auditory centers in echolocation tracking of a moving target by greater horseshoe bats. *Neirofiziologiya*, 16, 737–745.
Müller, R., Pannala, M., Reddy, O. P. K., & Meymand, S. Z. (2012). Design of a dynamic sensor inspired by bat ears. *Smart Materials and Structures*, 21, 094025.
Müller, R., Feng, L., & Pannala, M. (2013). Bat biosonar as an inspiration for dynamic sensing. *Proceedings of the Society of Photo-optical Instrumentation Engineers*, 8686, doi:10.1117/12.2011274
Neuweiler, G. (1962). Bau und Leistung des Flughundauges (*Pteropus giganteus*). *Zeitschrift für Vergleichende Physiologie*, 46, 13–56.
Neuweiler, G. (2000). *The biology of bats*. New York: Oxford University Press.
Neuweiler, G., Metzner, W., Heilmann, U., Rübsamen, R., Eckrich, M., & Costa, H. H. (1987). Foraging behaviour and echolocation in the rufous horseshoe bats, *Rhinolophus rouxi*, of Sri Lanka. *Behavioral Ecology and Sociobiology*, 20, 53–67.
Novick, A. (1963). Orientation in neotropical bats. II. Phyllostomatidae and Desmodontidea. *Journal of Mammalogy*, 44, 44–56.
Novick, A., & Griffin, D. R. (1961). Laryngeal mechanisms in bats for the production of orientation sounds. *Journal of Experimental Zoology*, 148, 125–145.
Pannala, M., Meymand, S. Z., & Müller, R. (2013). Interplay of static and dynamic features in biomimetic smart ears. *Bioinspired Biomimetics*, 8, 026008. doi:10.1088/1748-3182/8/4/049501
Pillat, J., & Schuller, G. (1998). Audiovocal behavior of Doppler-shift compensation in the horseshoe bat survives bilateral lesion of the paralemniscal tegmental area. *Experimental Brain Research*, 119, 17–26.
Pye, J. D. (1967). Synthesizing the waveforms of bats' pulses. In R-G. Busnel (Ed.), *Animal sonar systems, biology and bionics*. Volume I (pp. 43–64). Jouy-en-Josas: Laboratory of Physiological Acoustics, Institut National de la Recherche Agronomique – Centre National de la Recherche Scientific.
Pye, J. D. (1980). Echolocation signals and echoes in air. In R.-G. Busnel & J. F. Fish (Eds.), *Animal sonar systems* (pp. 309–333). New York: Plenum Press.
Pye, J. D., & Roberts, L. H. (1970). Ear movements in a hipposiderid bat. *Nature*, 225, 285–286.
Raghuram, H., Gopukumar, N., & Sripathi, K. (2007). Presence of single as well double clicks in the echolocation signals of a fruitbat, *Rousettus leschenaulti. Folia Zoologica*, 56, 33–38.
Raghuram, H., Thangadurai, C., Gopukumar, N., Nathar, K., & Sripathi, K. (2009). The role of olfaction and vision in the foraging behavior of an echolocating megachiropteran fruit bat, *Rousettus leschenaulti. Mammalian Biology*, 74, 9–14.
Ratcliffe, J. M. (2009) Predator-prey interaction in an auditory world. In R. Dukas & J. M. Ratcliffe (Eds.), *Cognitive ecology II* (pp. 201–225). Chicago: University of Chicago Press.
Ratcliffe, J. M., Elemans, C. P. H., Jakobsen, L., & Surlykke, A. (2013). How the bat got its buzz. *Biology Letters*, 9, 20121031.
Reger, J. F. (1978). A comparative study on the fine structure of tongue and cricothyroid muscle of the bat, *Myotis grisescens*, as revealed by thin section and freeze-fracture techniques. *Journal of Ultrastructural Research*, 63, 275–286.

Revel, J. P. (1962). The sarcoplasmic reticulum of the bat cricothyroid muscle. *Journal of Cell Biology*, 12, 571–688.
Roberts, L. H. (1973). Cavity resonances in the production of orientation cries. *Periodicum Biologorum*, 75, 27–32.
Roberts, L. H. (1975). Confirmation of the echolocation pulse production mechanism of *Rousettus*. *Journal of Mammalogy*, 56, 218–220.
Rübsamen, R., & Schuller, G. (1981). Laryngeal nerve activity during pulse emission in the CF-FM bat, *Rhinolophus ferrumequinum*. *Journal of Comparative Physiology A: Neuroethology, Sensory, Neural, and Behavioral Physiology*, 143, 323–327.
Rübsamen. R., & Schweizer, H. (1986). Control of echolocation pulses by neurons of the nucleus ambiguus in the rufous horseshoe bat, *Rhinolophus rouxi*. II. Afferent and efferent connections of the motor nucleus of the laryngeal nerves. *Journal of Comparative Physiology A: Neuroethology, Sensory, Neural, and Behavioral Physiology*, 159, 689–699.
Schneider, H., & Möhres, F. P. (1960). Die Ohrbewegungen der Hufeisenfledermäuse (Chiroptera, Rhinolophidae) und der Mechanismus des Bildhörens. *Zeitschrift für Vergleichende Physiologie*, 44, 1–40.
Schnitzler, H.-U. (1968). Die Ultraschallortungslaute der Hufeisennasen-Fledermäuse (Chiroptera, Rhinolophidae) in verschiedenen Orientierungssituationen. *Zeitschrift für Vergleichende Physiologie*, 57, 376–408.
Schnitzler, H.-U., & Denzinger, A. (2011). Auditory fovea and Doppler shift compensation, adaptations for flutter detection in echolocating bats using CF-FM signals. *Journal of Comparative Physiology A: Neuroethology, Sensory, Neural, and Behavioral Physiology*, 197, 541–559.
Schnitzler, H.-U., & Kalko, E. K. V. (2001). Echolocation by insect-eating bats. *Bioscience*, 51, 557–569.
Schuller, G. (1977). Echo delay and overlap with emitted orientation sounds and Doppler-shift compensation in the bat, *Rhinolophus ferrumequinum*. *Journal of Comparative Physiology A: Neuroethology, Sensory, Neural, and Behavioral Physiology*, 4, 103–114.
Schuller, G. (1986). Influence of echolocation pulse rate on Doppler-shift compensation control system in the greater horseshoe bat. *Journal of Comparative Physiology A: Neuroethology, Sensory, Neural, and Behavioral Physiology*, 158, 239–246.
Schuller G., & Moss, C. F. (2004). Vocal control and acoustically guided behavior in bats. In J. A. Thomas, C. F. Moss, & M. Vater (Eds.), *Echolocation in bats and dolphins* (pp. 3 –17). Chicago and London: The University of Chicago Press.
Schuller, G., & Pollak, G. D. (1979). Disproportionate frequency representation in the inferior colliculus of Doppler-compensating greater horseshoe bats, *Rhinolophus ferrumequinum*. *Journal of Comparative Physiology A: Neuroethology, Sensory, Neural, and Behavioral Physiology*, 132, 47–54.
Schuller, G., & Rübsamen, R. (1981). Laryngeal nerve activity during pulse emission in the CF-FM bat, *Rhinolophus ferrumequinum*. *Journal of Comparative Physiology A: Neuroethology, Sensory, Neural, and Behavioral Physiology*, 143, 317–321.
Schuller, G., & Radtke-Schuller, S. (1990). Neural control of vocalization in bats: Mapping of brainstem areas with electrical microstimulation. *Experimental Brain Research*, 79, 192–206.
Schuller, G., & Suga, N. (1976). Laryngeal mechanisms for the emission of CF-FM sounds in the Doppler-shift compensating bat, *Rhinolophus ferrumequinum*. *Journal of Comparative Physiology A: Neuroethology, Sensory, Neural, and Behavioral Physiology*, 107, 253–262.
Schuller, G., Fischer, S., & Schweizer, H. (1997). Significance of the paralemniscal tegmental area for audio-motor control in the moustached bat, *Pteronotus p. parnellii*: The afferent off efferent connections of the paralemniscal area. *European Journal of Neuroscience*, 9, 342–355.
Schwartz, C., & Smotherman, M. S. (2011). Mapping vocalization-related immediate early gene expression in echolocating bats. *Behavioral Brain Research*, 224, 358–368.
Schweizer, H., Rübsamen, R., & Rühle, C. (1981). Localization of brain stem motoneurons innervating the laryngeal muscles in the rufous horseshoe bat, *Rhinolophus rouxi*. *Brain Research*, 230, 41–50.

Simmons, J. A. (1974). Response of the Doppler echolocation system of the bat, *Rhinolophus ferrumequinum*. *Journal of the Acoustical Society of America*, 56, 672–682.
Simmons, J. A., & Stein, R. A. (1980). Acoustic imaging in bat sonar, echolocation signals and the evolution of echolocation. *Journal of Comparative Physiology A: Neuroethology, Sensory, Neural, and Behavioral Physiology*, 135, 61–84.
Simmons, N. B. (2005). *Mammal species of the World: a taxonomic and geographic reference.* Vol. 1. 3rd Ed. Baltimore: Johns Hopkins University Press.
Smotherman, M. S. (2007). Sensory feedback control of mammalian vocalization. *Behavioral Brain Research*, 182, 315–326.
Smotherman, M., & Metzner, W. (2005). Auditory-feedback control of temporal call patterns in echolocating horseshoe bats. *Journal of Neurophysiology*, 93, 1295–1303.
Smotherman, M., Zhang, S., & Metzner, W. (2003). A neural basis for auditory feedback control of vocal pitch. *Journal of Neuroscience*, 23, 1464–1477.
Smotherman, M., Kobayasi, K., Ma, J., Zhang, S., & Metzner, W. (2006). A mechanism for vocal-respiratory coupling in the mammalian parabrachial nucleus. *Journal of Neuroscience*, 26, 4860–4869.
Smotherman, M. S., Schwartz, C., & Metzner, W. (2010). Vocal-respiratory interactions in the parabrachial nucleus. In S. M. Brudzynski (Ed.), *Handbook of mammalian vocalization* (pp. 383–392). New York: Academic Press.
Speakman, J. R., & Racey, P. A. (1991). No cost of echolocation for bats in flight. *Nature*, 350, 421–423.
Speakman, J. R., Anderson, M. E., & Racey, P. A..(1989). The energy cost of echolocation in pipistrelle bats (*Pipistrellus pipistrellus*). *Journal of Comparative Physiology A: Neuroethology, Sensory, Neural, and Behavioral Physiology*, 165, 679–685.
Suga, N., Schlegel, P., Shimozawa, T., & Simmons, J. (1973). Orientation sounds evoked from echolocating bats by electrical stimulation of the brain. *Journal of the Acoustical Society of America*, 54, 793–797.
Surlykke, A., & Kalko, E.K.V. (2008). Echolocating bats cry out loud to detect their prey. *PLoS ONE*, 3, e2036.
Suthers, R. A. (2004). Vocal mechanisms in birds and bats: A comparative view. *Annals of the Brazilian Academy of Sciences*, 76, 247–252.
Suthers, R. A., & Fattu, J. M. (1973). Mechanism of sound production in echolocating bats. *American Zoologist*, 13, 1215–1226.
Suthers, R. A., & Fattu, J.M. (1982). Selective laryngeal neurotomy and the control of phonation by the echolocating bat, *Eptesicus*. *Journal of Comparative Physiology A: Neuroethology, Sensory, Neural, and Behavioral Physiology*, 145, 529–537.
Suthers, R. A., Thomas, S. P., & Suthers, B. J. (1972). Respiration, wing-beat and ultrasonic pulse emission in an echolocating bat. *Journal of Experimental Biology*, 56, 37–48.
Suthers, R. A., Hartley, D. J., & Wenstrup, J. J. (1988). The acoustic role of tracheal chambers and nasal cavities in the production of sonar pulses by the horseshoe bat, *Rhinolophus hildebrandti*. *Journal of Comparative Physiology A: Neuroethology, Sensory, Neural, and Behavioral Physiology*, 162, 799–813.
Suthers, R. A., Narins, P. M., Lin, W. Y., Schnitzler, H.-U., Denzinger, A., Xu, C. H., & Feng, A. S. (2006). Voices of the dead: Complex nonlinear vocal signals from the larynx of an ultrasonic frog. *Journal of Experimental Biology*, 209, 4984–4993.
Thomas, J. A., Moss, C. F., & Vater, M., Eds. (2004). *Echolocation in bats and dolphins*. Chicago: University of Chicago Press.
Tressler, J., & Smotherman, M. S. (2011). Regulation of bat echolocation pulse acoustics by striatal dopamine. *Journal of Experimental Biology*, 214, 3238–3247.
Vanderelst, D., Reijniers, J., & Peremans, H. (2014). Comment on "Ear deformations give bats a physical mechanism for fast adaptation of ultrasonic beam patterns". *Physical Review Letters*, 112, 079401.
von Herbert, H. (1985). Echoortungsverhalten des Flughundes *Rousettus aegyptiacus*. *Zeitschrift für Säugetierkunde*, 50, 141–152.

Wong, J. G., & Waters, D. A. (2001). The synchronisation of signal emission with wingbeat during the approach phase in soprano pipistrelles (*Pipistrellus pygmaeus*). *Journal of Experimental Biology*, 204, 575–583.

Yovel, Y., Falk, B., Moss, C. F., & Ulanovsky, N. (2010). Optimal localization by pointing off axis. *Science*, 327, 701–704.

Yovel, Y., Geva-Sagiv, M., & Ulanovsky, N. (2011a). Click-based echolocation in bats, not so primitive after all. *Journal of Comparative Physiology A: Neuroethology, Sensory, Neural, and Behavioral Physiology*, 197, 515–530.

Yovel, Y., Falk, B., Moss, C. F., & Ulanovsky, N. (2011b). Active control of acoustic field-of-view in a biosonar system. *PLoS Biology*, 9, e1001150.

Zhuang, Q., & Müller, R. (2006). Noseleaf furrows in a horseshoe bat act as resonance cavities shaping the biosonar beam. *Physical Review Letters*, 97, 218701.

Zhuang, Q., & Müller, R. (2007). Numerical study of the effect of the noseleaf on biosonar beam-forming in a horseshoe bat. *Physical Reviews E*, 76, 051902.

Chapter 4
To Scream or to Listen? Prey Detection and Discrimination in Animal-Eating Bats

Patricia L. Jones, Rachel A. Page, and John M. Ratcliffe

Keywords Aerial hawking • Bat echolocation • Eavesdropping • Signal intensity • Substrate gleaning

4.1 Introduction

Around 1940, Donald Griffin (Griffin 1944) and Sven Dijkgraaf (Dijkgraaf 1943, 1946) independently demonstrated that some bats use biological sonar while in flight. Soon thereafter, Griffin coined the term "echolocation" to describe how bats use the returning echoes from their vocalizations to orient and to locate objects in their path (Griffin 1944, 1958). Their experiments revealed a previously unconfirmed way of sensing. As a result, the concept of an animal's internal representations of the outside world was changed forever, and echolocation soon became part of popular discourse. In an essay published in 1974, the philosopher Thomas Nagel made famous the question, "What is it like to be a bat?" (Nagel 1974). Nagel used the question rhetorically as a metaphor for the position that understanding another person's experience of reality and, thus another's mind, is reductionist and ultimately absurd. Specifically, he argued that if one cannot fathom how bats "see" the world through their ears, how can one hope to describe and quantify another person's subjective experience? A reader of this volume, however, will be given the

P.L. Jones
Department of Ecology and Evolutionary Biology, Cornell University,
Corson Hall, Ithaca, NY 14853, USA
e-mail: plj6@cornell.edu

R.A. Page
Smithsonian Tropical Research Institute, Balboa, Panama
e-mail: pager@si.edu

J.M. Ratcliffe (✉)
Department of Ecology and Evolutionary Biology, University of Toronto,
25 Willcocks Street, Toronto, ON, Canada M5S 3B2
e-mail: j.ratcliffe@utoronto.ca

M.B. Fenton et al. (eds.), *Bat Bioacoustics*, Springer Handbook of Auditory Research 54, DOI 10.1007/978-1-4939-3527-7_4

opportunity to appreciate the extraordinary progress researchers have made since 1974 in understanding what it might be like to be an echolocating bat.

Throughout this volume, the chapters describe the sensory ecology and neurobiology of bat echolocation, from signal design, reception, and processing, to the role of echolocation in niche divergence, and the secondary social uses of biosonar signals for communication. For example, Surlykke, Simmons, and Moss (Chapter 10) discuss the latest ideas about how a bat "sees" with its ears through a well-developed and experimentally supported concept of auditory scene analysis. In this chapter, the research discussed concentrates specifically on the production and processing of auditory information in predatory (i.e., animal-eating) echolocating bats in nature and under conditions meant to simulate nature, especially in those species that hunt airborne prey (hawkers) and take prey from surfaces (gleaners).

Once undetected, and for the most part undetectable by human ears, advances in high-frequency recording technology now make echolocation one of the most measurable and accessible of senses. Simply put, call design indicates which sound frequencies will be (and will not be) in the returning echoes, call rate approximates how often bats update their auditory scene, and call directionality indicates where bats are directing their acoustic gaze. Similarly, one can record and analyze the sounds generated by would-be prey and assess behaviorally which kinds of sounds bat most easily detect. Researchers can also identify the sounds that are more or less attractive to bats or sounds that can even deter hunting bats.

This chapter concentrates on how animal-eating bats, including those that consume exclusively arthropods, use echolocation and prey-generated sounds to detect, discriminate, and locate prey. Most of today's more than 1,000 echolocating bat species are animal-eating (Simmons 2005; Fenton and Ratcliffe 2010), taking their prey either from the air, from surfaces, or both (Schnitzler and Kalko 2001). A small number of echolocating bats, perhaps a dozen species from a number of families, take some of their prey from water surfaces. These trawling species are not discussed in this chapter, other than to say that they use prey detection strategies similar to those used by aerially hawking bats (Siemers et al. 2001). "To scream or to listen," a component of this chapter's title, refers to a relationship first proposed by Griffin (1958). Specifically, aerially hawking bats tend to use calls of much higher intensity than those of bats gleaning prey from terrestrial surfaces. While researchers now know that there are many bats that both glean and hawk prey, this relationship between call intensity and foraging strategy still holds (Hackett et al. 2014). The calls of gleaning bats are less intense because they can be. At shorter distances between a bat and its would-be prey, the echoes returning from even faint calls will be as loud as or louder at the bat's ears than those returning from flying insects at longer distances.

Aerial hawking bats tend to rely on echolocation for prey detection and for all, or most, of the information about prey position and identity. Most substrate gleaning bats, in contrast, are listening for prey-generated sounds to determine prey position (Bell 1985). These two foraging behaviors are often distinguished as active listening (echolocation) and passive listening (using prey-generated cues). Among extant bats, and throughout evolutionary time, locating prey in the air by echolocation and finding prey on substrates by listening for prey cues probably have been the two important strategies for prey localization (Simmons and Geisler 1998). Whether gleaning bats can use echolocation alone to detect perched prey has been an area of lively debate for some time (Arlettaz et al. 2001).

Aerial hawking and substrate gleaning require different echolocation and flight behaviors and, perhaps, differential reliance on spatial memory and learning (Hulgard and Ratcliffe 2014). Our chapter has three sections focused on these two foraging strategies. In the first we discuss the evolution of echolocation and the foraging behaviors likely used by ancestral bats to set the stage for our examination of foraging in extant animal-eating bats. The second section describes the behavior of aerial hawkers and their complex relationship with insect prey, especially between aerial hawking bats and eared moths that respond to bat echolocation calls. The third section examines substrate gleaning bats and their reliance on prey-generated cues, with particular emphasis on eavesdropping bats that listen in on prey mating calls. Throughout we also discuss the effects of each of these foraging strategies on prey and prey adaptations to thwart bat detection. Together, these sections emphasize the diverse array of behaviors used by foraging bats to identify and localize their animal prey, highlighting the progress that bat researchers have made in answering Nagel's once rhetorical question, "What is it like to be a bat?"

4.2 Evolution of Echolocation

One current hypothesis proposes that echolocation first evolved for orientation and obstacle avoidance as the ancestors of modern bats jumped, glided, or flew between trees while foraging (Fenton et al. 1995; Simmons and Geisler 1998). If correct, this early form of echolocation was presumably quite rudimentary and not functional for prey localization; therefore, these proto-bats most likely located prey by listening for prey-generated cues such as the rustling sounds of arthropod prey moving through leaf litter or for insects flapping their wings against tree bark as they warmed up their muscles for flight (Schnitzler et al. 2003). In this scenario, echolocation was later co-opted for use in prey localization, especially in bats that hunt airborne prey on the wing, an evolutionary trajectory that has likely resulted in the sophisticated echolocators of today (Schnitzler and Kalko 2001). Whether early echolocating bats used their larynxes (like most extant bats) or their tongues (like pteropodid fruit bats in the genus *Rousettus*) to produce echolocation signals is not known. While we cannot be certain how many times laryngeal echolocation has evolved in bats, we favor the hypothesis that laryngeal echolocation evolved once in proto-bats, concurrently with powered flight, and was subsequently lost in the Pteropodidae (Figure 4.1) (Jones and Teeling 2006). The other equally parsimonious hypothesis is that laryngeal echolocation evolved twice: once in the suborder Yangochiroptera, and then a second time in the suborder Yinpterochiroptera after the divergence of the pteropodid bats, a group that includes the flying foxes (Teeling 2009).

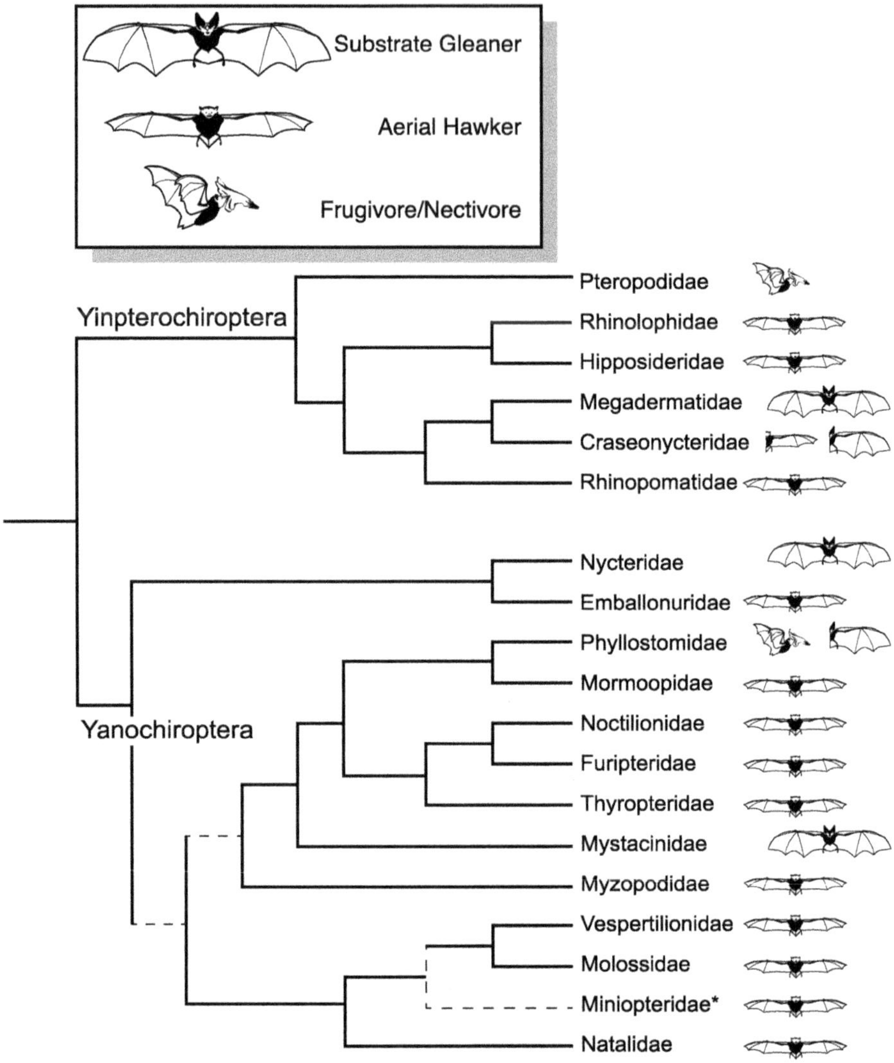

Fig. 4.1 Current phylogeny for bats (Jones and Teeling 2006). To the right of each family name, a Substrate Gleaner icon indicates that, in our opinion, this family is characterized by bat species that rely primarily on a gleaning strategy; all or most of which also take some prey by aerial hawking. An Aerial Hawker icon indicates that the family consists of species most of which primarily use a hawking strategy but includes behaviorally flexible species. Crasoenycteridae comprises a single behaviorally flexible species. A Frugivore/Nectivore icon indicates a family comprised solely or partially of frugivorous and nectivorous species

Each scenario, however, still suggests the same general story (Figure 4.1). That is, early laryngeal echolocating bats used powered flight, hunted animals, and took them in the air. The latter supposition is supported by the fact that most early fossil bats (~50 million years old) had wing designs suited for aerially hawking and not like those of modern gleaners (Simmons and Geisler 1998; Safi et al. 2005). Therefore, we suppose that while something akin to gleaning may have characterized proto-bats and the very earliest of bats, this trait may have been subsequently lost, at least as a primary means of prey capture, as bats evolved more sophisticated laryngeal echolocation and longer, narrower wings, and then gleaning evolved independently again multiple times (Simmons and Geisler 1998) (Figure 4.1). However, while this idea is supported by fossil evidence (Simmons and Geisler 1998) and phylogenetic trait reconstructions (Safi et al. 2005), it is not clear from a pure parsimony perspective (Figure 4.1), and deserves further attention.

4.3 Aerial Hawking

Echolocation has been well established as the primary means of detecting and locating prey in aerial hawking bats (Griffin 1958; Schnitzler and Kalko 2001). On an interception course with airborne prey, laryngeal echolocating bats from both infraorders (the Yangochiroptera and Yinpterochiroptera) emit calls and listen for echoes arriving at their ears to determine the size, shape, and texture of the would-be target, as well as to update information on the target's 3-D coordinates. The echolocation calls bats use for locating aerial prey can be as loud as 140 dB, measured 10 cm in front of the bat's mouth, and are typically greater than 120 dB, which is louder than a jet engine or a machine gun, and among the loudest biological signals to be described (Holderied and von Helversen 2003; Holderied et al. 2005; Surlykke and Kalko 2008). Griffin's early estimates put these values closer to 100 dB, likely because echolocation calls were assumed to be omnidirectional and measures of intensity using a single-directional microphone were therefore thought to be accurate. We now know that echolocation calls are highly directional, producing a narrow forwardly directed beam (Surlykke et al. 2009a; Jakobsen et al. 2013). As a result, a call must be measured along its acoustic axis to accurately estimate maximum intensity. The recent advent of multi-microphone arrays, made possible through multi-channel, high sampling rate analog-to-digital converters, have revealed how directional (Surlykke et al. 2009a) and loud bats' echolocation signals truly are and also allowed for the identification of the acoustic axis (i.e., the bat's acoustic gaze) (Ghose et al. 2006; Surlykke et al. 2009b). As a result of this 20–40 dB increase in our estimate of call intensity in aerial hawking bats, target detection distance estimates have dramatically increased, from 3–5 m to more than 20 m (Surlykke and Kalko 2008).

From target detection to capture, bats produce echolocation calls at emission rates starting out at 20 calls/s, and culminating at >150 calls/s just before contact (Griffin et al. 1960; Simmons et al. 1979). Griffin and colleagues (Griffin et al. 1960) were the first to describe the phases of vespertilionid attack sequences based on little brown bats, *Myotis lucifugus*. First in the sequence is the search phase, where the bat is often producing fewer than 20 calls/s. Second is the approach phase (between 20 and 100 calls/s), in which the bat has detected its prey and plots a course for its interception. The third and final phase is the "terminal buzz," the brief period of extremely high calling rate (>100 calls/s) just before contact with the prey

(Griffin et al. 1960; Simmons et al. 1979). The call rate that characterizes the buzz is thought to reflect the necessity for rapid auditory updates on the prey's position as the bat closes in on its moving target (Griffin et al. 1960).

Since the discovery of bat biosonar, it has been enigmatic how bats can produce buzz calls so quickly and, also, how the bats are then able to accurately process the returning echoes. It has now been shown that "superfast" muscles are responsible for call production during the buzz (Elemans et al. 2011). Superfast muscles can produce power at contraction/relaxation rates greater than 90 cycles/s and are rare in vertebrates (see Rome 2006 for review). Those found in bats are the first to be identified in mammals. How the bat's sonar receiver processes the echoes returning from a target during the buzz is also not well understood. However, echoes from each buzz call should have returned to the bat's ears before the bat produces its next buzz call (Kalko and Schnitzler 1989; Elemans et al. 2011). Researchers also have shown that some species of vespertilionids broaden their biosonar beam, and thus their acoustic field of view, as they close in on their target (Jakobsen and Surlykke 2010). This is largely accomplished by dropping the peak frequency of calls by roughly an octave during the buzz. By widening their field of view and rapidly updating their auditory scene during this terminal phase close to prey, the bats should reduce the chances that the target escapes off to the side of their sonar beam (Ratcliffe et al. 2013).

It is the approach phase, however, that it is characterized by different decision-making processing and, in fact, may be most interesting with respect to better understanding sensorimotor integration and guidance systems in flying animals. The approach phase begins when the bat has detected the target and assessed it as being worthy of further investigation. Kick and Simmons (1984) further classified the approach phase of Griffin et al. (1960) into approach and tracking, the latter term meaning that the bat is actively plotting a course for the target's interception based on estimates of the target's predicted flight path. Plotting a path to interception with an erratically moving target poses a particular challenge. In the laboratory, Ghose et al. (2006) explored this in greater detail and proposed that the big brown bat, *Eptesicus fuscus*, uses a constant absolute target direction strategy for prey tracking, which minimizes time to interception. In this strategy, the bat maintains the absolute direction to the target as a constant by shifting their flight direction and speed in response to changes in prey trajectory. Open space aerial hawking bat species combine long thin wings that enable fast flight (Norberg and Rayner 1987) with loud, directional echolocation calls that allow detection of small prey in open environments (Surlykke et al. 2009a). In many cases the loud echolocation calls that aerial hawking bats produce also provide important information for the hunted prey, as is discussed in the section below.

4.3.1 Echolocating Bats and Insects with Bat-Detecting Ears

Echolocation is an active spatial sensory system where the sender and intended receiver of the echolocation call is one and the same individual. Gillam and Fenton (Chapter 5) discuss the use of echolocation calls by other bats as a means of identifying bat species and, perhaps more often, for identifying potential profitable patches of prey through listening for other bats' feeding buzzes. Many insects, however, have also evolved ears (e.g., moths, mantids, lacewings) or co-opted preexisting ears (katydids and crickets) to detect the echolocation calls of bats (Miller and Surlykke 2001; Fullard et al. 2005). In response to a bat's echolocation calls, these insects typically exhibit evasive flight maneuvers, sometimes simply flying toward (or dropping to) the ground and, in other instances, initiating erratic flight to make themselves more difficult to track and more likely to escape the bat's acoustic field of view (Roeder 1967; Miller and Surlykke 2001). Within the moths alone, ears appear to have evolved multiple times (Hoy 1992; Fullard 1998; Yack et al. 1999). There is evidence that echolocating bats have been the selective force driving the evolution of ears in moths because different populations of moths have ears that are tuned to the frequencies of the echolocation calls of sympatric bats (Fullard 1998; ter Hofstede et al. 2013). The arms race between hearing moths and bats has long fascinated biologists, and this interaction is best explored in the sound-producing tiger moths (Arctiidae).

4.3.2 Case Study: Vespertilionid Bats and Sound-Producing Tiger Moths

The chemically defended tiger moths (Rothschild et al. 1970) are among those groups of moths with ears. Many tiger moths produce high-frequency sounds of their own in response to other high-frequency signals (e.g., bat echolocation calls) as well as in response to tactile stimulation (Blest et al. 1963; Fullard and Fenton 1977). These sounds are clicks: brief (less than 1 ms), broadband (often covering a frequency range from below 10 kHz to well above 100 kHz), and loud (approximately 70–90 dB in intensity) (Blest et al. 1963; Conner 1999). Sound-producing tiger moths produce these clicks (or more accurately, click modulation cycles) using a pair of modified metathoracic episternites, called tymbals. In most, but not all, sound-producing tiger moths, the tymbals buckle asynchronously. Depending on whether the tymbals are striated (i.e., scored by individual grooves called "microtymbals") or not, moths can produce as few as 20 clicks/s to more than 1,000 clicks/s (Barber and Conner 2007; Corcoran et al. 2009).

The adaptive function of tiger moth defensive sound production remains the source of some debate, with three competing hypotheses for the mechanism of bat deterrence. The first hypothesis suggests that the clicks are aposematic warning signals (Dunning 1968). There is strong evidence from naïve bats that the clicks function aposematically. Bats only avoid moths that make clicks if the moths are distasteful and then only after having experienced this negative pairing of cue and consequence (Hristov and Conner 2005). After being muted, noxious tiger moths are attacked by experienced bats (Ratcliffe and Fullard 2005). Moreover, within a given moth community, there are acoustic mimics in that some sound-producing tiger moths are palatable (Barber and Conner 2007; Barber et al. 2009).

The second hypothesis is that of acoustic startle, which suggests that clicks serve to startle approaching bats (Bates and Fenton 1990). Acoustic startle has received less support in recent years, mostly because bats quickly habituate to startle (Bates and Fenton 1990; Miller 1991), which would require clicking moths to be rare for startle to be effective, and this does not seem to be the case (Dunning et al. 1992).

The third major hypothesis suggests that the clicks act to jam bat echolocation (Fullard et al. 1979). Initially, this hypothesis lacked strong experimental support. However, it has now been confirmed that the incredibly high click rates of the tiger moth *Betholdia trigona* (a relatively palatable species) do jam bat biosonar and prevent the bat from accurately localizing the moth during the approach and terminal phases of attack (Corcoran et al. 2009). Field experiments have shown that bats capture ten times as many of these otherwise edible tiger moths if their sound-producing organs have been ablated than tiger moths with intact organs that are able to jam bat echolocation calls (Corcoran and Conner 2012). Clicking may save a moth's life, but it is energetically costly. The dogbane tiger moth, *Cycnia tenera*, preferentially produces clicks at bats when they are in the middle of their approach phase (Fullard et al. 2007), and they do so based on the activity of a single sensory neuron (Ratcliffe et al. 2009). Producing clicks only during the period of the bat's approach (in which the bat makes its final decision to pursue its would-be prey or to abort its attack) may save the moths energy (Ratcliffe and Fullard 2005; Ratcliffe et al. 2011).

Aerially hawking bats exhibit echolocation and pursuit behaviors that are assumed to be specialized to hunt flying prey in open air. Would-be prey species, in turn, have evolved a variety of strategies to avoid capture, including erratic escape maneuvers and sound production. These behaviors have made aerial hawking bats and their eared insect prey, especially moths, a textbook example of predator–prey interactions. The recent discoveries in moth sound production have highlighted the complexity of this interaction.

4.4 Substrate Gleaning

Based on the hypothesized scenario described at the beginning of this chapter, proto-bat ancestors likely were substrate gleaners that used echolocation for orientation and obstacle avoidance (Fenton et al. 1995; Simmons and Geisler 1998). These proto-bats probably relied on prey-generated cues, such as rustling sounds, to locate food. According to strong fossil evidence, however, the most recent laryngeal echolocating bat species from which all others are thought to have evolved would have been an aerial hawking species (Simmons and Geisler 1998; Safi et al. 2005). Therefore, many bat species and groups of species appear to have independently

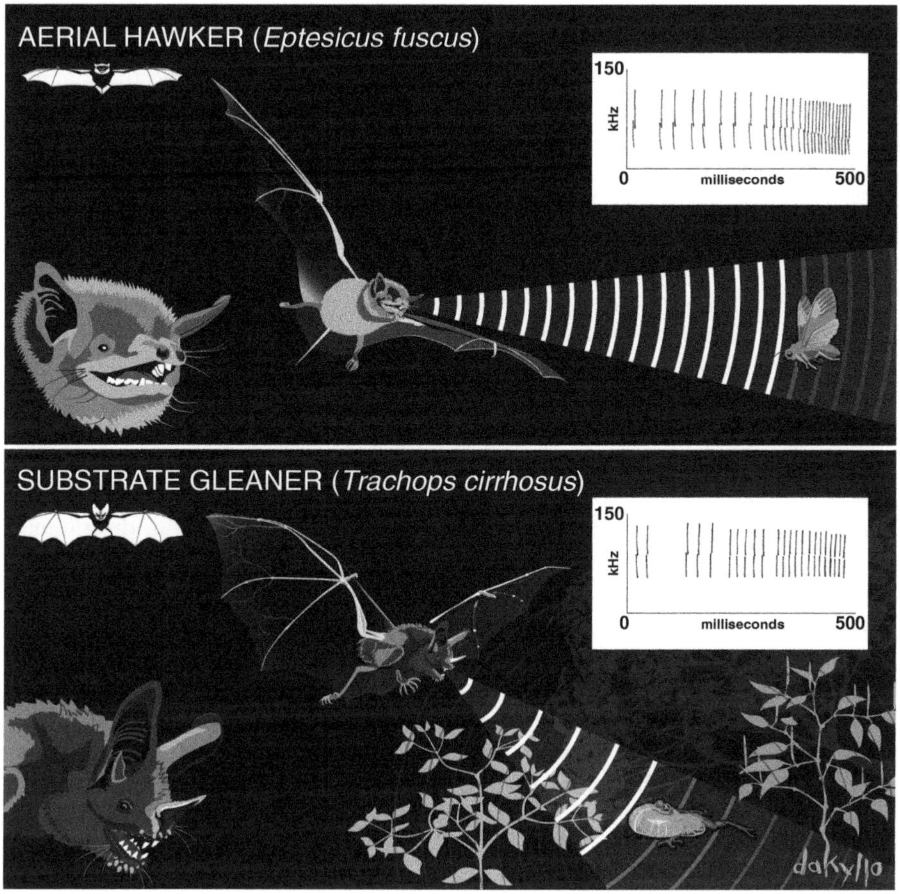

Fig. 4.2 Illustrations depicting two strategies for prey localization in bats. *Above*: Aerial hawking (capturing flying prey in the air) is exemplified here by the big brown bat, *Eptesicus fuscus*, capturing a moth on the wing. *Below*: Substrate gleaning (capturing prey on surfaces, e.g., on the forest floor or on vegetation) is exemplified here by the fringe-lipped bat, *Trachops cirrhosus*, capturing a frog calling in the forest understory. Aerial hawkers tend to rely on echolocation alone for prey detection (active listening); in contrast, substrate gleaners often eavesdrop on prey-emitted sounds to find their prey (passive listening), presumably using echolocation primarily for orientation. The divergent wing morphologies associated with these two hunting strategies are illustrated by the silhouettes in the upper left of each panel: Aerial hawkers tend to have long, thin wings for fast flight in open spaces (high wing loading), while substrate gleaners have short, broad wings, which make the bats slower but improve maneuverability for flight in the forest understory (low wing loading). In the upper right of each panel, spectrogram traces depict the echolocation calls emitted by each bat during an attack approach (time on the *x*-axis, frequency on the *y*-axis). As described in the text, the calls of hawking bats are typically louder than those of gleaning bats, which often use calls of shorter duration and broader bandwidth. Gleaning bats often, but not always, drop the terminal buzz from their attacks on substrate-borne prey (Drawings by D. Kyllo)

re-evolved the proto-bat behavior of substrate gleaning (Schnitzler et al. 2003; Figure 4.1). However, the use of specialized echolocation signals (short, high frequency, and broadband) coupled with highly maneuverable powered flight almost certainly sets today's gleaning bat species apart from the earliest progenitors of bats.

It is apparent that specialized gleaning bats differ from specialized aerial hawking species in both wing shape and echolocation behavior (Figure 4.2). Gleaning bats tend to have shorter and broader wings, which should confer greater maneuverability (Norberg and Rayner 1987). Gleaning bats also make much quieter echolocation calls (reviewed in Hackett et al. 2014), and their attack sequences often, but not always, lack the "terminal buzz" phase seen in aerial hawking species (Ratcliffe and Dawson 2003). The echolocation calls of gleaning bats also tend to be shorter in duration, with a broader bandwidth and a higher peak frequency (Schnitzler and Kalko 2001).

When taking prey from the ground or from vegetation, it had been supposed that instead of using echolocation, gleaning bats listen for prey-generated sounds to detect and localize prey (Schnitzler and Kalko 2001). Indeed, for most bat species that glean some or all of their prey, echolocation is thought to play at most a secondary role, used to orient in 3-D space and avoid obstacles, but not to discriminate their target from background vegetation. Arlettaz et al. (2001) used the phrase "acoustically blind" to describe the problem bats face in resolving prey on complex backgrounds like terrestrial vegetation.

To our minds, the subject is still open to debate for many gleaning species, particularly in light of more recent evidence demonstrating that most, perhaps all, gleaners do not stop echolocating as they close on prey. Instead, they use calls of lower intensity, missed by early recording equipment (Ratcliffe et al. 2005; Russo et al. 2007). Later in this section, a remarkable phyllostomid bat, the common big-eared (leaf-nosed) bat, *Micronycteris microtis*, is discussed. This species uses echolocation alone to detect and discriminate still and silent prey perched on vegetation (Geipel et al. 2013a). While previous work on gleaning bats from a number of families suggests that *M. microtis* may be the exception to the rule, further work is warranted.

4.4.1 Gleaning Bats Use Prey-Generated Cues

Among the prey cues used by gleaning bats are the incidental sounds of prey movement (Arlettaz 1996) and the advertisement signals prey produce to attract mates (Tuttle and Ryan 1981). Incidental sounds may include the rustling sound of prey moving through leaf litter or the sound of prey wing-beats against a hard surface. These incidental sounds are relatively generic to many types of prey, and, therefore, may be particularly useful as they encompass a wide variety of potential prey. Most rustling sounds, such as those created by beetles walking on leaf litter, are broadband, spanning a frequency range of 3–30 kHz with some energy up to 50 kHz and occasionally up to 100 kHz (Goerlitz et al. 2008). The broad range of frequencies encompassed by rustling sounds show substantial overlap with the hearing range of most bats, and their broadband components may make them particularly localizable (Bell 1982; Page and

Ryan 2008). Although generic to many types of prey, rustling sounds can also provide some information as they differ depending on the arthropod's size (Goerlitz and Siemers 2006) and the substrate type (Goerlitz et al. 2008), both of which could be useful for bats in determining which rustling sounds to approach and which to pass over.

Many bat species use prey-generated rustling sounds to locate prey (Bell 1982; Russo et al. 2007). One example is the greater mouse-eared bat, *Myotis myotis*, which forages by flying low over the ground and listening for prey-generated sounds. When *M. myotis* detects rustles, it briefly hovers and then lands on the ground with wings outstretched, often trapping the insect with its body and then picking it up in its mouth (Arlettaz 1996). This foraging strategy appears to be widespread for bats that glean insects from vegetation or the ground and requires specialized wing morphology that allows them to take off from the ground quickly to avoid terrestrial predators (Jones et al. 2003).

Most gleaning bats also flexibly employ aerial hawking strategies (Bell 1982; Ratcliffe et al. 2006). Many are predators on moths and other eared insects and use echolocation when these prey are aloft in open air but listen for the fluttering sounds of their warm-up wingbeats to locate them on surfaces (Ratcliffe and Dawson 2003). While we generally think of moths as prey to aerial hawkers and safe from predation when close to vegetation, this may not always be the case. For example, the western long-eared bat, *Myotis evotis*, can capture moths either by aerial hawking or by gleaning from the ground. When gleaning, *M. evotis* uses moth fluttering sounds to locate prey (Faure and Barclay 1994). This behavior also has been demonstrated in other myotids, the most species-rich genus of bats (Ratcliffe and Dawson 2003). At the extreme, the pallid bat, *Antrozous pallidus*, predominantly consumes arthropods but appears to attack them almost exclusively on the ground in response to prey-generated sounds (Bell 1982). Some predatory gleaning bats have expanded beyond arthropods to consume vertebrate prey. The African heart-nosed bat, *Cardioderma cor*, uses passive listening for prey rustling sounds to localize frog as well as insect prey (Ryan and Tuttle 1987). These and numerous other examples demonstrate that prey localization by passive listening for prey-emitted cues is widespread across gleaning bats from multiple families, as it is likely to be an excellent strategy for locating prey among clutter when echolocation is not as effective. As noted above, however, with the possible exception of *Macrotus californicus*, which has been reported to stop echolocating under bright moonlit conditions in the laboratory (Bell 1985), all gleaning species once thought to be silent have been demonstrated to produce echolocation calls throughout gleaning attacks (Schmidt et al. 2000; Russo et al. 2007), some species even while on the ground (Jones et al. 2003; Ratcliffe et al. 2005) but at lower intensities than when hawking.

Simultaneously listening to prey-generated sounds and echolocating raises some challenges. Barber et al. (2003) showed that for the gleaning vespertilionid pallid bat, *Antrozous pallidus*, attempting to process both forms of auditory information simultaneously may result in lower foraging performance overall. While this bat preferentially uses prey-generated sounds over echolocation to detect prey, its reliance on one modality or the other may reflect processing constraints limiting the ability to effectively integrate both at once. It has been proposed that the inferior colliculus is larger in gleaning species than aerial hawking species as a response to the challenge of processing both echolocation information and prey-generated sounds simultaneously (Baron et al. 1996; Reep and Bhatnagar 2000). However, a more recent comparative study did not find support for this hypothesis (Ratcliffe et al. 2006), suggesting, as do the results of Barber et al. (2003), that dual processing may be constrained by limited attention.

4.4.2 Gleaning Bats That Eavesdrop on Signaling Prey

Some bat species not only use generic rustling sounds to locate prey but also eavesdrop on species-specific mating calls of their prey. Eavesdropping is defined by Peake (2005) as "the use of information in signals by individuals other than the primary target" and differs from the use of inadvertent prey-generated cues, such as rustling sounds, in its reliance on signals rather than cues. Signals are defined as traits that evolved to communicate information (Karlson and Lüscher 1959). Cues are defined as incidental features present in the environment (Seeley 1995). Both signals and cues can be produced by a given prey species and can be used by predators. Signals, unlike cues, are specific to particular groups and even species and sexes of prey. This specificity may generate selection on the morphology and cognition of eavesdropping bats in order to successfully locate and identify calling prey.

Eavesdropping gleaning bats from a number of families prey on katydids (Orthoptera: Tettigoniidae) (Belwood and Morris 1987; ter Hofstede et al. 2008; Jones et al. 2011; Falk et al. 2015), which produce high-frequency and broadband calls (Korsunovskaya 2008). Some eavesdropping bats are reliant on katydid calls to locate them. The northern long-eared bat, *Myotis septentrionalis*, eavesdrops on the calls of the sword-bearing conehead katydid *Neoconocephalis ensiger* (ter Hofstede et al. 2008) and in captivity will attack speakers broadcasting this katydid's song. If the song is stopped before a bat lands on the speaker, the bat will abort its attack even if there is a katydid on the speaker. This demonstrates the dependence of *M. septentrionalis* on prey-emitted cues for localization and arguing against a role for echolocation in prey localization, at least in this species (ter Hofstede et al. 2008). As is the case with many eavesdropping predators, the response of *M. septentrionalis* is specific to the prey cue. For example, *M. septentrionalis* does not respond to the calls of the allopatric oceanic field cricket *Teleogryllus oceanicus* (ter Hofstede et al. 2008). Katydids are widespread throughout the world and likely are prey to other bat species whose foraging behavior has not yet been documented.

4.4.3 *Case Study: The Fringe-Lipped Bat*

Insects are not the only potential prey making loud, conspicuous signals. One of the most studied of the eavesdropping bats is the fringe-lipped bat, *Trachops cirrhosus*, that hunts frogs by approaching their calls (Tuttle and Ryan 1981). *T. cirrhosus* also eavesdrops on the calls of katydids (Falk et al. 2015) but preferentially approaches frog calls over katydid calls (Tuttle et al. 1985). During attacks on prey, *T. cirrhosus* echolocates throughout (Barclay et al. 1981). *T. cirrhosus* in the area surrounding the Panama Canal hunt the túngara frog, *Physalaemus pustulosus*, which calls from small puddles. Male túngara frogs can produce simple calls consisting of a ~350 ms frequency sweep ("whine"), and they can also make complex calls consisting of a whine followed by 1–7 broadband, approximately 40–80 ms, harmonic bursts ("chucks") (Ryan 1980). The foraging behavior of *T. cirrhosus* is worth discussing in depth for two reasons: it is the eavesdropping bat for which we have the best morphological, cognitive, and evolutionary explanations for response to prey cues, and it is a species that has been demonstrated to have remarkable learning capabilities.

Female túngara frogs (Ryan 1980) and *T. cirrhosus* (Ryan and Tuttle 1982) preferentially approach complex calls over simple calls. All male túngara frogs can produce complex calls, but complex call production is not correlated with male size (Ryan 1980). In female frogs, the preference for complex calls has apparently emerged from a perceptual bias for call complexity (Ryan et al. 1990; Phelps and Ryan 1998; but see Ron 2008). For *T. cirrhosus*, there are a number of potential factors that may produce this preference for call complexity. Complex calls contain more broadband frequency components than simple calls, which, in addition to their greater overall duration and the sharp onset and offset of the chucks, may make the calls easier signals to localize (Ryan 1985). Flight cage experiments confirm that complex calls are easier for bats to localize under certain conditions (Page and Ryan 2008). Another possibility is that bats prefer complex calls because frogs are more likely to make complex calls when there are other frogs calling nearby (Bernal et al. 2007); thus complex calls are indicative of higher prey densities. Bats may learn to associate complex calls with higher capture success because there are more available prey items when bats approach complex calls than when bats approach simple calls. *T. cirrhosus* have been shown to learn associations between acoustic stimuli and high prey rewards extremely quickly (Page and Ryan 2005; Jones et al. 2013), so it is likely that preference for call complexity has a learning component.

It is unclear whether, like female frogs, *T. cirrhosus* have a sensory bias for complex calls. In Ecuador the sister species of the túngara frog, Peter's dwarf frog (*Physalaemus petersi*), has populations that make complex calls and populations in which males only make simple calls (Boul and Ryan 2004). *T. cirrhosus* in Ecuador exhibit more passes over field playback speakers broadcasting complex calls than speakers broadcasting simple calls, even in populations where the local frogs do not make complex calls, suggesting a perceptual bias for complex calls (Trillo et al. 2012). In contrast, at the La Selva biological station in Costa Rica where túngara frogs are absent, captured *T. cirrhosus* individuals show no preferential response to complex versus simple calls of this allopatric frog in flight cage playback experiments (Jones et al. 2014).

Flight cage experiments in Panama also have tested for perceptual bias in bat preference for complex túngara calls over simple ones (Fugère et al. 2015). Captive bats were offered a choice between an unmodified simple túngara frog call versus a simple call modified to possess one of several acoustic properties differentiating complex calls from simple ones (e.g., longer duration, increased bandwidth, quick onset/offset amplitude modulation). With the exception of weak evidence suggesting that increased call duration could contribute to the bats' preference, there was no support for the perceptual bias hypothesis in *T. cirrhosus*. Rather, it is postulated that extreme flexibility and rapid learning abilities bypass any potential perceptual biases in shaping this bat's foraging behavior (Fugère et al. 2015). To further understand the mechanics underlying this bat's preference for call complexity, it may be particularly fruitful to examine the development of this preference in juvenile bats.

4.4.4 Auditory and Behavioral Adaptations to Eavesdropping

Hunting by eavesdropping may produce particular selective pressures on the hearing of predatory bats to improve detection and localization of prey calls. This has not been examined for eavesdropping bats that hunt insects, as insect calls generally have higher frequency components that overlap with the hearing range of most bats. *T. cirrhosus*, on the other hand, hunts frogs with <5 kHz calls (Tuttle and Ryan 1981). *T. cirrhosus* has auditory specializations for hearing low frequency sound, characterized by an unusual cochlear structure with peak neuron density in the apical portion of the cochlea, that part of the cochlea proposed to detect low frequency sound (Bruns and Burda 1989). It is unknown whether other eavesdropping bats have similar specializations for locating prey calls.

Eavesdropping on prey cues may also drive changes in behavior. As mentioned previously, *M. septentrionalis* eavesdrops on calls of the katydid *N. ensiger* but does not respond to the calls of the allopatric cricket species, *T. oceanicus* (ter Hofstede et al. 2008). Such specificity in eavesdropping also characterizes *T. cirrhosus*, which responds to the calls of palatable frog species but not to the calls of poisonous toads (Tuttle and Ryan 1981). Also, *T. cirrhosus* exhibits population variation in response to some prey calls depending on the availability of that species (Jones et al. 2014). Although bats respond to some prey calls and not to others, *T. cirrhosus* do generalize their responses to include similar-sounding novel calls (Ryan and Tuttle 1983). *T. cirrhosus* also can learn novel associations between prey cues and prey quality very quickly through both individual learning (Page and Ryan 2005) and social learning (Page and Ryan 2006; Jones et al. 2013). The learning capabilities of *T. cirrhosus* may be a solution to the challenge posed by eavesdropping as a foraging strategy, where each prey item is associated with a specific signal.

4.4.5 Sensory Niche Partitioning in Gleaning Bats

The two forms of passive listening for prey-generated cues (listening for the rustling sounds of prey moving across the substrate and eavesdropping on prey mating calls) can enable niche partitioning between closely related species of gleaning bats through different sensory ecologies. Bats that listen for prey-emitted rustling sounds do not necessarily respond to prey calls, as evidenced by *A. pallidus* and their lack of a response to orthopteran calls (Bell 1982) and *Cardioderma cor* (Megadermatidae) and their lack of an attack response to frog calls (Ryan and Tuttle 1987).

Sensory niche partitioning has been demonstrated for the European greater and lesser mouse-eared bats, *Myotis myotis* and *M. blythii oxygnathus*, which are morphologically very similar and roost together in the same caves. *M. myotis* predominantly eats carabid beetles, which it locates using the rustling sounds of beetles moving through leaf litter (Russo et al. 2007), and *M. b. oxygnathus* predominantly eats katydids, which it locates by eavesdropping on the katydid calling songs (Jones et al. 2011). Differences in the types of prey cues used by these two closely related species may therefore enable niche partitioning. Similarly, while *T. cirrhosus* preferentially approaches frog calls over katydid calls, a closely related species, the white-throated round-eared bat, *Lophostoma sylvicolum*, only approaches katydid calls and does not respond to frog calls (Tuttle et al. 1985). Further research compared two other closely related phyllostomid bats, *Tonatia saurophila* and *Micronycteris microtis*, and demonstrated that each of these four gleaning species differs in which acoustic features of katydid calls best predict their attacks (Falk et al. 2015). Such differences in foraging behavior potentially partition niches for these four often sympatric Neotropical phyllostomid bats.

Passive listening versus active prey localization through echolocation can also enable niche partitioning, as appears to be the case for the closely related *Myotis bechstennii* and *M. nattereri*. *M. bechstennii* relies more on prey-generated cues while *M. nattereri* relies more on echolocation to locate prey (Siemers and Swift 2006). These studies suggest that the means by which bats locate prey can have important consequences for species divergence. Interestingly, these and other European vespertilionid bats also exhibit signs of sensory niche partitioning with respect to aerial hawking, with those species able to produce calls containing very high frequencies best able to exploit airborne prey close to background vegetation (Siemers and Schnitzler 2004). Dietary evidence for these different niches is somewhat lacking, however, and DNA barcode analysis is required to better document who eats what and even when (Clare et al. 2009). Niche partitioning between sympatric bat species is discussed further in Chapter 6 (Denzinger, Kalko, Tschapka, Grinnell, and Schnitzler).

4.4.6 Challenges in Relying on the Use of Prey-Emitted Acoustic Cues

For prey hunted by eavesdropping predators, the best recourse to avoid being eaten is to stop calling. Many katydid species stop calling when they hear bat echolocation calls (ter Hofstede et al. 2008, 2010), and túngara frogs stop calling when they see a bat fly overhead (Tuttle et al. 1982). As the bat approaches the prey, therefore, the prey-generated cue stops. Bats that hunt prey by listening for their rustling sounds may face a similar challenge if the rustling prey can hear bat echolocation calls and cease moving. In order to detect the motionless and silent prey, these gleaning bats may require echolocation. It is possible that the use of prey-generated cues gives bats general information on the location of the prey, and they then use echolocation in the final stage of attack. This has been demonstrated for *T. cirrhosus*, as they continue to use echolocation as they approach calling prey and update their information about prey size if the prey that is available is manipulated (Page et al. 2012).

T. cirrhosus can obtain echoacoustic information not only from the prey per se but from perturbations made by prey in the surrounding environment. When túngara frogs call, the repeated inflation and deflation of the vocal sac produces ripples on the water surface. *T. cirrhosus* uses echolocation to key in on these ripples and prefers calls coupled with ripples to calls with no ripples present (Halfwerk et al. 2014). Ripples are a particularly salient cue because even if a calling frog detects an approaching bat and goes silent, the frog cannot immediately stop the trail of ripples that continue for another several seconds, thus leaving an unavoidable "footprint" of the frog's presence. The continual use of echolocation while approaching prey-generated cues has been demonstrated for a number of bat species (Schmidt et al. 2000; Russo et al. 2007) and indicates the importance of echolocation even for gleaning bats that do not require echolocation to locate prey.

Bats that locate prey using prey-generated cues may be particularly susceptible to interference from background noise in prey detection. *Myotis myotis* avoids areas with anthropogenic traffic noise, presumably because of the interference between such background noise and prey-generated rustling sounds (Schaub et al. 2008). Similarly, *T. cirrhosus* does not approach the calls of the pug-nosed treefrog, *Smilisca sila*, when they are masked by the water sounds that are common at the bubbling stream sites where *S. sila* chooses to call (Tuttle and Ryan 1982). Although aerial foraging bats that rely on echolocation to locate prey also may be negatively impacted by background noise, they are likely less impacted than passive-listening bats due to the high frequency of echolocation calls.

4.4.7 *Case Study: The Common Big-Eared Bat Defies Categories*

For decades it was thought that bats hunting in highly cluttered conditions, such as the dense forest understory, could not use echolocation alone to glean silent, motionless prey. Foraging in such conditions is a very difficult sensory task. Weaker echoes from prey overlap with stronger echoes from the surrounding vegetation, resulting in backward masking, while the bats' emitted signals and their returning echoes can also become confounded, resulting in forward masking (Neuweiler 1989). Due to both these masking effects, it was argued that bats could not use echolocation alone to find prey when gleaning in dense clutter (Schnitzler and Kalko 1998; Arlettaz et al. 2001; Schnitzler et al. 2003). Geipel et al. (2013a), however, made a striking discovery with the Neotropical common big-eared bat, *Micronycteris microtis*, that proved this assumption wrong.

Micronycteris microtis seems do the "impossible": They take silent, motionless dragonflies and other insect prey, including stick insects, off of the surface of leaves in the dense rainforest understory at night. *M. microtis* fly up and down individual plants, hover in front of resting prey, and then move in for capture. By offering *M. microtis* manipulated dragonfly prey in a small flight cage and recording their hunting behavior with high-speed video, Geipel et al. (2013a) showed that the strategy used by *M. microtis* for hunting prey consists of two main components. First, their stereotypical 3-D hovering likely enables them to shift the angle between the target and background, reducing interference between returning echoes and enabling the bat to pinpoint the prey on the leaf precisely and to assess its quality. Second, the bat's short, broadband, high-frequency echolocation calls (broadcast at a high repetition rate for a gleaning bat) are likely short enough in duration that a bat that stays a sufficient distance from its prey can discriminate outgoing calls from incoming echoes. Like some other gleaners, *M. microtis* lacks a terminal phase buzz.

The combined strategy of 3-D hovering and emitting short, high-frequency echolocation calls allows these bats to be quite selective in their assessment of prey. In flight cage experiments the bats successfully rejected dummies that had incorrect shape, surface structure, or material. *M. microtis* apparently has an echo-acoustic search image for its prey and has the ability to detect, localize, and assess prey among leaves using echolocation alone. This skill may take some time to perfect, as evidenced by the observations that mothers continue to feed young well after weaning, when the young are already foraging for themselves (Geipel et al. 2013b). Whether other gleaning bat species have similar capabilities remains to be seen, as does the possibility that many gleaning bats possess the more modest ability to track prey moving on substrate based on changes in echo timing and structure from one echolocation call to the next (Ratcliffe et al. 2005; Ratcliffe 2009).

4.5 Summary

Animal-eating bats fall predominantly in three groups: aerial hawkers, substrate gleaners, and those that use both strategies. The latter may include the majority of species. All appear to be descended from a laryngeal echolocating bat that fossils indicate was an aerial hawking species. Over the course of the 50 million years or more since the evolution of echolocation in modern bats, this extraordinary sensory system has been highly modified for different tasks and challenges. One of the drivers of this evolution is the arms race between echolocating bats and their would-be prey, which have evolved a number of strategies to avoid capture and have likely required bats to develop the remarkable echolocation abilities that characterize aerial hawkers (for example, the low-intensity aerial hawking bat, *Barbastella barbastellus*; Goerlitz et al. 2010).

However, there are many bats, from many different families, that have reverted to the proto-bat ancestor's hunting strategy of gleaning prey from terrestrial surfaces. At the two extremes, specialized substrate gleaning species differ from open space aerial hawkers in wing shape, echolocation call design, and perhaps even cognitive abilities (Norberg and Rayner 1987; Fenton 1990; Schnitzler and Kalko 2001; Ratcliffe 2009; Hulgard and Ratcliffe 2014). Another important difference between substrate gleaning bats and aerial hawking species is the use of prey-generated cues. The echo-acoustic challenge of locating prey among leaves on a surface makes listening for the rustling sounds of prey moving or the calls that prey make to attract mates, particularly advantageous for gleaning bats. In turn, listening for prey cues drives a number of sensory adaptations, such as the large ears characteristic of substrate gleaners, as well as different auditory and cognitive processing. This foraging behavior also generates selective pressure on prey mating signals and behaviors (Tuttle and Ryan 1981).

While some animal-eating bats fall squarely into these two categories of aerial hawkers that rely on echolocation to locate prey and substrate gleaners that use prey-generated cues, it would appear that most species flexibly and opportunistically recruit the two strategies (Ratcliffe and Dawson 2003; Ratcliffe et al. 2006). However, there are also some remarkable exceptions, such as the gleaning common big-eared bat (*Micronycteris microtis*) that can find silent stationary prey amongst leaves using echolocation alone (Geipel et al. 2013a) and the aerial hawking species *B. barbastellus* that uses calls of intensities similar to those used by gleaning bats (Goerlitz et al. 2010). In this chapter we have attempted to succinctly summarize the auditory information and foraging behaviors used by animal-eating bats to find their prey, and we have demonstrated some of the progress researchers have made toward providing an answer to Nagel's question, "What is it like to be a bat?"

References

Arlettaz, R. (1996). Feeding behaviour and foraging strategy of free-living mouse-eared bats, *Myotis myotis* and *Myotis blythii*. *Animal Behaviour*, 51(1), 1–11.

Arlettaz, R., Jones, G., & Racey, P. A. (2001). Effect of acoustic clutter on prey detection by bats. *Nature*, 414(6865), 742–745. doi:10.1038/414742a

Barber, J. R., & Conner, W. E. (2007). Acoustic mimicry in a predator prey interaction. *Proceedings of the National Academy of Sciences of the USA*, 104(22), 9331–9334. doi:10.1073/pnas.0703627104

Barber, J. R., Razak, K. A., & Fuzessery, Z. M. (2003). Can two streams of auditory information be processed simultaneously? Evidence from the gleaning bat *Antrozous pallidus*. *Journal of Comparative Physiology A: Neuroethology, Sensory, Neural, and Behavioral Physiology*, 189(11), 843–855. doi:10.1007/s00359-003-0463-6

Barber, J. R., Chadwell, B. A., Garrett, N., Schmidt-French, B., & Conner, W. E. (2009). Naïve bats discriminate arctiid moth warning sounds but generalize their aposematic meaning. *Journal of Experimental Biology*, 212(14), 2141–2148. doi:10.1242/jeb.029991

Barclay R. M. R., Fenton M. B., Tuttle M. D., & Ryan M. J. (1981). Echolocation calls produced by *Trachops cirrhosus* (Chiroptera: Phyllostomatidae) while hunting for frogs. *Canadian Journal of Zoology*, 59, 750–753. doi:10.1139/z81-107

Baron, G., Stephan, H., & Frahm, H. D. (1996). *Comparative neurobiology in Chiroptera*. Basel: Birkhäuser.

Bates, D. L., & Fenton, M. B. (1990). Aposematism or startle? Predators learn their responses to the defenses of prey. *Canadian Journal of Zoology*, 68(1), 49–52. doi:10.1139/z90-009

Bell, G. (1982). Behavioral and ecological aspects of gleaning by a desert insectivorous bat *Antrozous pallidus* (Chiroptera: Vespertilionidae). *Behavioral Ecology and Sociobiology*, 10(3), 217–223.

Bell, G. (1985). The sensory basis of prey location by the California leaf-nosed bat *Macrotus californicus* (Chiroptera: Phyllostomidae). *Behavioral Ecology and Sociobiology*, 16, 343–347.

Belwood, J., & Morris, G. (1987). Bat predation and its influence on calling behavior inneotropical katydids. *Science*, 238(4823), 64.

Bernal, X. E., Page, R. A., Rand, A. S., & Ryan, M. J. (2007). Cues for eavesdroppers: Do frog calls indicate prey density and quality? *American Naturalist*, 169(3), 409–415. doi:10.1086/510729

Blest, A. D., Collett, T. S., & Pye, J. D. (1963). The generation of ultrasonic signals by a New World arctiid moth. *Proceedings of the Royal Society of London B: Biological Sciences*, 158(971), 196–207. doi:10.1098/rspb.1963.0042

Boul, K. E., & Ryan, M. J. (2004). Population variation of complex advertisement calls in *Physalaemus petersi* and comparative laryngeal morphology. *Copeia*, 3, 624–631. doi:10.1643/CH-03-153R2

Bruns, V., & Burda, H. (1989). Ear morphology of the frog-eating bat (*Trachops cirrhosus*, family: Phyllostomidae): Apparent specializations for low-frequency hearing. *Journal of Morphology*, 199, 103–118.

Clare, E. L., Fraser, E. E., Braid, H. E., Fenton, M. B., & Hebert, P. D. N. (2009). Species on the menu of a generalist predator, the eastern red bat (*Lasiurus borealis*): Using a molecular approach to detect arthropod prey. *Molecular Ecology*, 18(11), 2532–2542.

Conner, W. E. (1999). 'Un chant d'appel amoureux': Acoustic communication in moths. *Journal of Experimental Biology*, 202, 1711–1723.

Corcoran, A. J., & Conner, W. E. (2012). Sonar jamming in the field: Effectiveness and behavior of a unique prey defense. *Journal of Experimental Biology*, 215, 4278–4287. doi:10.1242/jeb.076943

Corcoran, A. J., Barber, J. R., & Conner, W. E. (2009). Tiger moth jams bat sonar. *Science*, 325(5938), 325–327. doi:10.1126/science.1174096

Dijkgraaf, S. (1943). Over een merkwaardige functie van den gehoorsin bij vleermuizen. *Verslagen Nederlandische Akademie van Wetenschappen Afd. Naturkunde*, 52, 622–627.

Dijkgraaf, S. (1946) Die sinneswelt der fledermäuse. *Experientia*, 2, 438–448.
Dunning, D. C. (1968). Warning sounds of moths. *Zeitschrift für Tierpsychologie*, 25(2), 129–138.
Dunning, D. C., Acharya, L., Merriman, C. B., & Ferro, L. D. (1992). Interactions between bats and arctiid moths. *Canadian Journal of Zoology*, 70(11), 2218–2223. doi:10.1139/z92-298
Elemans, C. P. H., Mead, A. F., Jakobsen, L., & Ratcliffe, J. M. (2011). Superfast muscles set maximum call rate in echolocating bats. *Science*, 333(6051), 1885–1888. doi:10.1126/science.1207309
Falk, J. J., ter Hofstede, H. M., Jones, P. L., Dixon, M. M., Faure, P. A., Kalko, E. K. V., & Page, R. A. (2015). Sensory-based niche partitioning in a multiple predator–multiple prey community. *Proceedings of the Royal Society of London B: Biological Sciences*, 282(1808), 20150520. doi: 10.1098/rspb.2015.0520
Faure, P., & Barclay, R. (1994). Substrate-gleaning versus aerial-hawking: Plasticity in the foraging and echolocation behaviour of the long-eared bat, *Myotis evotis*. *Journal of Comparative Physiology A: Neuroethology, Sensory, Neural, and Behavioral Physiology*, 174, 651–660.
Fenton, M. B. (1990). The foraging behaviour and ecology of animal-eating bats. *Canadian Journal of Zoology*, 68(3), 411–422.
Fenton, M. B., & Ratcliffe, J. M. (2010). Bats. *Current Biology*, 20, R1060-R1062.
Fenton, M., Audet, D., Orbrist, M. K., & Rydell, J. (1995). Signal strength, timing, and self-deafening: The evolution of echolocation in bats. *Paleobiology*, 21(2), 229–242.
Fugère, V., O'Mara, T. M., & Page, R.A. (2015). Perceptual bias does not explain preference for prey call adornment in the frog-eating bat. *Behavioral Ecology and Sociobiology*, doi: 10.1007/s00265-015-1949-2
Fullard, J. H. (1998). The sensory coevolution of moths and bats. In R. R. Hoy, A. N. Popper, & R. R. Fay (Eds), *Comparative hearing: Insects* (pp. 279–326). New York: Springer. doi:10.1007/978-1-4612-0585-2_8
Fullard, J. H., & Fenton, M. (1977). Acoustic and behavioural analyses of the sounds produced by some species of Nearctic Arctiidae (Lepidoptera). *Canadian Journal of Zoology*, 55(8), 1213–1224. doi:10.1139/z77-160
Fullard, J. H., Fenton, M. B., & Simmons, J. A. (1979). Jamming bat echolocation: The clicks of arctiid moths. *Canadian Journal of Zoology*, 57(3), 647–649. doi:10.1139/z79-076
Fullard, J. H., Ratcliffe, J. M., & Guignion, C. (2005). Sensory ecology of predator–prey interactions: Responses of the AN2 interneuron in the field cricket, *Teleogryllus oceanicus* to the echolocation calls of sympatric bats. *Journal of Comparative Physiology A: Neuroethology, Sensory, Neural, and Behavioral Physiology*, 191(7), 605–618.
Fullard, J. H., Ratcliffe, J. M., & Christie, C. G. (2007). Acoustic feature recognition in the dogbane tiger moth, *Cycnia tenera*. *Journal of Experimental Biology*, 210(14), 2481–2488. doi:10.1242/jeb.001909
Geipel, I., Jung, K., & Kalko, E. K. V. (2013a). Perception of silent and motionless prey on vegetation by echolocation in the gleaning bat *Micronycteris microtis*. *Proceedings of the Royal Society of London B: Biological Sciences*, 280(1754), 20122830. doi:10.1098/rspb.2012.2830
Geipel, I., Kalko, E. K. V., Wallmeyer, K., & Knörnschild, M. (2013b). Postweaning maternal food provisioning in a bat with a complex hunting strategy. *Animal Behaviour*, 85, 1435–1441.
Ghose, K., Horiuchi, T. K., Krishnaprasad, P. S., & Moss, C. F. (2006). Echolocating bats use a nearly time-optimal strategy to intercept prey. *PLoS Biology*, 4(5), e108. doi:10.1371/journal.pbio.0040108
Goerlitz, H. R., & Siemers, B. M. (2006). Sensory ecology of prey rustling sounds: Acoustical features and their classification by wild grey mouse lemurs. *Functional Ecology*, 21, 143–153.
Goerlitz, H. R., Greif, S., & Siemers, B. M. (2008). Cues for acoustic detection of prey: Insect rustling sounds and the influence of walking substrate. *Journal of Experimental Biology*, 211(17), 2799–2806.
Goerlitz, H. R., ter Hofstede, H. M., Zeale, M. R. K., Jones, G., & Holderied, M. W. (2010). An aerial-hawking bat uses stealth echolocation to counter moth hearing. *Current Biology*, 20(17), 1568–1572.
Griffin, D. R. (1944). How bats guide their flight by supersonic echoes. *American Journal of Physics*, 12(6), 342–345. doi: 10.1119/1.1990634

Griffin, D. R. (1958). *Listening in the dark: The acoustic orientation of bats and men.* New Haven, CT: Yale University Press.

Griffin, D. R., Webster, F. A., & Michael, C. R. (1960). The echolocation of flying insects by bats. *Animal Behaviour*, 111, 141–154.

Hackett, T. D., Korine, C., & Holderied, M. W. (2014). A whispering bat that screams: Bimodal switch of foraging guild from gleaning to aerial hawking in the desert long-eared bat. *Journal of Experimental Biology*, 217(17), 3028–3032. doi:10.1242/u200Bjeb.100362

Halfwerk, W., Jones, P. L., Taylor, R. C., Ryan, M. J., & Page, R. A. (2014). Risky ripples allow bats and frogs to eavesdrop on a multisensory sexual display. *Science*, 343(6169), 413–416. doi:10.1126/science.1244812

Holderied, M. W., & von Helversen, O. (2003). Echolocation range and wingbeat period match in aerial-hawking bats. *Proceedings of the Royal Society of London B: Biological Sciences*, 270(1530), 2293–2299.

Holderied, M. W., Korine, C., Fenton, M. B., Parsons, S., Robson, S., & Jones, G. (2005). Echolocation call intensity in the aerial hawking bat *Eptesicus bottae* (Vespertilionidae) studied using stereo videogrammetry. *Journal of Experimental Biology*, 208(7), 1321–1327.

Hoy, R. R. (1992). The evolution of hearing in insects as an adaptation to predation from bats. In D. B. Webster, R. R. Fay, & A. N. Popper (Eds), *The evolutionary biology of hearing* (pp. 115–129). New York: Springer.

Hristov, N. I., & Conner, W. E. (2005). Sound strategy: Acoustic aposematism in the bat–tiger moth arms race. *Naturwissenschaften*, 92(4), 164–169.

Hulgard, K., & Ratcliffe, J. M. (2014). Niche-specific cognitive strategies: Object memory interferes with spatial memory in the predatory bat, *Myotis nattereri. Journal of Experimental Biology*, 217, 3293–3300.

Jakobsen, L., & Surlykke, A. (2010). Vespertilionid bats control the width of their biosonar sound beam dynamically during prey pursuit. *Proceedings of the National Academy of Sciences of the USA*, 107(31), 13930–13935. doi:10.1073/pnas.1006630107

Jakobsen, L., Ratcliffe, J. M., & Surlykke, A. (2013). Convergent acoustic field of view in echolocating bats. *Nature*, 493, 93–96. doi:10.1038/nature11664

Jones, G., & Teeling, E. C. (2006). The evolution of echolocation in bats. *Trends in Ecology and Evolution*, 21(3), 149–156. doi:10.1016/j.tree.2006.01.001

Jones, G., Webb, P. I., Sedgeley, J. A., & O'Donnell, C. F. J. (2003). Mysterious *Mystacina*: How the New Zealand short-tailed bat (*Mystacina tuberculata*) locates insect prey. *Journal of Experimental Biology*, 206, 4209–4216.

Jones, P., Page, R., Hartbauer, M., & Siemers, B. M. (2011). Behavioral evidence for eavesdropping on prey song in two Palearctic sibling bat species. *Behavioral Ecology and Sociobiology*, 65, 333–340.

Jones, P., Ryan, M., & Page, R. (2014). Population and seasonal variation in response to prey calls by an eavesdropping bat. *Behavioral Ecology and Sociobiology*, 608, 605–615. doi:10.1007/s00265-013-1675-6

Jones, P. L., Ryan, M. J., Flores, V., & Page, R. A. (2013). When to approach novel prey cues? Social learning strategies in frog-eating bats. *Proceedings of the Royal Society of London B: Biological Sciences*, 280(1772), 20132330. doi:10.1098/rspb.2010.1562

Kalko, E. K. V., & Schnitzler, H.-U. (1989). The echolocation and hunting behavior of Daubenton's bat, *Myotis daubentoni. Behavioral Ecology and Sociobiology*, 24(4), 225–238. doi:10.1007/BF00295202

Karlson, P., & Lüscher, M. (1959). "Pheromones": A new term for a class of biologically active substances. *Nature*, 183(4653), 55–56. doi:10.1038/183055a0

Kick, S. A., & Simmons, J. A. (1984). Automatic gain control in the bat's sonar receiver and the neuroethology of echolocation. *Journal of Neuroscience*, 4(11), 2725–2737.

Korsunovskaya, O. (2008). Acoustic signals in katydids (Orthoptera, Tettigonidae). *Entomological Review*, 88(9), 1032–1050.

Miller, L. (1991). Arctiid moth clicks can degrade the accuracy of range difference discrimination in echolocating big brown bats, *Eptesicus fuscus. Journal of Comparative Physiology A: Neuroethology, Sensory, Neural, and Behavioral Physiology*, 168, 571–579.

Miller, L. A., & Surlykke, A. (2001). How some insects detect and avoid being eaten by bats: Tactics and countertactics of prey and predator. *BioScience*, 51, 571–582.

Nagel, T. (1974). What is it like to be a bat? *Philosophical Review*, 83(4), 435–450.

Neuweiler, G. (1989). Foraging ecology and audition in echolocating bats. *Trends in Ecology and Evolution*, 4(6), 160–166. doi:10.1016/0169-5347(89)90120-1

Norberg, U. M., & Rayner, J. (1987). Ecological morphology and flight in bats (Mammalia; Chiroptera): Wing adaptations, flight performance, foraging strategy and echolocation. *Philosophical Transactions of the Royal Society B*, 316(1179), 335–427.

Page, R. A., & Ryan, M. J. (2005). Flexibility in assessment of prey cues: Frog-eating bats and frog calls. *Proceedings of the Royal Society of London B: Biological Sciences*, 272(1565), 841–847. doi:10.1098/rspb.2004.2998

Page, R. A., & Ryan, M. J. (2006). Social transmission of novel foraging behavior in bats: Frog calls and their referents. *Current Biology*, 16, 1201–1205.

Page, R. A., & Ryan, M. J. (2008). The effect of signal complexity on localization performance in bats that localize frog calls. *Animal Behaviour*, 76(3), 761–769.

Page, R. A., Schnelle, T., Kalko, E. K. V., Bunge, T., & Bernal, X. E. (2012). Sequential assessment of prey through the use of multiple sensory cues by an eavesdropping bat. *Naturwissenschaften*, 99(6), 505–509. doi:10.1007/s00114-012-0920-6

Peake, T. M. (2005). Eavesdropping in communication networks. In P. K. McGregor (Ed.), *Animal communication networks* (pp. 13–37). Cambridge: Cambridge University Press.

Phelps, S. M., & Ryan, M. J. (1998). Neural networks predict response biases of female túngara frogs. *Proceedings of the Royal Society of London B: Biological Sciences*, 265(1393), 279–285.

Ratcliffe, J. M. (2009). Predator-prey interaction in an auditory world. In R. Dukas, & J. M. Ratcliffe, *Cognitive ecology II* (pp. 201–225). Chicago: University of Chicago Press.

Ratcliffe, J. M., & Dawson, J. W. (2003). Behavioural flexibility: The little brown bat, *Myotis lucifugus*, and the northern long-eared bat, *M. septentrionalis*, both glean and hawk prey. *Animal Behaviour*, 66, 847–856.

Ratcliffe, J. M., & Fullard, J. H. (2005). The adaptive function of tiger moth clicks against echolocating bats: An experimental and synthetic approach. *Journal of Experimental Biology*, 208, 4689–4698.

Ratcliffe, J. M., Raghuram, H., Marimuthu, G., Fullard, J. H., & Fenton, M. B. (2005). Hunting in unfamiliar space: Echolocation in the Indian false vampire bat, *Megaderma lyra*, when gleaning prey. *Behavioral Ecology and Sociobiology*, 58, 157–164.

Ratcliffe, J. M., Fenton, M. B., & Shettleworth, S. J. (2006). Behavioral flexibility positively correlated with relative brain volume in predatory bats. *Brain, Behavior and Evolution*, 67(3), 165–176. doi:10.1159/000090980

Ratcliffe, J. M., Fullard, J. H., Arthur, B. J., & Hoy, R. R. (2009). Tiger moths and the threat of bats: Decision-making based on the activity of a single sensory neuron. *Biology Letters*, 5, 368–371.

Ratcliffe, J. M., Fullard, J. H., Arthur, B. J., & Hoy, R. R. (2011). Adaptive auditory risk assessment in the dogbane tiger moth when pursued by bats. *Proceedings of the Royal Society of London B: Biological Sciences*, 278, 364–370.

Ratcliffe, J. M., Elemans, C. P. H., Jakobsen, L., & Surlykke, A. (2013). How the bat got its buzz. *Biology Letters*, 9(2), 1–5.

Reep, R. L., & Bhatnagar, K. P. (2000). Brain ontogeny and ecomorphology in bats. In R. A. Adams, & S. C. Pedersen (Eds), *Ontogeny, functional ecology, and evolution of bats* (pp. 93–136). Cambridge: Cambridge University Press.

Roeder, K. D. (1967). *Nerve cells and insect behavior.* Cambridge, MA: Harvard University Press.

Rome, L. C. (2006). Design and function of superfast muscles: New insights into the physiology of skeletal muscle. *Annual Review of Physiology*, 68, 193–221.

Ron, S. R. (2008). The evolution of female mate choice for complex calls in túngara frogs. *Animal Behaviour*, 76, 1783–1794.

Rothschild, M., Reichstein, T., Euw, J. von, Aplin, R., & Harman, R. R. M. (1970). Toxic lepidoptera. *Toxicon*, 8(4), 293–296. doi:10.1016/0041-0101(70)90006-1

Russo, D., Jones, G., & Arlettaz, R. (2007). Echolocation and passive listening by foraging mouse-eared bats *Myotis myotis* and *M. blythii*. *Journal of Experimental Biology*, 210(1), 166–176. doi:10.1242/jeb.02644

Ryan, M. J. (1980). Female mate choice in a neotropical frog. *Science*, 209(4455), 523–525.

Ryan, M. J. (1985). *The túngara frog: A study in sexual selection and communication*. Chicago: University of Chicago Press.

Ryan, M. J., & Tuttle, M. D. (1982). Bat predation and sexual advertisement in a neotropical anuran. *American Naturalist*, 119(1), 136–139.

Ryan, M. J., & Tuttle, M. D. (1983). The ability of the frog-eating bat to discriminate among novel and potentially poisonous frog species using acoustic cues. *Animal Behaviour*, 31, 827–833.

Ryan, M. J., & Tuttle, M. D. (1987). The role of prey-generated sounds, vision, and echolocation in prey localization by the African bat *Cardioderma cor* (Megadermatidae). *Journal of Comparative Physiology A: Neuroethology, Sensory, Neural, and Behavioral Physiology*, 161(1), 59–66.

Ryan, M. J., Fox, J. H., Wilczynski, W., & Rand, A. S. (1990). Sexual selection for sensory exploitation in the frog *Physalaemus pustulosus*. *Nature*, 343(6253), 66–67. doi:10.1038/343066a0

Safi, K., Seid, M. A., & Dechmann, D. K. N. (2005). Bigger is not always better: When brains get smaller. *Biology Letters*, 1(3), 283–286. doi:10.1098/rsbl.2005.0333

Schaub, A., Ostwald, J., & Siemers, B. M. (2008). Foraging bats avoid noise. *Journal of Experimental Biology*, 211(19), 3174–3180. doi:10.1242/jeb.022863

Schmidt, S., Hanke, S., & Pillat, J. (2000). The role of echolocation in the hunting of terrestrial prey – new evidence for an underestimated strategy in the gleaning bat, *Megaderma lyra*. *Journal of Comparative Physiology A: Neuroethology, Sensory, Neural, and Behavioral Physiology*, 186(10), 975–988. doi:10.1007/s003590000151

Schnitzler, H.-U., & Kalko, E. K. V. (1998). How echolocating bats search and find food. In T. H. Kunz, & P. A. Racey (Eds), *Bat biology and conservation* (pp. 183–196). Washington, DC: Smithsonian Institution Press.

Schnitzler, H.-U., & Kalko, E. K. V. (2001). Echolocation by insect-eating bats. *BioScience*, 51(7), 557- 569.

Schnitzler, H.-U., Moss, C. F., & Denzinger, A. (2003). From spatial orientation to food acquisition in echolocating bats. *Trends in Ecology and Evolution*, 18(8), 386–394. doi:10.1016/S0169-5347(03)00185-X

Seeley, T. D. (1995). *The wisdom of the hive*. Cambridge, MA: Belknap Press of Harvard University Press.

Siemers, B. M., & Schnitzler, H.-U. (2004). Echolocation signals reflect niche differentiation in five sympatric congeneric bat species. *Nature*, 429, 657–661.

Siemers, B. M., & Swift, S. (2006). Differences in sensory ecology contribute to resource partitioning in the bats *Myotis bechsteinii* and *Myotis nattereri* (Chiroptera: Vespertilionidae). *Behavioral Ecology and Sociobiology*, 59(3), 373–380.

Siemers, B. M., Stilz, P., & Schnitzler, H.-U. (2001). The acoustic advantage of hunting at low heights above water: Behavioural experiments on the European trawling bats *Myotis capaccinii*, *M. dasycneme* and *M. daubentonii*. *Journal of Experimental Biology*, 204(22), 3843–3854.

Simmons, J. A., Fenton, M. B., & O'Farrell, M. J. (1979). Echolocation and pursuit of prey by bats. *Science*, 203(4375), 16–21.

Simmons, N. B. (2005). Order Chiroptera. In D. E. Wilson, & D. M. Reeder (Eds), *Mammal species of the World: A taxonomic and geographic reference, 3rd ed.* Volume 1 (pp. 312–529). Baltimore: Johns Hopkins University Press.

Simmons, N. B., & Geisler, J. H. (1998). Phylogenetic relationships of Icaronycteris, Archaeonycteris, Hassianycteris, and Palaeochiropteryx to extant bat lineages, with comments on the evolution of echolocation and foraging strategies in Microchiroptera. *Bulletin of the American Museum of Natural History*, 235, 2–182.

Surlykke, A., & Kalko, E. (2008). Echolocating bats cry out loud to detect their prey. *PLoS ONE*, 3(4), e2036.
Surlykke, A., Pedersen, S. B., Jakobsen, L. (2009a). Echolocating bats emit a highly directional sonar sound beam in the field. *Proceedings of the Royal Society of London B: Biological Sciences*, 276, 853–860.
Surlykke, A., Ghose, K., & Moss, C. M. (2009b). Acoustic scanning of natural scenes by echolocation in the big brown bat, *Eptesicus fuscus*. *Journal of Experimental Biology*, 212, 1011–1020.
Teeling, E. C. (2009). Hear, hear: The convergent evolution of echolocation in bats? *Trends in Ecology and Evolution*, 24(7), 351–354. doi:10.1016/j.tree.2009.02.012
ter Hofstede, H. M., Ratcliffe, J. M., & Fullard, J. H. (2008). The effectiveness of katydid (*Neoconocephalus ensiger*) song cessation as antipredator defence against the gleaning bat *Myotis septentrionalis*. *Behavioral Ecology and Sociobiology*, 63(2), 217–226.
ter Hofstede, H. M., Kalko, E. K. V., & Fullard, J. H. (2010). Auditory-based defense against gleaning bats in neotropical katydids (Orthoptera: Tettigoniidae). *Journal of Comparative Physiology A: Neuroethology, Sensory, Neural, and Behavioral Physiology*, 196(5), 349–358. doi:10.1007/s00359-010-0518-4
ter Hofstede, H. M., Goerlitz, H. R., Ratcliffe, J. M., Holderied, M. W., & Surlykke, A. (2013). The simple ears of noctuid moths are tuned to the calls of their sympatric bat community. *Journal of Experimental Biology*, 216, 3954–3962.
Trillo, P. A., Athanas, K. A., Goldhill, D. H., Hoke, K. L., & Funk, W. C. (2012). The influence of geographic heterogeneity in predation pressure on sexual signal divergence in an Amazonian frog species complex. *Journal of Evolutionary Biology*, 26(1), 216–222. doi:10.1111/jeb.12041
Tuttle, M. D., & Ryan, M. J. (1981). Bat predation and the evolution of frog vocalizations in the Neotropics. *Science*, 214(4521), 677–678.
Tuttle, M. D., & Ryan, M. J. (1982). The role of synchronized calling, ambient light, and ambient noise, in anti-bat-predator behavior of a treefrog. *Behavioral Ecology and Sociobiology*, 11(2), 125–131.
Tuttle, M. D., Taft, L. K., & Ryan, M. J. (1982). Evasive behaviour of a frog in response to bat predation. *Animal Behaviour*, 30, 393–397.
Tuttle, M. D., Ryan, M. J., & Belwood, J. J. (1985). Acoustical resource partitioning by two species of phyllostomid bats (*Trachops cirrhosus* and *Tonatia sylvicola*). *Animal Behaviour*, 33(4), 1369–1371.
Yack, J. E., Scudder, G., & Fullard, J. H. (1999). Evolution of the metathoracic tympanal ear and its mesothoracic homologue in the Macrolepidoptera (Insecta). *Zoomorphology*, 119, 93–103.

Chapter 5
Roles of Acoustic Social Communication in the Lives of Bats

Erin Gillam and M. Brock Fenton

Keywords Agonistic interactions • Contact calls • Distress calls • Eavesdropping • Echolocation • Isolation calls • Mating • Mother-pup recognition • Personal information • Social calls • Territoriality

5.1 Introduction

The diversity in animal communication systems highlights the importance of signals in a wide variety of behavioral interactions. Furthermore, these systems are generally under strong selective pressures and shaped by a variety of ecological, behavioral, and social factors. For example, an animal's habitat provides needed resources in the form of food and cover, while also serving as the signaling background for communication. Environmental characteristics, such as light levels and vegetative structure, can strongly impact how efficiently signals travel between senders and receivers. In addition, optimal signal structure typically is a trade-off between maximizing responses from desired receivers, such as potential mates, and minimizing responses from eavesdroppers, such as predators (Endler 1987). While a discussion of the suite of behavioral and social influences on signaling is beyond the scope of this chapter, it is important to keep in mind the array of factors that can drive the evolution of animal communication systems.

From a sensory ecology perspective, bats are particularly interesting for studying the evolution of communication systems and the role signaling plays in behavior. The approximately 1,260 living species of bats show considerable

E. Gillam (✉)
Department of Biological Sciences, North Dakota State University,
218 Stevens Hall, Dept 2715, PO Box 6050, Fargo, ND 58108, USA
e-mail: Erin.Gillam@ndsu.edu

M.B. Fenton
Department of Biology, University of Western Ontario,
1151 Richmond St., London, ON, Canada N6A 5B7
e-mail: bfenton@uwo.ca

M.B. Fenton et al. (eds.), *Bat Bioacoustics*, Springer Handbook of Auditory Research 54, DOI 10.1007/978-1-4939-3527-7_5

variation in ecology and sociality. Many species of bats live in social groups, setting the stage for communication among conspecifics. Bat social systems exhibit broad diversity both within and among species in group size, social structure, and mating system (McCracken and Wilkinson 2000). Bats also exhibit extensive variability in ecological characteristics, such as diet and roost use (Kunz and Fenton 2006), which can potentially influence the evolution of communication systems.

While bats actively use a variety of modalities for communication, most is known about acoustic signaling systems and their role in bat behavior. This bias likely reflects the nocturnal activity of bats and their tendency to live in confined, dark places. In these settings, light levels may be low or absent, reducing the efficacy of visual signaling. Alternatively, sound can be adapted to a wide variety of environmental conditions and behavioral situations. Acoustic signals travel rapidly, permitting efficient information exchange between senders and receivers. Aspects of call structure, such as amplitude, duration, and frequency, affect other characteristics of signal transmission, such as the distance traveled in the environment and the ability to localize the position of the sender (Wiley and Richards 1978, 1982). This ultimately means that bat acoustic signals are designed to optimize transmission in reference to the surrounding environment and the behavioral function of the signal. Bats directly control a range of signal features, including general signal design (Mora et al. 2011), the use of harmonics (Bates et al. 2011), call intensity (Surlykke and Kalko 2008), and patterns of sound radiation away from the mouth (Jakobsen and Surlykke 2010; Jakobsen et al. 2013).

The first objective of this chapter is to provide an overview of what has been learned about social calls of bats in the last 20 years. Focus is placed on the various functions of social calls and the diversity of structure and usage observed across species, as well as the continuum between echolocation and communication. Calls are categorized based on their presumed function, while recognizing that any one signal can simultaneously serve more than one function (e.g., courtship signals and territorial defense).

The second objective of this chapter is to examine the artificial line between echolocation and social calls. Echolocation signals are primarily used for orientation in the environment, as well as detection, tracking and identification of prey. Yet, there is a growing body of evidence that echolocation calls are rich in social information, sometimes encoding information about age, sex, and other individual features. Furthermore, bats are highly attentive to the echolocation calls of conspecifics, deriving information about the availability of foraging and roosting opportunities from these signals; this form of interception further emphasizes the potential communicative function of echolocation signals.

Finally, promising directions for future research are identified, particularly how future research can test hypotheses about the evolution of signaling systems in bats by applying signaling theory and asking questions that have been investigated in other taxa.

5.2 Social Calls

Since 1960, a staggering amount of research (>6,000 publications) has focused on echolocation in bats, ranging from the neurophysiological bases of signal processing to measures of call flexibility and signal optimization under different behavioral and ecological conditions (see reviews in Popper et al. 1995; Surlykke et al. 2014). Significantly less research has focused on social calls of bats, although the number of studies in this area has greatly increased in the last 15 years. Such studies examine not only the structure of social calls but also the behavioral context in which they are emitted, which provides clues to signal function. It is important to note that while acoustic signals appear to dominate the bat communication world, such sounds are often coupled with visual and chemical signals that can also play critical roles in communicating the correct signal to the receiver (Voigt et al. 2008).

5.2.1 Mate Attraction

In spite of the diversity of bats, details about the mating behavior and associated signaling systems are known for approximately 50 species or less (4 % of all bats). The lack of data reflects a combination of sampling method limitations, nocturnal activity patterns, small body size, and tendency to roost in inaccessible locations. Most available data come from species such as the greater sac-winged bat (*Saccopteryx bilineata*) that roosts in exposed areas (Voigt et al. 2008) and performs courtship behaviors outside of the roost (Lundberg and Gerell 1986). Other data come from species studied in captivity (i.e., Brazilian free-tailed bats, *Tadarida brasiliensis*; Bohn et al. 2008). What is known suggests that acoustic and olfactory signals can be central to courtship and mating. Like most mammals, female bats bear the brunt of the costs of reproduction (pregnancy and lactation), and females are the choosy sex.

The most extensive data about mating and acoustic signaling comes from the greater sac-winged bat, a Neotropical insectivore that roosts in small groups on the sides of trees and in other similarly open structures. These bats exhibit resource-defense polygyny in which males guard groups of females (i.e., harems; Bradbury and Vehrencamp 1977; Voigt et al. 2008) and defend small, often adjacent, roosting territories (Bradbury and Emmons 1974). While females tend to stay in the same harem group, there is extensive extra-pair paternity, with up to 70 % of offspring not sired by the harem male (Heckel et al. 1999). This shows that males are unable to control the movements of females within and between groups, indicating that attraction of females for mating opportunities is critical for male fitness.

In addition to chemical and visual signals (Bradbury and Emmons 1974; Voigt and von Helversen 1999), displaying male greater sac-winged bats also sing. In general, a "song" differs from a "call" in that songs are composite signals that include multiple unique calls or phrases (Catchpole and Slater 2008). In this species, males produce a highly variable song containing four call types, one of which encodes information about individual identity and, potentially, colony identity (Behr and von Helversen 2004; Davidson and Wilkinson 2004). The structure of the courtship song affects sexual selection, as males with more composite syllables in their songs have more females in their harems than males with simpler songs (Davidson and Wilkinson 2004). Males also produce long (66.7 ± 44.1 ms) tonal whistles that are directed toward females during courtship displays (Behr and von Helversen 2004; Davidson and Wilkinson 2004). Production of such long whistles while hovering is likely to be energetically expensive, suggesting that whistling rate could reflect male quality (Behr and von Helversen 2004; Voigt et al. 2008).

Complex multi-syllabic acoustic signals that function in male advertisement occur in at least 11 bat species. Most species that produce songs or calls associated with courtship exhibit a mating system of resource-defense polygyny, although such signaling also occurs in species exhibiting lek polygyny and promiscuity. As with the greater sac-winged bat, Nathusius's pipistrelle (*Pipistrellus nathusii*) and the Brazilian free-tailed bat exhibit multimodal courtship displays that include a song composed of multiple motifs and phrases (Figure 5.1). While these songs are variable, they clearly follow basic syntactical rules while encoding information about the identity of the signaler (Russ and Racey 2007; Bohn et al. 2009). Courtship signals have also been described in a variety of other species, including little brown bats (*Myotis lucifugus*) (Barclay and Thomas 1979), common noctules (*Nyctalus noctula*) (Kozhurina 1996) and hammer-headed bats (*Hypsignathus monstrosus*) (Bradbury 1977), among others. Figure 5.1 demonstrates the variation that can be seen in such mating signals, ranging from the complex multisyllabic songs of greater sac-winged bats to the motif-containing signals of Brazilian free-tailed bats (motifs are common in birdsong), to the simpler songs of Seba's short-tailed bats that contain a single, repeated call type (Knörnschild et al. 2014).

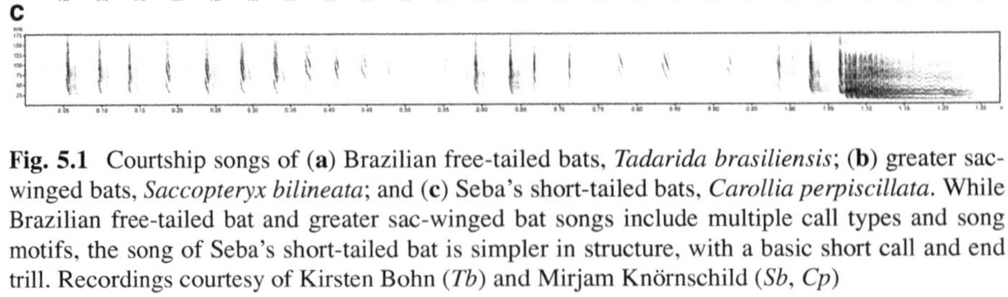

Fig. 5.1 Courtship songs of (**a**) Brazilian free-tailed bats, *Tadarida brasiliensis*; (**b**) greater sac-winged bats, *Saccopteryx bilineata*; and (**c**) Seba's short-tailed bats, *Carollia perpiscillata*. While Brazilian free-tailed bat and greater sac-winged bat songs include multiple call types and song motifs, the song of Seba's short-tailed bat is simpler in structure, with a basic short call and end trill. Recordings courtesy of Kirsten Bohn (*Tb*) and Mirjam Knörnschild (*Sb*, *Cp*)

5.2.2 Antagonistic Interactions

During aggressive encounters, communication signals convey information about the competitive ability of the opponents, perhaps allowing individuals to avoid the potentially high costs of fighting (Bradbury and Vehrencamp 2011). Calls specifically emitted during agonistic encounters have been described in several bat species, the earliest report coming from Vaughan (1976) who described the singing behavior of male heart-nosed bats (*Cardioderma cor*) during territorial displays. In some species, agonistic calls are emitted as a general response to negative/stressful encounters, whether this involves manipulation by a researcher or aggressive interactions with conspecifics (Bastian and Schmidt 2008; Bohn et al. 2008). Research on big brown bats (*Eptesicus fuscus*) (Gadziola et al. 2012) and greater false vampire bats (*Megaderma lyra*) (Gadziola et al. 2012) revealed that the effect/arousal state of the sender is encoded in the structure of agonistic calls. Bats in a higher state of aggression exhibit different temporal patterns of emission and syllable diversity compared to bats in less aggressive behavioral states.

Agonistic calls used for repelling conspecifics from foraging sites have been reported in multiple European *Pipistrellus* species (Barlow and Jones 1997; Georgiakakis and Russo 2012) and the greater false vampire bat (McWilliam 1987). Such calls are most commonly produced during periods of low insect activity, presumably when the presence of other foraging bats is likely to reduce an individual's foraging success. In *Pipistrellus* species, social calls do not appear to be used for long-term defense of feeding sites, likely because insect densities are highly variable in time and space, making territory establishment unprofitable. The feeding sites of male greater false vampire bats overlap with long-term roosting territories, hence social calls could serve a dual purpose, but calling activity is negatively correlated with insect density (McWilliam 1987), suggesting a strong role for feeding territory defense.

Greater sac-winged bats engage in extensive male-male competition, which primarily consists of territorial songs that are structurally different from courtship songs (Figure 5.2) (Behr and von Helversen 2004). Territorial songs are most often produced during times when females are away from the roost, indicating that they are involved in competition for mating territories (Behr and von Helversen 2004). Territorial songs encode information about caller identity (Behr et al. 2006; Eckenweber and Knörnschild 2013), and calling rates increase during periods of greater male-male competition (Eckenweber and Knörnschild 2013). Playback experiments have revealed that low-frequency songs elicit stronger vocal responses from territorial males, indicating that call structure may provide information about male competitive ability (Behr et al. 2009). Males producing longer songs with more composite syllables and lower frequency terminal buzzes also have larger harems and greater reproductive success, suggesting that these signals also play a role in female choice (Davidson and Wilkinson 2004; Behr et al. 2006). Overall, these data suggest that honest signaling about male quality in greater sac-winged bats may be enforced by the significant cost of producing "high quality" territorial signals, such that lower quality males lack the energy to produce such costly songs (Zahavi 1975).

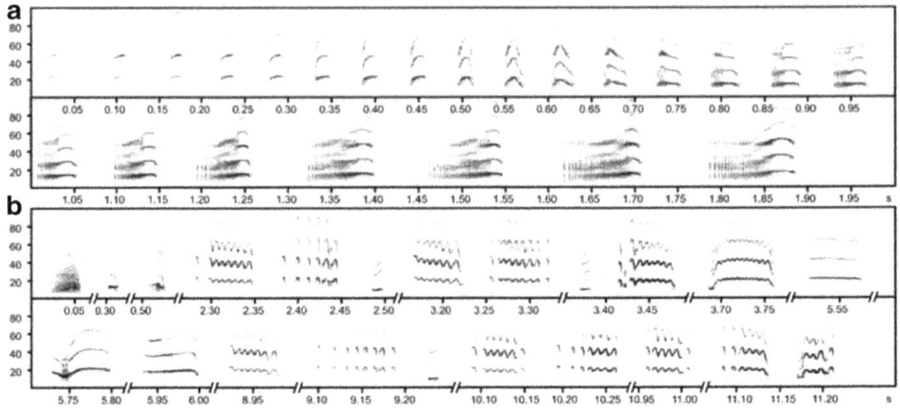

Fig. 5.2 Differences between the (**a**) territorial song and (**b**) courtship song of *Saccopteryx bilineata*. Courtship song is a 12-s fraction taken from a song that was 4 min 25 s in duration (Reproduced from Behr and von Helversen 2004)

5.2.3 Locating Conspecifics

Many species of bats live in social groups, suggesting that one or more mechanisms permit individuals to find social partners after periods of separation. The need for effective location of roost mates often reflects the fission-fusion society known from many species of bats (e.g., Wilkinson 1985; Rhodes 2007). Specific communication signals can allow bats to find roosting conspecifics (e.g., Vaughan and O'Shea 1976), and roost mates are often important to minimize the costs of thermoregulation (e.g., Willis and Brigham 2004, 2007). Locating conspecifics, whether on the wing or at roosts, can provide benefits such as passive information transfer about food (Wilkinson 1992) and roosts (Ruczyński et al. 2007) or maintenance of social relationships (Willis and Brigham 2004). While spatial memory and olfactory cues certainly play a role in relocating conspecifics (Gustin and McCracken 1987; McCracken 1993), acoustic communication can also be a key mechanism in maintaining/affecting group cohesion. Such signals, known as "contact calls," have been documented in a variety of taxa, especially among birds and primates (Kondo and Watanabe 2009). Among bats, contact calls can be categorized into two groups: those exchanged specifically between females and their offspring and those exchanged among individuals that are part of the same social group.

5.2.3.1 Mother-Pup Recognition

In many bats, pups produce "isolation calls" when separated from their mother. All statistical analyses of isolation calls to date have found evidence for individual signatures (e.g., Gelfand and McCracken 1986; Scherrer and Wilkinson 1993). Further, all studies that have presented mothers with isolated infants or playbacks of isolation calls have revealed that females specifically respond to the calls of their pup, indicating discrimination solely based on infant call structure (e.g., Balcombe 1990; Knörnschild and von Helversen 2008). This is especially impressive given that many studies also report that pup isolation calls change with age, indicating that for successful discrimination to occur, mothers must have an evolving template of their infant's isolation call (de Fanis and Jones 1995; Knörnschild et al. 2007). Further, mothers are generally able to identify their own offspring among hundreds, if not thousands, of other pups (Balcombe 1990).

Females also vocalize when approaching pups. In some species, these are echolocation pulse trains associated with orientation; in others, a specialized social call, known as a "directive call," is produced. Older pups orient toward vocalizing mothers (Balcombe and McCracken 1992; de Fanis and Jones 1995) and antiphonal calling between mother and pup appears to play a role in reunions in some species (Matsumura 1979; Brown et al. 1983). Directive calls and/or echolocation pulses have statistically unique individual signatures in some species (Brown 1976; Esser and Schmidt 1989). Yet, discrimination tests have yielded mixed results. In some species, pups do not appear to discriminate, actively approaching and attempting to nurse with any female (Balcombe 1990; de Fanis and Jones 1995). In others, there is mutual recognition of mother and pup (de Fanis and Jones 1995; Esser and Lud 1997), although olfactory signals likely are involved. In general, selection on female recognition of pups should be strong because of the cost of giving milk to unrelated young rather than related young. Alternatively, pups that can obtain resources from unrelated mothers will benefit, generating selective pressure for young to approach any female, while selection for the cognitive ability to discriminate their mother from other adult females likely is absent (Knörnschild and von Helversen 2008).

5.2.3.2 Group Formation and Cohesion

Many bats use contact calls for maintaining contact with conspecifics. This often means bouts of antiphonal calling between two or more individuals. Female greater spear-nosed bats, *Phyllostomus hastatus*, produce group-specific "screech" calls when exiting a roost. Playback studies have demonstrated that females preferentially respond to the calls of group members compared to non-group members, indicating that discrimination based on call structure is occurring (Boughman and Wilkinson 1998). Screech calls attract social group members to the caller's location, presumably facilitating group foraging (Boughman and Wilkinson 1998; Wilkinson and Boughman 1998). Antiphonal exchanges of individual-specific social calls have been documented in common vampire bats (*Desmodus rotundus*) (Carter et al. 2012) and white-winged vampire bats (*Diaemus youngi*) (Carter et al. 2008, 2009) that are physically separated from conspecifics. Pallid bats (*Antrozous pallidus*) also exchange consistent, individual-specific contact calls, primarily during the early morning hours when selecting day roosts (Arnold and Wilkinson 2011). Playback experiments revealed that individuals preferentially respond to the contact calls of familiar over unfamiliar bats, suggesting that discrimination is occurring in this species (Arnold and Wilkinson 2011). Spix's disk-winged bats (*Thyroptera tricolor*) use two social calls for recruiting groupmates to ephemeral leaf roosts, with flying bats producing an "inquiry" call and roosting bats answering with a "response" call (Chaverri et al. 2010); both call types encode information about individual identity (Gillam and Chaverri 2012). Interestingly, choice experiments have demonstrated that flying bats preferentially come to the response calls of group members but that roosting bats indiscriminately reply to the inquiry calls of any flying conspecific (Chaverri et al. 2013).

5.2.4 Distress

In many taxa, individuals emit alarm calls when a predator is detected. Distress calls are a specialized type of alarm call emitted by animals in especially dire circumstances, such as when captured by a predator or researcher (Bradbury and Vehrencamp 2011). Distress calls are known from many bat species, including phyllostomids (August 1979; Ryan et al. 1985), vespertilionids (Fenton et al. 1976; Russ et al. 2004), and rhinolophids (Luo et al. 2013), among others. While several functions of distress calls have been proposed (Bradbury and Vehrencamp 2011), in bats the most common hypothesis is that these signals attract conspecifics that will mob the predator, helping the captured individual to escape. Yet, only one definitive observation of mobbing has ever been reported in bats (Knörnschild and Tschapka 2012). Bat distress calls generally attract other bats to the location of the sender (Fenton et al. 1976; Hill and Greenaway 2005). In captive animal and playback studies, free-flying bats respond to distress calls by altering their flight behavior and circling close to the captive bat/speaker (Ryan et al. 1985; Russ et al. 2005). These responses can be quite intense. For example, Russ et al. (1998) documented an 80-fold increase in bat activity when a caged individual producing distress calls was placed within 5 m of a soprano pipistrelle (*Pipistrellus pygmaeus*) roost. In some cases, bats only respond readily to the distress calls of their own species (Fenton et al. 1976); in others, the broadcasts of conspecific and heterospecific distress signals elicit similar response intensities (August 1979; Russ et al. 2004). Russ et al. (2004) even demonstrated that three European *Pipistrellus* species responded strongly to distress calls of sucker-footed bats (*Myzopoda aurita*), which are native to Madagascar.

Distress calls exhibit striking structural similarities across bat species (Figure 5.3) (Fenton et al. 1976; August 1979), even those that are distantly related (Russ et al. 2004). Distress calls tend to be short, high-intensity, low-frequency signals with high repetition rates and extensive harmonic structure (August 1985; Luo et al. 2013). In a broader context, Morton (1977) proposed a set of motivational-structural rules in relation to avian and mammalian vocalizations, stating that distress calls should be broadband with increasing frequencies over time. Based on data from four species, Russ et al. (2004) documented agreement with this rule. Experimental manipulations of call structures that exaggerate this pattern generate the strongest

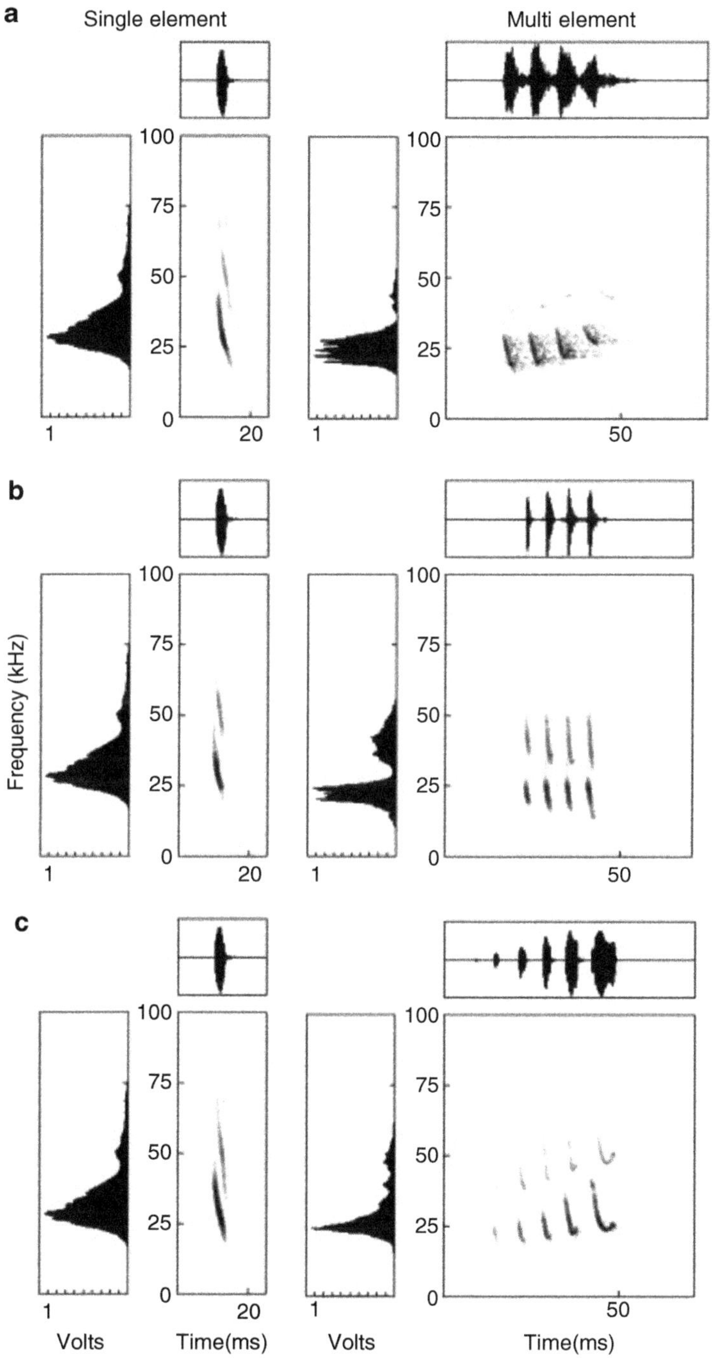

Fig. 5.3 Sonograms, oscillograms, and power spectra of typical single element and multiple element distress calls from: (**a**) *Pipistrellus nathusii*, (**b**) *Pipistrellus pipistrellus*, and (**c**) *Pipistrellus pygmaeus* (Reproduced from Russ et al. 2004)

responses (Russ et al. 2005). Sounds of low frequency prevail in many distress calls, suggesting that these calls travel long distances in the environment. This means that they are heard by more receivers than similar, higher frequency signals (Morton 1975). Finally, the high repetition rate of distress calls should not only increase detection by other individuals due to redundancy (Shannon and Weaver 1949) but may also provide information about the arousal state of the caller (August 1985; Russ et al. 2005).

5.3 The Communicative Role of Echolocation

Unlike social calls that involve actively broadcasting information, communication can also be passive when unintentional cues produced by an individual are detected and convey information to other animals about the location, identity, or activities of the sender. Eavesdropping receivers can use such cues in many different ways, such as assessing the quality of a foraging patch or sizing up the fighting ability of competitors for mates (Valone 2007), among others.

The intensities, numbers, cadences, and patterns of calls used in echolocation make the system vulnerable to eavesdroppers, whether by conspecifics, other species of bats, potential prey, or predators. As usual when writing about the echolocation behavior of bats, one ends up back at the classic book *Listening in the Dark* (Griffin 1958). Here, Griffin's description of red bats (*Lasiurus borealis*) responding to the echolocation calls, particularly feeding buzzes of conspecifics, was an early demonstration of the continuum between echolocation and communication. Möhres (1967) provided the first clues that echolocation might encode information about the personal characteristics of the sender, documenting that captive greater horseshoe bats (*Rhinolophus ferrumequinum*) use echolocation to locate preferred roost mates. Suthers (1967) illustrated how greater bulldog bats (*Noctilio leporinus*) on a collision course with a conspecific modified their echolocation to include a low-frequency "honk" that resulted in one bat changing its flight path. These results further blurred the line between echolocation as a signaling system exclusively used for orientation/prey capture and one that potentially conveys information to others. Examination of the metabolic costs of echolocation in lesser bulldog bats (*N. albiventris*) indicate that calls produced outside of a foraging context may reflect active signaling, as such signal production carries an energetic cost (Dechmann et al. 2013). In short, the distinction between echolocation and social calls of bats is arbitrary, as either signal may serve a communication role (although the long durations of many social calls preclude their serving a role in echolocation).

Echolocation calls of aerial hawking bats are generally very intense, with normal broadcast amplitudes of 125–140 dB SPL at 10 cm from the mouth (Surlykke and Kalko 2008). While higher frequency echolocation calls are subject to strong atmospheric attenuation due to their mostly high frequencies, they can still be perceived by other individuals within several tens of meters of the calling bat. If echolocation call structure relates to characteristics of the individual or their behavioral state, eavesdropping individuals within earshot could assess such information about the sender and, potentially, use it to their advantage. The types of information passively conveyed via echolocation can be broadly classified into two categories: (1) activity information, and (2) personal information (Jonker et al. 2010; Knörnschild et al. 2012).

5.3.1 Activity Information

Bats change their echolocation calls in stereotypical ways as they detect and approach insect prey (Griffin 1958). It has been shown that the fine-scale structure of the final terminal phase, or "feeding buzz," can indicate whether a bat successfully captured a prey item or not (Surlykke et al. 2003). If individuals attend to the calls of nearby conspecifics, information about feeding success can be revealed, and eavesdroppers can potentially assess the quality of nearby feeding areas. Such use of so-called "public information" has been documented in a wide variety of avian and mammalian species (Danchin et al. 2004; Valone 2007). Attention to public information may be especially valuable for echolocating bats, as the need to detect weak returning echoes severely limits target detection range but attending to conspecific cues could greatly expand the area over which a bat can assess foraging patch quality. Dechmann et al. (2010) illustrate this idea nicely with velvety free-tailed bats (*Molossus molossus*), calculating that an individual's maximum detection distance of 0.5–2 m using echolocation alone can potentially be increased up to 54 m by incorporating passive eavesdropping on conspecific calls.

Experimental work involving broadcasts of echolocation in the field have shown that bats attend and respond to the echolocation calls of other bats. Studies by Barclay (1982), Leonard and Fenton (1984), and Balcombe and Fenton (1988) demonstrated how four bat species (little brown bats; red bats; Yuma myotis, *M. yumanensis*; and spotted bats, *Euderma maculatum*) are attracted to playbacks of the echolocation calls of conspecifics (and even heterospecifics in one case). Gillam (2007), Dechmann et al. (2009), and Übernickel et al. (2013) showed in other species (Brazilian free-tailed bats, greater bulldog bats, and lesser bulldog bats) that individuals are particularly attracted to the feeding buzz section of echolocation call sequences, presumably as these signals provide the most information about capture success and foraging patch quality. Radio tracking of lesser bulldog bats revealed that bats forage within earshot of social group mates 94 % of the time, suggesting that eavesdropping may be an important mechanism driving the formation of foraging groups in this species (Dechmann et al. 2009). In other words, while some species, like greater spear-nosed bats, may actively recruit conspecifics into foraging groups (Boughman and Wilkinson 1998), foraging groups can also be formed passively due to the benefits that individuals attain by attending to the echolocation cues of conspecifics. Safi and Kerth (2007) further extended this idea by conducting a comparative phylogenetic analysis and concluded that eavesdropping is a key factor driving male sociality in temperate bat species.

Interestingly, eavesdropping on nearby individuals is not universal in bats, as greater sac-winged bats (Übernickel et al. 2013) and common pipistrelles (*P. pipistrellus*) (Jonker et al. 2010) are not attracted to the calls of conspecifics. Differences in the use of eavesdropping are likely related to the behavioral and ecological characteristics of a given species. Specifically, bats that exploit rich, highly ephemeral patches of insects should benefit most from eavesdropping, as such prey are patchily distributed throughout the environment yet sufficiently rich in density that the presence of multiple bats feeding in the same area is unlikely to reduce individual foraging success (Barta and Szep 1992; Wilkinson 1992). Greater sac-winged bats forage within the forest interior where insect prey are relatively uniform in distribution; hence it follows that eavesdropping may not be an effective strategy for increasing prey detection rates in this species (Übernickel et al. 2013). While common pipistrelles exploit ephemeral patches of prey, these vary in their richness such that individuals are known to establish feeding territories, emit feeding defense calls, and actively chase other bats from foraging grounds (Barlow and Jones 1997).

Eavesdropping can also be involved in finding another limited resource: roosts. Ruczyński et al. (2007) explored how noctule bats locate new cavity roosts in trees; this should be a challenge when relying on echolocation alone, as cavity openings are small and difficult to acoustically distinguish from the complex and irregular structure of a tree trunk. The authors assessed the ability of bats to locate cavity openings using echolocation alone in the presence of non-social cues (visual information and temperature differences) and social cues (conspecific echolocation calls and odors). Broadcast of echolocation calls from the cavity was the only cue that significantly decreased the time for bats to locate the roost opening, suggesting that eavesdropping plays an important role in finding roosts and potentially plays a role in group formation. It has also been shown that eavesdropping on social calls can facilitate location of roosts in the common noctule (Furmankiewicz et al. 2011). Ruczyński et al. (2009) explored roost-finding behavior in the presence of cues in two additional species, brown long-eared bats and Daubenton's bats (*M. daubentonii*). Interestingly, the presence of echolocation cues reduced the search time for roost openings in Daubenton's bats but not brown long-eared bats, presumably because the low-amplitude calls of this species cannot be effectively detected from within a roost except at very short distances.

5.3.2 Personal Information

The primary function of echolocation is orientation in the environment and, for many species, detection and localization of insect prey. Hence, it would be expected that the strongest selection pressures on signal structure would relate to optimizing echo detection based on the behavioral, ecological, and climactic conditions of the signaling environment. Echolocation structure can be influenced by phylogeny, with closely related species exhibiting higher degrees of call similarity due to phylogenetic inertia (Jones and Teeling 2006). Despite these constraining factors, several studies suggest that personal information, such as sex, age, or individual identity, are at least statistically discriminable in the echolocation calls of some species (thoroughly reviewed in Jones 1995 and Jones and Siemers 2011).

Sex differences in echolocation call structure have been most thoroughly documented in bats using high duty cycle echolocation reflected by calls dominated with a constant frequency (CF), particularly rhinolophids and hipposiderids. In some species, females produce higher frequency calls than males. This especially interesting relationship is surprising because females are generally larger than males. In this situation, body scaling rules would predict the opposite relationship with call frequency (Jones et al. 1992; Chen et al. 2009). In other cases, males emit higher frequencies (Guillén et al. 2000) or there appear to be no differences between the sexes (Heller and von Helversen 1989). Despite these average differences, there is often extensive overlap in the range of frequencies used by males and females (Chen et al. 2009), suggesting that call frequency may not be a particularly useful signal for bats to discriminate the sex of nearby conspecifics (although see Neuweiler et al. 1987). Among low duty cycle echolocating bats, such as those whose calls are dominated by frequency-modulated signals, big brown bats emit sex-specific echolocation calls. This has been documented in a context associated with mating and reproduction (not foraging), suggesting that bats may actively adjust their call structure during times of mating to permit discrimination based on echolocation (Grilliot et al. 2009). Alternatively, other species such as Daubenton's bats (Jones and Kokurewicz 1994) do not exhibit sex-specific differences in echolocation call structure; it is likely that this is the case in many species but that non-significant differences between sexes are simply not highlighted or included in publications.

Relatively few studies have moved beyond examining statistical differences in call parameters between the sexes to assess if bats can actually use this information to make effective discriminations. Three species—big brown bats (Kazial and Masters 2004), Mehely's horseshoe bats (*R. mehelyi*), and Mediterranean horseshoe bats (*R. euryale*) (Schuchmann et al. 2012)—differentially respond to echolocation calls based on the sex of the caller, providing some of the first evidence that bats can actually use sex differences in call structure to assess information about other bats within eavesdropping range. Knörnschild et al. (2012) found that male greater sac-winged bats at roosts differentially respond to the sexually dimorphic echolocation calls of this species, producing courtship calls in response to the echolocation of an approaching female and aggressive territorial calls toward the echolocation of approaching males. These results highlight the need for further experimental work examining the discriminatory ability of bats in reference to personal information encoded in echolocation.

Other life history characteristics that have been investigated for echolocation call signatures include age, reproductive state, and body size/condition. In all but one species studied to date (big brown bats; Masters et al. 1995; Kazial et al. 2001), call frequencies are lower among juveniles compared to adults. Ontogenetic studies of call development have shown that change in frequency, along with other call features, is a gradual process, suggesting an association with the development of the structures and musculature of the laryngeal system (Gould 1975; Moss et al. 1997). In greater horseshoe bats, maternal learning plays a role in determining the frequency of calls produced by juveniles (Jones and Ransome 1993). It is not clear if maternal contributions to juvenile echolocation are the exception or the rule among bats. In reference to reproductive condition, only one study has demonstrated that lactating and non-lactating females (little brown bats) produce statistically different echolocation call structures (Kazial et al. 2008a). Few studies have reported differences in echolocation call structure in relation to body size (see reviews in Jones 1995; Jones and Siemers 2011), although those species in which differences have been found primarily align with body scaling rules in that larger animals produce lower frequency calls (Jones and Kokurewicz 1994; although see Jones et al. 1993). Body condition, generally calculated in bats as "forearm length/body mass," has also not been shown to be commonly encoded within echolocation, the exception being some CF echolocators (Jones et al. 1994; Guillén et al. 2000; Siemers et al. 2005). Overall, the statistical differences among bats falling into the described categories (sex, age, etc.) provide little information about how bats actually use this information on the wing when detecting the calls of conspecifics. The substantial degree of overlap generally observed between groups (i.e., juvenile versus adult) suggests that at least some of these differences may be of limited value in real-life discriminatory tasks (Jones and Siemers 2011).

Information about specific group or individual identity has been reported to be discriminable in the echolocation calls of some bat species. Signals can generally be viewed as having an "information capacity," which refers to the potential to encode unique signatures within the available parameter space of the signal (Beecher 1989). The primary use of echolocation for orientation and prey detection means that calls are constrained by certain parameter boundaries. Yet, there is substantial flexibility within that parameter space for effectively adapting to local ecological and behavioral conditions. Hence, it is somewhat surprising that unique signatures needed to identify groups or individuals can still be maintained (Siemers and Kerth 2006). Colony-level differences in echolocation have been reported in little brown bats (Pearl and Fenton 1996; Jameson and Hare 2009), and these differences do not reflect differing levels of clutter at roost entrances (Jameson and Hare 2009). Family-level differences (mother and offspring) in echolocation call structure, presumably driven by genetic relatedness, have been reported in big brown bats (Masters et al. 1995). Hiryu et al. (2006) found that individual Taiwanese leaf-nosed bats (*Hipposideros terasensis*) joining a new colony actively adjusted their resting echolocation call frequency to the frequency used by other colony members. Such social learning of a common group signal structure has been reported for the social calls of at least one bat (Boughman and Wilkinson 1998) as well as in other taxa (e.g., Nousek et al. 2006). Recognition of group versus non-group members via echolocation may also be valuable when an individual is making decisions about whether to produce potentially costly social/behavioral responses (Voigt-Heucke et al. 2010).

Many studies have assessed individual variation in echolocation calls. Most studies record bats in only one behavioral situation, although Burnett et al. (2001) found that when testing seven big brown bats, 73 % of calls could be assigned to the correct individual in any of three laboratory scenarios (perched, held in hand, or flying). Kazial et al. (2008b) tested for discriminatory ability by habituating little brown bats to the calls of one individual (Bat A) and assessing behavioral changes when the playback was switched to another call from the same bat (Bat A) or to the call of a different bat (Bat B). Bats were able to detect the change in caller identity, increasing their own calling rates only when the playback was switched to a different individual. Yovel et al. (2009) conducted a related discriminatory experiment with *M. myotis*, documenting that bats not only discriminated among known individuals but also discriminated the calls of new individuals to which the listener had not previously been exposed. Comparing experimental results with statistical call classifiers indicated that greater mouse-eared bats use a simple classification system by which they primarily make discriminations based on knowledge of the average call structure of a given individual's echolocation. Interestingly, Yovel et al. (2009) also documented that the spectral structure of echolocation appears to provide sufficient information to encode individual differences, with temporal characters contributing relatively less to this process.

5.3.3 Interspecific Differences in Echolocation: Evidence for Acoustic Communication?

When examining communities of bats, frequency partitioning across species appears to occur. There have been efforts to understand the factors driving this divergence in call structure among sympatric species. Jones and Siemers (2011) nicely summarize the hypotheses and associated literature that have been put forth to date to explain interspecific differences in echolocation, including allometric scaling effects (Aldridge and Rautenbach 1987; Jones 1999), dietary niche partitioning (Jones and Barlow 2003), and interspecific communication (Schuchmann and Siemers 2010). Jung et al. (2014) analyzed echolocation call structure in 18 Neotropical molossids to examine what factors contribute to differences in echolocation between species. Despite high levels of individual flexibility in call structure, there were clear differences among species and genera, suggesting a strong role for phylogeny in shaping echolocation. Evidence for allometric scaling and impacts of call structure on prey perception were also documented, although the authors note that these factors do not explain all of the variation observed within these species and that the high plasticity in call structure may be key in allowing animals to encode social information into their calls. Such findings stress the importance of conducting large-scale analyses involving suites of species to better understand the factors driving divergence in echolocation, specifically when trying to understand the communicative role of echolocation signals.

5.4 Future Directions

Looking to the future, three research directions are identified here that may be particularly fruitful to pursue when studying acoustic communication in bats.

1. *Energetic costs of signaling displays*: Producing social signals can be energetically expensive, especially the complex songs and/or displays that males produce to attract females. For example, displaying male hammer-headed bats produce loud display calls at leks. When a female approaches, a male roughly doubles his rate of display calling, resulting in a buzz-like sound (Bradbury 1977) that must be associated with a major increase in energy use. Such high-intensity courtship songs have been documented in other species as well, including non-echolocating pteropodids (Wickler and Seibt 1976). Dechmann et al. (2013) provide a nice example of such a study, examining how the use of echolocation in a communicative, non-foraging context can have a non-trivial metabolic cost. Further work looking at acoustic signaling in bats, especially high-amplitude social calls, would provide insight into the cost-benefit trade-off of signaling, potentially allowing us to understand under what conditions such communicative systems would be expected to evolve.
2. *Differential fitness based on signaling behavior*: As mating behavior and sexual selection in bats continues to be explored, the focus should move beyond delineating call function and move toward understanding the role that such signals play in affecting male reproductive success and fitness. If females can detect differences in male signals and such signals are indicative of overall male vigor, then females may base much of their decision about who to mate with on information provided in the songs of males.
3. *Evolution of social calling behavior in a phylogenetic context*: As the availability of molecular tools has increased in recent decades, so have attempts to interpret behavioral patterns in a phylogenetic context. Within bats, some studies have already begun to look at patterns of echolocation divergence in terms of phylogenetic relatedness (i.e., Jones and Teeling 2006; Jung et al. 2014). The potential for using molecular and phylogenetic analyses to understand the evolution of social calls in bats remains, to our knowledge, an untapped area of research.

5.5 Summary

The gregarious nature of bats, along with extensive interspecific variation in ecology, behavior, and social organization, make them a fascinating taxon for examining communication systems. In several species, males produce social calls and/or songs to attract females for mating purposes; such signals may encode information about male quality, as in at least one bat, song complexity is correlated with male reproductive success. Specific social calls are often produced during aggressive encounters with other individuals, often associated with access to feeding sites or to females. Contact calls help to maintain group cohesion and are exchanged between mothers and non-volant offspring (called isolation and directive calls in bats) or more generally between adult group members. In many cases, contact calls encode information about the identity of the sender, permitting discrimination of an individual's own offspring, parent, or group member from other signalers in the environment. Distress calls, which are produced when animals are being attacked, have been reported in many bat species. Interestingly, these calls almost always have an attractive function, drawing other nearby individuals to the location of the sender. As a result, distress calls in bats may primarily function to elicit mobbing of a predator, although more research is necessary to confirm this hypothesis.

While bat echolocation is primarily viewed as a signaling system for orientation and prey detection and tracking, increasingly it is clear that echolocation calls also encode information about the transmitter that is readily available to other individuals. Echolocation conveys information about the foraging activity of the sender; hence, eavesdropping individuals can gather information about the feeding success of nearby bats. Bats are particularly attracted to the feeding buzzes of conspecifics, which are emitted when an individual attempts to capture prey. Such eavesdropping is most valuable in species that exploit patchy, ephemeral food resources, since detection of many feeding buzzes should indicate the location of a high-quality foraging area. Eavesdropping on echolocation calls also has been shown to be the primary mechanism by which some bats locate tree roosts, which are difficult to detect without such cues. Personal information, including species, sex, age, body condition, reproductive condition, group identity and/or individual identity, can be encoded into echolocation. While most research has focused on the statistical discriminability of these features, more studies are now assessing if bats can actually assess personal information in echolocation calls and make behavioral decisions based on that knowledge. In many cases, such discriminatory abilities have been demonstrated.

Many exciting and potentially rich avenues of research remain relatively unexplored among aspects of communication in bats. Studies assessing the energetic costs of social calls will provide clues about the costs and benefits of signaling. Research focused on assessing the impacts of signaling behavior on fitness will be valuable for understanding the role of acoustic communication in sexual selection of bats. Finally, couching information about social calls within a phylogenetic framework will provide insight into the major factors driving the evolution of communication systems in bats.

References

Aldridge, H., & Rautenbach, I. L. (1987). Morphology, echolocation and resource partitioning in insectivorous bats. *Journal of Animal Ecology*, 56(3), 763–778.

Arnold, B. D., & Wilkinson, G. S. (2011). Individual specific contact calls of pallid bats (*Antrozous pallidus*) attract conspecifics at roosting sites. *Behavioral Ecology and Sociobiology*, 65(8), 1581–1593.

August, P. V. (1979). Distress calls in *Artibeus jamaicensis*: Ecology and evolutionary implications. In J. F. Eisenberg (Ed.), *Vertebrate ecology in the northern neotropics* (pp. 151–159). Washington, DC: Smithsonian Institution Press.
August, P. V. (1985). Acoustical properties of the distress calls of *Artibeus jamaicensis* and *Phyllostomus hastatus* (Chiroptera: Phyllostomidae). *The Southwestern Naturalist*, 30(3), 371–375.
Balcombe, J. P. (1990). Vocal recognition of pups by mother Mexican free-tailed bats, *Tadarida brasiliensis mexicana*. *Animal Behaviour*, 39(5), 960–966.
Balcombe, J. P., & Fenton, M. B. (1988). Eavesdropping by bats—the influence of echolocation call design and foraging strategy. *Ethology*, 79(2), 158–166.
Balcombe, J. P., & McCracken, G. F. (1992). Vocal recognition in Mexican free-tailed bats—do pups recognize mothers? *Animal Behaviour*, 43(1), 79–87.
Barclay, R. M. R. (1982). Interindividual use of echolocation calls—eavesdropping by bats. *Behavioral Ecology and Sociobiology*, 10(4), 271–275.
Barclay, R. M. R., & Thomas, D. W. (1979). Copulation call of *Myotis lucifugus*: A discrete situation-specific communication signal. *Journal of Mammalogy*, 60(3), 632–634.
Barlow, K. E., & Jones, G. (1997). Function of pipistrelle social calls: Field data and a playback experiment. *Animal Behaviour*, 53(5), 991–999.
Barta, Z., & Szep, T. (1992). The role of information transfer under different food patterns: A simulation study. *Behavioral Ecology*, 3(4), 318–324.
Bastian, A., & Schmidt, S. (2008). Affect cues in vocalizations of the bat, *Megaderma lyra*, during agonistic interactions. *Journal of the Acoustical Society of America*, 124(1), 598–608.
Bates, M. E., Simmons, J. A., & Zorikov, T. V. (2011). Bats use echo harmonic structure to distinguish their targets from background clutter. *Science*, 333(6042), 627–630.
Beecher, M. D. (1989). Signalling systems for individual recognition: An information theory approach. *Animal Behaviour*, 38(2), 248–261.
Behr, O., & von Helversen, O. (2004). Bat serenades—complex courtship songs of the sac-winged bat (*Saccopteryx bilineata*). *Behavioral Ecology and Sociobiology*, 56(2), 106–115.
Behr, O., von Helversen, O., Heckel, G., Nagy, M., Voigt, C. C., & Mayer, F. (2006). Territorial songs indicate male quality in the sac-winged bat *Saccopteryx bilineata* (Chiroptera, Emballonuridae). *Behavioral Ecology*, 17(5), 810–817.
Behr, O., Knörnschild, M., & von Helversen, O. (2009). Territorial counter-singing in male sac-winged bats (*Saccopteryx bilineata*): Low-frequency songs trigger a stronger response. *Behavioral Ecology and Sociobiology*, 63(3), 433–442.
Bohn, K. M., Schmidt-French, B., Ma, S. T., & Pollak, G. D. (2008). Syllable acoustics, temporal patterns, and call composition vary with behavioral context in Mexican free-tailed bats. *Journal of the Acoustical Society of America*, 124(3), 1838–1848.
Bohn, K. M., Schmidt-French, B., Schwartz, C., Smotherman, M., & Pollak, G. D. (2009). Versatility and stereotypy of free-tailed bat songs. *PLoS ONE*, 4(8), e6746.
Boughman, J. W., & Wilkinson, G. S. (1998). Greater spear-nosed bats discriminate group mates by vocalizations. *Animal Behaviour*, 55(6), 1717–1732.
Bradbury, J. W. (1977). Lek mating behavior in the hammer-headed bat. *Zeitschrift für Tierpsychologie*, 45(3), 225–255.
Bradbury, J. W., & Emmons, L. H. (1974). Social organization of some Trinidad bats. *Zeitschrift für Tierpsychologie*, 36(1–5), 137–183.
Bradbury, J. W., & Vehrencamp, S. L. (1977). Social organization and foraging in emballonurid bats. III. Mating systems. *Behavioral Ecology and Sociobiology*, 2(1), 1–17.
Bradbury, J. W., & Vehrencamp, S. L. (2011). *Principles of animal communication*. Sunderland, MA: Sinauer Associates, Inc.
Brown, P. (1976). Vocal communication in pallid bat, *Antrozous pallidus*. *Zeitschrift für Tierpsychologie*, 41(1), 34–54.
Brown, P. E., Brown, T. W., & Grinnell, A. D. (1983). Echolocation, development, and vocal communication in the lesser bulldog bat, *Noctilio albiventris*. *Behavioral Ecology and Sociobiology*, 13(4), 287–298.

Burnett, S. C., Kazial, K. A., & Masters, W. M. (2001). Discriminating individual big brown bat (*Eptesicus fuscus*) sonar vocalizations in different recording situations. *Bioacoustics*, 11(3), 189–210.
Carter, G. G., Skowronski, M. D., Faure, P. A., & Fenton, M. B. (2008). Antiphonal calling allows individual discrimination in white-winged vampire bats. *Animal Behaviour*, 76(4), 1343–1355.
Carter, G. G., Fenton, M. B., & Faure, P. A. (2009). White-winged vampire bats (*Diaemus youngi*) exchange contact calls. *Canadian Journal of Zoology*, 87(7), 604–608.
Carter, G. G., Logsdon, R., Arnold, B. D., Menchaca, A., & Medellin, R. A. (2012). Adult vampire bats produce contact calls when isolated: Acoustic variation by species, population, colony, and individual. *PLoS ONE*, 7(6), e38791.
Catchpole, C., & Slater, P. (2008). *Bird song: Biological themes and variations*. Cambridge, UK: Cambridge University Press.
Chaverri, G., Gillam, E. H., & Vonhof, M. J. (2010). Social calls used by a leaf-roosting bat to signal location. *Biology Letters*, 6(4), 441–444.
Chaverri, G., Gillam, E. H., & Kunz, T. H. (2013). A call-and-response system facilitates group cohesion among disc-winged bats. *Behavioral Ecology*, 24(2), 481–487. doi: 10.1093/beheco/ars188
Chen, S.-F., Jones, G., & Rossiter, S. J. (2009). Determinants of echolocation call frequency variation in the Formosan lesser horseshoe bat (*Rhinolophus monoceros*). *Proceedings of the Royal Society of London B: Biological Sciences*, 276(1674), 3901–3909.
Danchin, E., Giraldeau, L.-A., Valone, T. J., & Wagner, R. H. (2004). Public information: From nosy neighbors to cultural evolution. *Science*, 305(5683), 487–491.
Davidson, S. M., & Wilkinson, G. S. (2004). Function of male song in the greater white-lined bat, *Saccopteryx bilineata*. *Animal Behaviour*, 67, 883–891.
de Fanis, E., & Jones, G. (1995). Post-natal growth, mother-infant interactions and development of vocalizations in the vespertilionid bat *Plecotus auritus*. *Journal of Zoology*, 235(1), 85–97.
Dechmann, D. K. N., Heucke, S. L., Giuggioli, L., Safi, K., Voigt, C. C., & Wikelski, M. (2009). Experimental evidence for group hunting via eavesdropping in echolocating bats. *Proceedings of the Royal Society of London B: Biological Sciences*, 276(1668), 2721–2728.
Dechmann, D. K. N., Kranstauber, B., Gibbs, D., & Wikelski, M. (2010). Group hunting: A reason for sociality in molossid bats? *PLoS ONE*, 5(2), e9012.
Dechmann, D. K. N., Wikelski, M., van Noordwijk, H. J., Voigt, C. C., & Voigt-Heucke, S. L. (2013). Metabolic costs of bat echolocation in a non-foraging context support a role in communication. *Frontiers in Physiology*, doi: 10.3389/fphys.2013.00066
Eckenweber, M., & Knörnschild, M. (2013). Social influences on territorial signaling in male greater sac-winged bats. *Behavioral Ecology and Sociobiology*, 67(4), 639–648.
Endler, J. A. (1987). Predation, light intensity and courtship behaviour in *Poecilia reticulata* (Pisces: Poeciliidae). *Animal Behaviour*, 35(5), 1376–1385.
Esser, K.-H., & Schmidt, U. (1989). Mother-infant communication in the lesser spear-nosed bat *Phyllostomus discolor* (Chiroptera, Phyllostomidae)—evidence for acoustic learning. *Ethology*, 82(2), 156–168.
Esser, K.-H., & Lud, B. (1997). Discrimination of sinusoidally frequency-modulated sound signals mimicking species-specific communication calls in the FM-bat *Phyllostomus discolor. Journal of Comparative Physiology A: Neuroethology, Sensory, Neural, and Behavioral Physiology*, 180(5), 513–522.
Fenton, M. B., Belwood, J. J., Fullard, J. H., & Kunz, T. H. (1976). Responses of *Myotis lucifugus* (Chiroptera-Vespertilionidae) to calls of conspecifics and to other sounds. *Canadian Journal of Zoology*, 54(9), 1443–1448.
Furmankiewicz, J., Ruczyński, I., Urban, R., & Jones, G. (2011). Social calls provide tree-dwelling bats with information about the location of conspecifics at roosts. *Ethology*, 117(6), 480–489.
Gadziola, M. A., Grimsley, J. M. S., Faure, P. A., & Wenstrup, J. J. (2012). Social vocalizations of big brown bats vary with behavioral context. *PLoS ONE*, 7(9), e44550.
Gelfand, D. L., & McCracken, G. F. (1986). Individual variation in the isolation calls of Mexican free-tailed bat pups (*Tadarida brasiliensis mexicana*). *Animal Behaviour*, 34(4), 1078–1086.

Georgiakakis, P., & Russo, D. (2012). The distinctive structure of social calls by Hanak's dwarf bat *Pipistrellus hanaki*. *Acta Chiropterologica*, 14(1), 167–174.

Gillam, E. H. (2007). Eavesdropping by bats on the feeding buzzes of conspecifics. *Canadian Journal of Zoology*, 85(7), 795–801.

Gillam, E. H., & Chaverri, G. (2012). Strong individual signatures and weaker group signatures in contact calls of Spix's disc-winged bat, *Thyroptera tricolor. Animal Behaviour*, 83(1), 269–276.

Gould, E. (1975). Experimental studies of the ontogeny of ultrasonic vocalizations in bats. *Developmental Psychobiology*, 8(4), 333–346.

Griffin, D. R. (1958). *Listening in the dark*. New Haven, CT: Yale University Press.

Grilliot, M. E., Burnett, S. C., & Mendonça, M. T. (2009). Sexual dimorphism in big brown bat (*Eptesicus fuscus*) ultrasonic vocalizations is context dependent. *Journal of Mammalogy*, 90(1), 203–209.

Guillén, A., Juste, J., & Ibáñez, C. (2000). Variation in the frequency of the echolocation calls of *Hipposideros ruber* in the Gulf of Guinea: An exploration of the adaptive meaning of the constant frequency value in rhinolophoid CF bats. *Journal of Evolutionary Biology*, 13(1), 70–80.

Gustin, M. K., & McCracken, G. F. (1987). Scent recognition between females and pups in the bat *Tadarida brasiliensis mexicana. Animal Behaviour*, 35(1), 13–19.

Heckel, G., Voigt, C. C., Mayer, F., & von Helversen, O. (1999). Extra-harem paternity in the white-lined bat *Saccopteryx bilineata* (Emballonuridae). *Behaviour*, 136(9), 1173–1185.

Heller, K.-G., & von Helversen, O. (1989). Resource partitioning of sonar frequency bands in rhinolophoid bats. *Oecologia*, 80(2), 178–186.

Hill, D. A., & Greenaway, F. (2005). Effectiveness of an acoustic lure for surveying bats in British woodlands. *Mammal Review*, 35(1), 116–122.

Hiryu, S., Katsura, K., Nagato, T., Yamazaki, H., Lin, L.-K., Watanabe, Y., & Riquimaroux, H. (2006). Intra-individual variation in the vocalized frequency of the Taiwanese leaf-nosed bat, *Hipposideros terasensis*, influenced by conspecific colony members. *Journal of Comparative Physiology A: Neuroethology, Sensory, Neural, and Behavioral Physiology*, 192(8), 807–815.

Jakobsen, L., & Surlykke, A. (2010). Vespertilionid bats control the width of their biosonar sound beam dynamically during prey pursuit. *Proceedings of the National Academy of Sciences of the USA*, 107(31), 13930–13935.

Jakobsen, L., Ratcliffe, J. M., & Surlykke, A. (2013). Convergent acoustic field of view in echolocating bats. *Nature*, 493(7430), 93–96.

Jameson, J. W., & Hare, J. F. (2009). Group-specific signatures in the echolocation calls of female little brown bats (*Myotis lucifugus*) are not an artifact of clutter at the roost entrance. *Acta Chiropterologica*, 11(1), 163–172.

Jones, G. (1995). Variation in bat echolocation: Implications for resource partitioning and communication. *Le Rhinolophe*, 11, 53–59.

Jones, G. (1999). Scaling of echolocation call parameters in bats. *Journal of Experimental Biology*, 202(23), 3359–3367.

Jones, G., & Ransome, R. D. (1993). Echolocation calls of bats are influenced by maternal effects and change over a lifetime. *Proceedings of the Royal Society of London B: Biological Sciences*, 252(1334), 125–128.

Jones, G., & Kokurewicz, T. (1994). Sex and age variation in echolocation calls and flight morphology of Daubenton's bats, *Myotis daubentonii. Mammalia*, 58(1), 41–50.

Jones, G., & Barlow, K. E. (2003). *Cryptic species of echolocating bats*. Chicago, IL: University of Chicago Press.

Jones, G., & Teeling, E. C. (2006). The evolution of echolocation in bats. *Trends in Ecology & Evolution*, 21(3), 149–156.

Jones, G., & Siemers, B. (2011). The communicative potential of bat echolocation pulses. *Journal of Comparative Physiology A: Neuroethology, Sensory, Neural, and Behavioral Physiology*, 197(5), 447–457.

Jones, G., Gordon, T., & Nightingale, J. (1992). Sex and age differences in the echolocation calls of the lesser horseshoe bat, *Rhinolophus hipposideros. Mammalia*, 56(2), 189–194.

Jones, G., Morton, M., Hughes, P. M., & Budden, R. M. (1993). Echolocation, flight morphology and foraging strategies of some West African hipposiderid bats. *Journal of Zoology*, 230(3), 385–400.

Jones, G., Sripathi, K., Waters, D. A., & Marimuthu, G. (1994). Individual variation in the echolocation calls of three sympatric Indian hipposiderid bats and an experimental attempt to jam bat echolocation. *Folia Zoologica*, 43(4), 347–361.

Jonker, M. N., de Boer, W. F., Kurvers, R. H. J. M., & Dekker, J. J. A. (2010). Foraging and public information use in common pipistrelle bats (*Pipistrellus pipistrellus*): A field experiment. *Acta Chiropterologica*, 12(1), 197–203.

Jung, K., Molinari, J., & Kalko, E. K. V. (2014). Driving factors for the evolution of species-specific echolocation call design in New World free-tailed bats (Molossidae). *PLoS ONE*, 9(1), e85279.

Kazial, K. A., & Masters, W. M. (2004). Female big brown bats, *Eptesicus fuscus*, recognize sex from a caller's echolocation signals. *Animal Behaviour*, 67(5), 855–863.

Kazial, K. A., Burnett, S. C., & Masters, W. M. (2001). Individual and group variation in echolocation calls of big brown bats, *Eptesicus fuscus* (Chiroptera: Vespertilionidae). *Journal of Mammalogy*, 82(2), 339–351.

Kazial, K. A., Pacheco, S., & Zielinski, K. N. (2008a). Information content of sonar calls of little brown bats (*Myotis lucifugus*): Potential for communication. *Journal of Mammalogy*, 89(1), 25–33.

Kazial, K. A., Kenny, T. L., & Burnett, S. C. (2008b). Little brown bats (*Myotis lucifugus*) recognize individual identity of conspecifics using sonar calls. *Ethology*, 114(5), 469–478.

Knörnschild, M., & von Helversen, O. (2008). Nonmutual vocal mother-pup recognition in the greater sac-winged bat. *Animal Behaviour*, 76(3), 1001–1009.

Knörnschild, M., & Tschapka, M. (2012). Predator mobbing behaviour in the greater spear-nosed bat, *Phyllostomus hastatus. Chiroptera Neotropical*, 18(2), 1132–1135.

Knörnschild, M., von Helversen, O., & Mayer, F. (2007). Twin siblings sound alike: Isolation call variation in the noctule bat, *Nyctalus noctula. Animal Behaviour*, 74, 1055–1063.

Knörnschild, M., Jung, K., Nagy, M., Metz, M., & Kalko, E. (2012). Bat echolocation calls facilitate social communication. *Proceedings of the Royal Society of London B: Biological Sciences*, 279(1748), 4827–4835.

Knörnschild, M., Feifel, M., & Kalko, E. K. V. (2014). Male courtship displays and vocal communication in the polygynous bat, *Carollia perspicillata. Behaviour*, 151, 781–798.

Kondo, N., & Watanabe, S. (2009). Contact calls: Information and social function. *Japanese Psychological Research*, 51(3), 197–208.

Kozhurina, E. (1996). What may the "songs" of noctule bats disclose to an observer? *Myotis*, 34, 5–15.

Kunz, T. H., & Fenton, M. B. (2006). *Bat ecology.* Chicago: University of Chicago Press.

Leonard, M. L., & Fenton, M. B. (1984). Echolocation calls of *Euderma maculatum* (Vespertilionidae) — use in orientation and communication. *Journal of Mammalogy*, 65(1), 122–126.

Lundberg, K., & Gerell, R. (1986). Territorial advertisement and mate attraction in the bat *Pipistrellus pipistrellus. Ethology*, 71(2), 115–124.

Luo, B., Jiang, T. L., Liu, Y., Wang, J., Lin, A. Q., Wei, X. W., & Feng, J. (2013). Brevity is prevalent in bat short-range communication. *Journal of Comparative Physiology A: Neuroethology, Sensory, Neural, and Behavioral Physiology*, 199(4), 325–333.

Masters, W. M., Raver, K. A. S., & Kazial, K. A. (1995). Sonar signals of big brown bats, *Eptesicus fuscus*, contain information about individual identity, age, and family affiliation. *Animal Behaviour*, 50(5), 1243–1260.

Matsumura, S. (1979). Mother-infant communication in a horseshoe bat (*Rhinolophus ferrumequinum nippon*) — development of vocalization. *Journal of Mammalogy*, 60(1), 76–84.

McCracken, G. F. (1993). Locational memory and female-pup reunions in Mexican free-tailed bat maternity colonies. *Animal Behaviour*, 45(4), 811–813.

McCracken, G. F., & Wilkinson, G. S. (2000). Bat mating systems. In E. G. Crichton & P. H. Krutzsch (Eds.), *Reproductive biology of bats* (pp. 321–362). London, UK: Academic Press.

McWilliam, A. N. (1987). Territorial and pair behavior of the African false vampire bat, *Cardioderma cor* (Chiroptera, Megadermatidae), in coastal Kenya. *Journal of Zoology*, 213, 243–252.

Möhres, F. (1967). Communicative characters of sonar signals in bats. In R.-G. Busnel (Ed.), *Animal sonar systems: Biology and bionics* (pp. 939–945). Tome II, Jouy-en-Josas: Laboratoire de Physiologie Acoustique.
Mora, E. C., Ibáñez, C., Macías, S., Juste, J., López, I., & Torres, L. (2011). Plasticity in the echolocation inventory of *Mormopterus minutus* (Chiroptera, Molossidae). *Acta Chiropterologica*, 13(1), 179–187.
Morton, E. S. (1975). Ecological sources of selection on avian sounds. *The American Naturalist*, 109(965), 17–34.
Morton, E. S. (1977). On the occurrence and significance of motivation-structural rules in some bird and mammal sounds. *The American Naturalist*, 111(981), 855–869.
Moss, C. F., Redish, D., Gounden, C., & Kunz, T. H. (1997). Ontogeny of vocal signals in the little brown bat, *Myotis lucifugus*. *Animal Behaviour*, 54(1), 131–141.
Neuweiler, G., Metzner, W., Heilmann, U., Rübsamen, R., Eckrich, M., & Costa, H. H. (1987). Foraging behaviour and echolocation in the rufous horseshoe bat (*Rhinolophus rouxi*) of Sri Lanka. *Behavioral Ecology and Sociobiology*, 20(1), 53–67.
Nousek, A. E., Slater, P. J. B., Wang, C., & Miller, P. J. O. (2006). The influence of social affiliation on individual vocal signatures of northern resident killer whales (*Orcinus orca*). *Biology Letters*, 2(4), 481–484.
Pearl, D. L., & Fenton, M. B. (1996). Can echolocation calls provide information about group identity in the little brown bat (*Myotis lucifugus*)? *Canadian Journal of Zoology*, 74(12), 2184–2192.
Popper, A, N., & Fay, R. R. (Eds). (1995). *Hearing by bats*. New York, NY: Springer-Verlag.
Rhodes, M. (2007). Roost fidelity and fission–fusion dynamics of white-striped free-tailed bats (*Tadarida australis*). *Journal of Mammalogy*, 88(5), 1252–1260.
Ruczyński, I., Kalko, E. K. V., & Siemers, B. M. (2007). The sensory basis of roost finding in a forest bat, *Nyctalus noctula*. *Journal of Experimental Biology*, 210, 3607–3615.
Ruczyński, I., Kalko, E. K. V., & Siemers, B. M. (2009). Calls in the forest: A comparative approach to how bats find tree cavities. *Ethology*, 115(2), 167–177.
Russ, J. M., & Racey, P. A. (2007). Species-specificity and individual variation in the song of male Nathusius' pipistrelles (*Pipistrellus nathusii*). *Behavioral Ecology and Sociobiology*, 61(5), 669–677.
Russ, J. M., Racey, P. A., & Jones, G. (1998). Intraspecific responses to distress calls of the pipistrelle bat, *Pipistrellus pipistrellus*. *Animal Behaviour*, 55(3), 705–713.
Russ, J. M., Jones, G., Mackie, I. J., & Racey, P. A. (2004). Interspecific responses to distress calls in bats (Chiroptera: Vespertilionidae): A function for convergence in call design? *Animal Behaviour*, 67(6), 1005–1014.
Russ, J. M., Jones, G., & Racey, P. A. (2005). Responses of soprano pipistrelles, *Pipistrellus pygmaeus*, to their experimentally modified distress calls. *Animal Behaviour*, 70(2), 397–404.
Ryan, J. M., Clark, D. B., & Lackey, J. A. (1985). *Artibeus lituratus* (Chiroptera, Phyllostomidae) response to distress calls of conspecifics. *Journal of Mammalogy*, 66(1), 179–181.
Safi, K., & Kerth, G. (2007). Comparative analyses suggest that information transfer promoted sociality in male bats in the temperate zone. *American Naturalist*, 170(3), 465–472.
Scherrer, J. A., & Wilkinson, G. S. (1993). Evening bat isolation calls provide evidence for heritable signatures. *Animal Behaviour*, 46(5), 847–860.
Schuchmann, M., & Siemers, B. R. M. (2010). Variability in echolocation call intensity in a community of horseshoe bats: A role for resource partitioning or communication? *PLoS ONE*, 5(9), e12842.
Schuchmann, M., Puechmaille, S. J., & Siemers, B. M. (2012). Horseshoe bats recognise the sex of conspecifics from their echolocation calls. *Acta Chiropterologica*, 14(1), 161–166.
Shannon, C. E., & Weaver, W. (1949). *The mathematical theory of communication*. Urbana: University of Illinois Press.
Siemers, B., & Kerth, G. (2006). Do echolocation calls of wild colony-living Bechstein's bats (*Myotis bechsteinii*) provide individual-specific signatures? *Behavioral Ecology and Sociobiology*, 59(3), 443-454.

Siemers, B. M., Beedholm, K., Dietz, C., Dietz, I., & Ivanova, T. (2005). Is species identity, sex, age or individual quality conveyed by echolocation call frequency in European horseshoe bats? *Acta Chiropterologica*, 7(2), 259–274.

Surlykke, A., & Kalko, E. K. V. (2008). Echolocating bats cry out loud to detect their prey. *PLoS ONE*, 3(4), e2036.

Surlykke, A., Futtrup, V., & Tougaard, J. (2003). Prey-capture success revealed by echolocation signals in pipistrelle bats (*Pipistrellus pygmaeus*). *Journal of Experimental Biology*, 206(1), 93–104.

Surlykke, A., Nachtigall, P., Fay, R. R., & Popper, A. N. (Eds). (2014). *Biosonar.* New York, NY: Springer-Verlag.

Suthers, R. A. (1967). Comparative echolocation by fishing bats. *Journal of Mammalogy*, 48(1), 79–87.

Übernickel, K., Tschapka, M., & Kalko, E. K. V. (2013). Selective eavesdropping behaviour in three neotropical bat species. *Ethology*, 119(1), 66–76.

Valone, T. J. (2007). From eavesdropping on performance to copying the behavior of others: A review of public information use. *Behavioral Ecology and Sociobiology*, 62(1), 1–14.

Vaughan, T. A. (1976). Nocturnal behavior of the African false vampire bat (*Cardioderma cor*). *Journal of Mammalogy*, 57(2), 227–248.

Vaughan, T. A., & O'Shea, T. J. (1976). Roosting ecology of the pallid bat, *Antrozous pallidus*. *Journal of Mammalogy*, 57(1), 19–42.

Voigt, C. C., & von Helversen, O. (1999). Storage and display of odour by male *Saccopteryx bilineata* (Chiroptera, Emballonuridae). *Behavioral Ecology and Sociobiology*, 47(1–2), 29–40.

Voigt, C. C., Behr, O., Caspers, B., von Helversen, O., Knörnschild, M., Mayer, F., & Nagy, M. (2008). Songs, scents, and senses: Sexual selection in the greater sac-winged bat, *Saccopteryx bilineata*. *Journal of Mammalogy*, 89, 1401–1410.

Voigt-Heucke, S. L., Taborsky, M., & Dechmann, D. K. N. (2010). A dual function of echolocation: Bats use echolocation calls to identify familiar and unfamiliar individuals. *Animal Behaviour*, 80(1), 59–67.

Wickler, W., & Seibt, U. (1976). Field studies on the African fruit bat, *Epomophorus wahlbergi* (Sundevall), with special reference to male calling. *Zeitschrift für Tierpsychologie*, 40(4), 345–376.

Wiley, R. H., & Richards, D. G. (1978). Physical constraints on acoustic communication in the atmosphere: Implications for the evolution of animal vocalizations. *Behavioral Ecology and Sociobiology*. 3(1), 69–94.

Wiley, R. H., & Richards, D. G. (1982). Adaptations for acoustic communication in birds: Sound transmission and signal detection. In D. Kroodsma and E.A. Miller (Eds.), *Acoustic communication in birds I* (pp. 131–181). New York, NY: Academic Press.

Wilkinson, G. S. (1985). The social organization of the common vampire bat. *Behavioral Ecology and Sociobiology*, 17(2), 123–134.

Wilkinson, G. S. (1992). Information transfer at evening bat colonies. *Animal Behaviour*, 44(3), 501–518.

Wilkinson, G. S., & Boughman, J. W. (1998). Social calls coordinate foraging in greater spear-nosed bats. *Animal Behaviour*, 55(2), 337–350.

Willis, C. R., & Brigham, R. M. (2004). Roost switching, roost sharing and social cohesion: Forest-dwelling big brown bats, *Eptesicus fuscus*, conform to the fission–fusion model. *Animal Behaviour*, 68(3), 495–505.

Willis, C. R., & Brigham, R. M. (2007). Social thermoregulation exerts more influence than microclimate on forest roost preferences by a cavity-dwelling bat. *Behavioral Ecology and Sociobiology*, 62(1), 97–108.

Yovel, Y., Melcon, M. L., Franz, M. O., Denzinger, A., & Schnitzler, H.-U. (2009). The voice of bats: How greater mouse-eared bats recognize individuals based on their echolocation calls. *PLoS Computational Biology*, 5(6), e1000400.

Zahavi, A. (1975). Mate selection—a selection for a handicap. *Journal of Theoretical Biology*, 53(1), 205–214.

Chapter 6
Guild Structure and Niche Differentiation in Echolocating Bats

Annette Denzinger, Elisabeth K.V. Kalko, Marco Tschapka, Alan D. Grinnell, and Hans-Ulrich Schnitzler

Keywords Bat • Echolocation • Foraging • Guild • Habitat • Masking • Niche differentiation • Resource partitioning

6.1 Diversity in Bats

Bats (Order: Chiroptera) differ from all other mammals in their capacity for powered flight and, with the exception of Old World fruit bats (Family: Pteropodidae), in their use of laryngeal echolocation (Denzinger et al. 2004; Schnitzler et al. 2004; Jones and Teeling 2006). Most Old World fruit bats mainly rely on vision for

---

Author (Elisabeth K. V. Kalko) was deceased at the time of publication.

A. Denzinger (✉) • H.-U. Schnitzler
Animal Physiology, Institute for Neurobiology, University of Tübingen,
Auf der Morgenstelle 28, Tübingen 72076, Germany
e-mail: annette.denzinger@uni-tuebingen.de; hans-ulrich.schnitzler@uni-tuebingen.de

E.K.V. Kalko
Institute of Experimental Ecology, University of Ulm,
Albert-Einstein-Allee 11, Ulm 89069, Germany

Smithsonian Tropical Research Institute,
Apartado, P.O. Box 0843-03092, Balboa Ancón, Panamá, Republica de Panama
e-mail: marco.tschapka@uni-ulm.de

M. Tschapka
Institute for Evolutionary Ecology and Conservation Genomics,
Albert-Einstein-Allee 11, Ulm 89069, Germany

Smithsonian Tropical Research Institute,
Apartado, P.O. Box 0843-03092, Balboa Ancón, Panamá, Republica de Panama
e-mail: marco.tschapka@uni-ulm.de

A.D. Grinnell
Department of Physiology, David Geffen School of Medicine,
University of California, Los Angeles, CA 90095, USA
e-mail: adg@mednet.ucla.edu

M.B. Fenton et al. (eds.), *Bat Bioacoustics*, Springer Handbook of Auditory Research 54, DOI 10.1007/978-1-4939-3527-7_6

orientation and foraging; however, bats from the genus *Roussettus* use tongue clicks for lingual echolocation (Kulzer 1956). Since their echolocation systems and foraging behaviors differ in many aspects from that of laryngeal echolocators, *Roussettus* spp. are not considered in this chapter.

Bats with laryngeal echolocation are ecologically more diverse than any other group of mammals, especially in the tropics where bat assemblages comprise large numbers of species with highly diverse foraging areas, foraging modes, and diets (Kalko 1997). The high diversity in bats is based on a great variety of food sources, such as insects and other arthropods, small vertebrates including fish, fruit, nectar, pollen, and even blood (Gardner 1977; Rojas et al. 2012). When searching for food, bats have to perform a multitude of tasks depending on where they forage, what they eat, and how they acquire their food. In the course of evolution, the echolocation and hearing systems of the slightly more than 1,000 echolocating bat species have been adapted to gather behaviorally relevant information while performing species-specific tasks. Bats foraging under similar ecological conditions share similar adaptations in their sensory and motor systems and in their behaviors. Thus bats can be assigned to functional groups or guilds that indicate the ecological position of species within the local ecosystems (Schnitzler and Kalko 1998; Denzinger and Schnitzler 2013).

6.2 Sensory and Motor Tasks of Foraging Bats

When searching for food, bats are challenged in several ways. They have to orient in space, recognize biotopes that provide specific resources, and find and acquire their food.

6.2.1 Spatial Orientation

Foraging bats must determine their own position in relation to the world around them while navigating from the roost to their foraging areas and back and while searching for food. During migration they even travel over large distances through areas where they have never been before (e.g., Rojas-Martinez et al. 1999; McGuire et al. 2012). The different spatial orientation challenges facing bats have been subdivided into three levels of navigation (Schnitzler et al. 2003). Small-scale navigation encompasses all tasks where bats locate and identify nearby targets such as obstacles, landing sites, or landmarks by echolocation. Middle-scale navigation includes the ability of bats to fly on routes along landmarks to goals that are beyond their echolocation range. This planning of routes requires a topological representation of the home range. Finally, large-scale navigation includes migration and homing. Large-scale navigation is not based on echolocation (Hill and Smith 1984) but relies on other modalities, such as vision and magnetic sensitivity (Holland et al. 2006).

The sensory and motor tasks of navigating bats depend on their distance to background targets. Depending on signal frequency, dynamic range of the echolocation system, and atmospheric conditions, the maximum range for the detection of landmarks and other background targets, such as a forest edge or a meadow, can be nearly 70 m for a bat operating with low-frequency signals in temperate zones or down to only 2.4 m in a bat using high-frequency signals in the humid tropics (Stilz and Schnitzler 2012). The closer bats fly to the background, the less space they have for flight maneuvers and the greater is the risk of colliding with obstacles. Species-specific adaptations in wing morphology, which reflect maneuverability, indicate the preferred foraging distance to the background (Norberg and Rayner 1987; Fenton 1990; Norberg 1994). Bats flying in the vicinity of background targets change their echolocation behavior, thus indicating that they collect environmental information necessary for route planning and obstacle avoidance.

6.2.2 Biotope Recognition

Biotope recognition in bats is a multimodal process that includes all senses and is also affected by spatial memory. Nevertheless, echolocation always plays a key role. Background targets such as trees, forest edges, meadows, water surfaces, and rocks not only are effective as landmarks and guide lines for route following, but they also provide information about foraging biotopes that deliver specific prey. Biotope recognition, therefore, is just as important for foraging bats as landmark recognition. Statistical properties of echoes from vegetation allow the classification of typical biotope elements such as trees and bushes (Grunwald et al. 2004; Yovel et al. 2009, 2011).

6.2.3 Food Finding

The difficulty of finding food depends strongly on where the wanted resource is positioned. With decreasing distance between prey and background, it gets more difficult for bats to find and acquire a food item (Schnitzler and Kalko 1998). If food is found solely by echolocation, bats must be able to separate food echoes from background echoes or clutter, which is hampered by and may even be made impossible if the food echoes overlap with the clutter echoes. In this case, bats have to rely on other sensory cues such as odor or prey-generated sounds to find the food (Schmidt et al. 1988; Hessel and Schmidt 1994). According to their feeding habits, echolocating bats can be separated into two groups: animalivorous and phytophagous species. Animalivorous species feed on prey animals that do their best to avoid being eaten, whereas phytophagous species feed on fruits or nectar and pollen of flowers, which are readily advertised because bats provide essential seed dispersing and pollination services to the plants.

6.3 The Masking Problem

Foraging bats emit a continuous sequence of echolocation signals and analyze the complex of emitted signal and returning echoes to detect, localize, and classify echo-producing targets. The key challenge is distinguishing prey echoes from echoes returning from background targets. One of the main problems is the masking of a weak prey echo by a strong signal or by strong background echoes. Depending on the temporal position of a food item echo in relation to the emitted signal and to clutter echoes, two forms of masking exist. Forward masking occurs if the food echo is close in time or overlaps with the preceding emitted signal. Backward masking occurs if the food echo is close in time or overlaps with the succeeding clutter echoes. To a first approximation, therefore, it is assumed that relevant masking occurs if either the emitted signal or the clutter echoes (or both) start to overlap with the echo from the food item.

Insects are positioned in the signal overlap zone if they are so close to a bat that their echoes overlap with the emitted signal and in the clutter overlap zone if they are so close to the background that the clutter echoes from the background overlap with the prey echo (Figure 6.1). An undisturbed evaluation of echoes is possible only if targets are positioned in the overlap-free window between the signal and clutter overlap zones (Kalko and Schnitzler 1993; Schnitzler and Kalko 2001; Schnitzler et al. 2003). The width of the overlap zones and the overlap-free window depend on signal duration. At a sound speed of 340 m/s, the width of the overlap zones is calculated by multiplying the sound duration (in ms) by 0.17 m. A reduction of the signal duration by 1 ms reduces the width of the overlap zones by 0.17 m and increases the width of the overlap-free window by 0.34 m (Figure 6.1). Field and laboratory data from several species indicate that bats tend to avoid overlap between the emitted signal and the echoes from the target of interest (Kalko and Schnitzler 1989, 1993). The masking effects depend critically on the structure and sound pressure level (SPL) of the interfering sounds and their temporal relation to the target echo. Steeply modulated signals tolerate more overlap than shallowly modulated signals; hence their masking zones are smaller than the overlap zones calculated from sound duration (Siemers and Schnitzler 2004).

Bats that use signals with a long constant-frequency (CF) component followed by a short frequency-modulated (FM) component (CF-FM signal) use another principle to avoid masking. These bats compensate for Doppler shifts created by their own flight speed by lowering the frequency of the loud emitted signals. Thus they reduce masking by keeping the frequency of the weak echoes constant and exactly in the center of their auditory fovea (a highly expanded frequency representation on the basilar membrane in the range of the second harmonic of the echolocation signals, where the auditory threshold is very low), while the emission frequency is kept at a slightly lower frequency range with a higher threshold (for review see Schnitzler and Denzinger 2011). However, like other bats using FM signals alone, they need to avoid overlap of their terminal FM components (Schnitzler 1968). Fenton (1995) used the two different principles of masking avoidance to classify

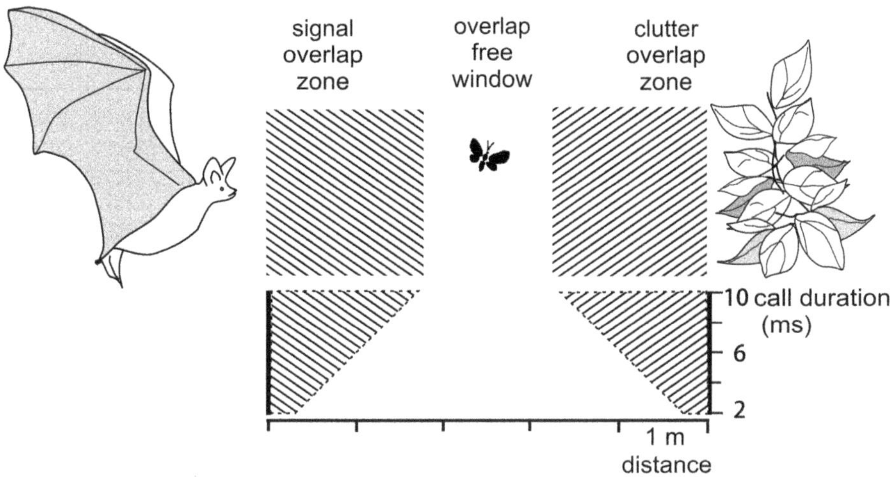

Fig. 6.1 Overlap situation in a bat foraging for an insect at a distance of 5 m from vegetation. If an insect flies in the signal overlap zone, its echo overlaps with the emitted signal. In the clutter overlap zone, prey echoes overlap with the echoes from the background. No overlap occurs in the overlap-free window. The width of the zones depends on signal duration. A reduction of the signal duration from 10 to 2 ms reduces the width of the signal and clutter overlap zone from 1.70 to 0.34 m. A reduction of signal duration by 1 ms increases the width of the overlap-free window by 0.34 cm (adapted from Denzinger and Schnitzler 2013)

bats into two groups. Based on the duty cycle of the echolocation signals he discriminated between non-overlap-tolerant low duty cycle bats, which use short echolocation signals so that pulse and echo are separated in time and overlap-tolerant high duty cycle bats, which achieve a separation of their long pulses from the overlapping echoes in the frequency range.

6.4 Habitat Types and Foraging Modes

6.4.1 Definitions

Comparative studies of the echolocation and foraging behavior of bats have shown that the position of a bat's preferred food relative to background sources of echoes has been the most important ecological constraint during evolution. Sensory and motor adaptations reflect where bats search for food and how they acquire it. Several attempts have been made to classify bats according to foraging habitats. All approaches distinguish three main foraging areas or habitat types, which Fenton (1990) named open, edge, and closed habitats (for a review see Schnitzler et al. 2003).

Habitat is not just the place where an animal lives. In agreement with Krausman (1999), a foraging habitat is defined by the resource availability and the conditions that a species encounters when searching for food (Denzinger and Schnitzler 2013). For bats, habitat types are defined by the echolocation conditions in which each species forages for its typical resources. This functional definition implies that bats exploiting similar resources under similar conditions use the same type of habitat. The spatial extent of such a functionally defined habitat type differs between species. Three habitat types that obviously contain elements of previous definitions are distinguished here: open, edge, and narrow space (Figure 6.2).

In "open spaces" bats exploit airborne prey flying so far from the background that clutter echoes reach the bats considerably later than the echoes from the prey, if at all (Figure 6.2). The criterion for open space is that bats do not react to background objects in their echolocation behavior. The flying prey is caught in aerial-hawking mode.

In "edge spaces" bats exploit airborne prey found near the edges of vegetation and buildings, in gaps, just above the ground, or drifting on water surfaces. Here the prey echo is always promptly followed by clutter echoes from the background. In edge space, bats react to background targets in their echolocation behavior. The echolocation situation differs depending on whether bats capture airborne prey in aerial-hawking mode or take prey from water surfaces in trawling mode. When foraging in aerial-hawking mode, each pair of emitted signal and prey echo is followed by background echoes (Figure 6.2). When foraging in trawling mode, bats fly low over water and emit signals at an oblique angle toward the water surface. The flat water surface acts like a mirror and reflects most of the signal away from the bat. Only sound waves that hit the water perpendicularly are reflected back to the bat, so that a low-flying trawling bat receives an echo from below immediately after signal emission (Greif and Siemers 2010). The time delay between signal emission and this ventral echo encodes the flight height of the bat above the water. All other emitted sound waves are reflected away from the bat due to the mirror properties of a flat surface. Clutter echoes are produced if the direct and mirrored sound waves hit background structures and are projected back to the bat. A prey item drifting on the water surface produces an additional echo after the echo from the water below, which delivers the information that the bats need for the detection, localization, and classification of the prey (Figure 6.2). The echoes from drifting prey always contain both a direct and a mirrored component, making the echo stronger than that from the same prey item suspended in air (Siemers et al. 2001, 2005). This increase in echo SPL is called the mirror effect (Siemers et al. 2005). It has to be acknowledged that trawling bats can also hunt successfully under conditions when the water surface is far from mirror smooth, so the situation is often more challenging (e.g., Blood and Clark 1998).

In "narrow spaces" bats exploit a wide variety of resources. Animalivorous species forage for animals that are found near or on surfaces such as vegetation or the ground. Phytophagous species forage for fruits or flowers offering nectar, both of which are part of the background. In both groups, the echoes from food items often overlap and interfere with echoes from the background. These bats have the difficult task of distinguishing the food item from background targets, which may not be possible with echolocation alone.

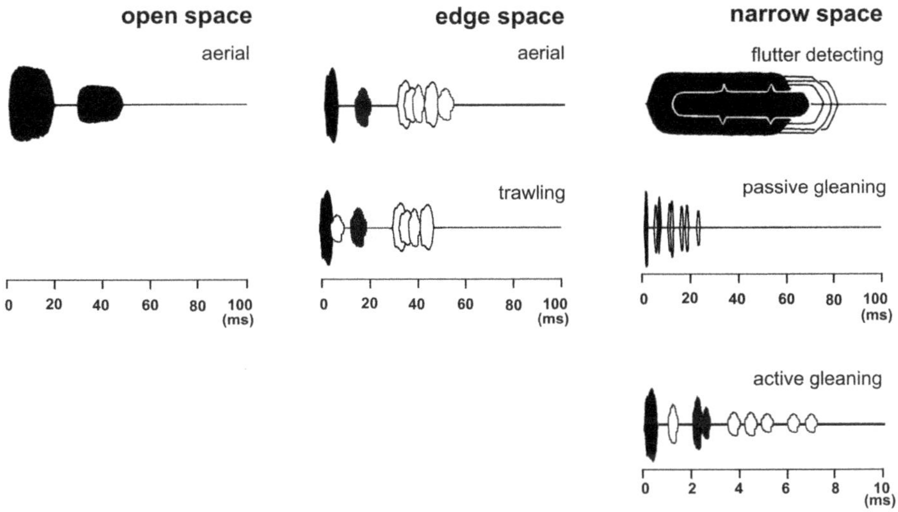

Fig. 6.2 Habitat types and foraging modes of bats. Diagrams of echolocation scenarios for bats searching for food in three types of foraging habitats with typical foraging modes. The emitted signal and the prey echo are depicted in *black*. Echoes from background targets are depicted in *white*. "Open space" bats perceive a pulse-echo pair of the emitted signal and the returning prey echo without clutter echoes. In "edge space," the pulse-echo pair is followed by clutter echoes from background. In trawling bats, the emitted signal is immediately followed by an echo from the water surface (in *white*). In "narrow space," the prey echoes are positioned in the clutter overlap zone. In flutter-detecting foragers, the long signals are modulated by the beating wings of insects and, therefore, can be distinguished from unmodulated clutter echoes. In passive-gleaning foragers, the prey echoes of the very short signals are buried in clutter echoes and cannot be distinguished. These bats have to rely on other senses for the detection of their prey. Active-gleaning foragers rely on favorable echolocation situations where the prey echoes are either isolated in time or so conspicuous that bats are able to find them between clutter echoes (adapted from Denzinger and Schnitzler 2013)

Different foraging strategies have evolved to solve this problem. Bats that are specialized to perceive the wing movements of fluttering insects use the *flutter-detecting mode*. The echoes of the long CF components of their signals are modulated in amplitude and frequency by the fluttering wings of prey insects, allowing them to be distinguished from unmodulated clutter echoes from the background (reviewed in Schnitzler and Denzinger 2011). Echoes from non-flying prey that are buried in clutter and are not specific enough to be discriminated from clutter echoes pose a detection and classification problem that cannot be solved by echolocation. Under these conditions, bats have to rely on other senses, and they use food-generated cues such as odors or prey-generated sounds to find the food (Laska 1990; Page and Ryan 2005). These bats use the *passive-gleaning mode*. Some species identify echoes from targets of interest with echolocation even amid clutter echoes from background targets. They forage in the *active-gleaning mode*. Either they are able to separate the food echoes from clutter echoes on a microsecond time scale (Denzinger and Schnitzler 2013; Geipel et al. 2013) or they search for conspicuous echoes. Such echoes are produced, for example, by dish-shaped leaves and are rather invariant over a larger approach angle. The leaves advertise the flowers and guide the bats toward them (von Helversen and von Helversen 1999; Simon et al. 2011).

6.4.2 Borders Between Habitats

The definition of habitat categories into open, edge, and narrow space poses the problem of defining the borders between them. Here, definitions are used that were first proposed by Schnitzler and Kalko (1998, 2001) based on the echolocation behavior of the bats.

Bats in open space do not react to the background whereas in edge space they do. This border between open and edge is species specific. Schaub and Schnitzler (2007) showed for the parti-color bat, *Vespertilio murinus*, that the border between the two habitat types is at 6 m from vertical background structures and 5 m above the ground. Closer than that in either dimension, the echolocation behavior (flight speed, signal duration, and signal bandwidth) is changed systematically in relation to the distance to background. The border between open and edge space was determined to be at a height of about 5 m in the common pipistrelle, *Pipistrellus kuhlii*; about 3 m in the soprano pipistrelle, *Pipistrellus pygmaeus*; and 8–10 m in the serotine bat, *Eptesicus serotinus*, and the northern bat, *Eptesicus nilssonii* (Kalko and Schnitzler 1993; Rydell 1993; Jensen and Miller 1999). Denzinger and Schnitzler (2013) concluded that the species-specific extensions of edge space reflect differences in the ability of bats to maneuver near background structures. Large and fast-flying bats need more space for their maneuvers than smaller bats that are better equipped for obstacle avoidance. The distance-dependent changes in echolocation behavior in edge space indicate that bats collect information that they need for successful navigation and maneuvering near background structures. However, background objects can be detected at distances greater than those at which bats begin to react (Stilz and Schnitzler 2012).

The border between edge and narrow space is defined as a function of the sensory demands of finding food that is positioned close to or on background structures (Schnitzler and Kalko 2001; Schnitzler et al. 2003; Denzinger and Schnitzler 2013). Narrow space begins when the food item echo is masked by background echoes. As already mentioned, however, masking effects depend on the echolocation behavior of the bats and are determined by the structure, SPLs, and temporal relationship between food and clutter echoes. Therefore the exact dimension of the masking zone cannot be determined. For practical purposes, the border between the overlap free window and clutter overlap zone is used as the border between edge and narrow space. The calculation of the overlap zone from the signal duration is only a first approximation for the real masking zone.

6.5 Bat Guilds

6.5.1 *Definition of Guilds*

A structuring of bat assemblages based on habitat types alone is not sufficient for the identification of distinct functional groups or guilds since it does not take into account the different adaptations for specific foraging modes. Root (1967) defined a guild as "a group of species that exploits the same class of environmental resources in a similar way." Bats from the same guild perform similar tasks and share similar adaptations, whereas bats from different guilds differ in the resources they use and in the way they obtain them.

Several approaches have been used to categorize bats into groups with similar feeding habits and foraging behaviors (for reviews see Fenton 1990; Kalko 1998; Schnitzler et al. 2003). For example, bats can be categorized by feeding behavior: aerial insectivory, gleaning insectivory, piscivory, sanguivory, nectarivory, frugivory, omnivory, and carnivory (McNab 1971; Hill and Smith 1984). Based on wing morphology and foraging behavior, bats have been classified as fast hawking, slow hawking, trawling, gleaning, hovering, fly-catching, and perch hunting (Norberg and Rayner 1987). In another approach, foraging behavior and habitat use were used to classify bats into groups (Aldridge and Rautenbach 1987; Neuweiler 1989; Fenton 1990). This approach has been developed further with a more precise definition of bat habitats and the assignment of bats into functional groups or guilds according to the ecological conditions and the motor and sensory tasks they encounter (Schnitzler and Kalko 2001; Schnitzler et al. 2003; Denzinger and Schnitzler 2013).

Here, following Schnitzler et al. (2003) and Denzinger and Schnitzler (2013), guilds are exclusively defined by habitat type and foraging mode. In a former approach, Schnitzler and Kalko (1998, 2001) additionally defined guilds based on the bats' diet, distinguishing aerial insectivores, gleaning insectivores, carnivores, piscivores, sanguivores, frugivores, nectarivores, and omnivores. However, comparative studies have revealed that the type of food had far less influence on the evolution of adaptive foraging and echolocation behavior than the foraging conditions, which are sufficiently described by habitat type and foraging mode (Schnitzler et al. 2003). For instance, gleaning insectivorous bats, as well as gleaning carnivorous bats, search in the passive mode for insects or mice moving on the ground and face similar tasks. Both have to find and approach sound sources that produce rustling noises within a cluttered background. Likewise, frugivorous and nectarivorous bats have to perform similar tasks when searching for fruits or flowers, both of which advertise their presence to the bats by odor and/or by specific reflective properties (Thies et al. 1998; von Helversen et al. 2000; Simon et al. 2011). Seven guilds defined by habitat type and foraging mode are sufficient to structure even the most diverse bat assemblages (Figs. 6.3 and 6.4) (Denzinger and Schnitzler 2013).

Fig. 6.3 The three foraging habitats of bats with typical representative species and examples of their echolocation signals. The borders between the habitat types are determined by the species-specific echolocation behavior

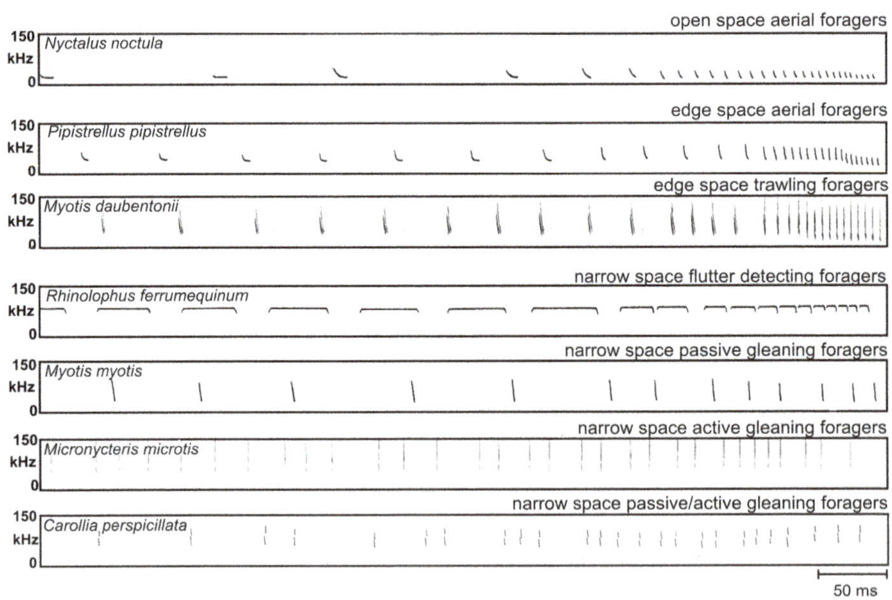

Fig. 6.4 Search and approach sequences from representative members of each of the seven bat guilds. The approach sequences of "open space" and "edge space" foragers end with a distinct terminal group consisting of buzz I and buzz II. The flutter-detecting forager *Rhinolophus ferrumequinum* maintains the CF component of the long CF-FM signals in all situations, even in the terminal group. All other "narrow space" foragers approach stationary targets and mostly emit less distinct terminal groups without buzz II (figure as originally published in Denzinger and Schnitzler 2013)

6.5.1.1 Open Space Aerial Foragers

Bats foraging in open space must locate prey that is often dispersed over a large volume of space and thus is difficult to find. These bats have evolved echolocation systems adapted for long-range detection of prey. In search flight, they emit shallowly frequency-modulated signals with call durations of 8–25 ms (Figure 6.4) that facilitate the detection of weak prey echoes since the signal energy is concentrated for a substantial time in the corresponding neural filters of the auditory system (Schnitzler and Kalko 1998). Additionally, long signals increase the chance of perceiving glints (short amplitude peaks in echoes from a moving reflector at the instant of perpendicular sound incidence) in the echo generated by insect wingbeats. The probability of receiving such a glint depends on the duty cycle and the wingbeat rate of the insect. For instance, a bat with a duty cycle of 10 % perceives an average of 6 glints/s from a moth with a wingbeat rate of 60 Hz (Schnitzler and Kalko 1998). The frequency of the narrowband signal is usually below 35 kHz and can reach values below 10 kHz (e.g., in the European free-tailed bat, *Tadarida teniotis*). Frequencies are species specific but may overlap between species. Bats use either the first or second harmonic predominantly, a characteristic that is related to phylogeny. Intervals between search pulses are long because calls are often emitted only every second or third wingbeat. Even pulse intervals of more than 1 s are possible (e.g., in the genus *Tadarida* and *Eumops*).

The average source levels measured in open space bats range from 104 to 111 dB SPL at 1 m (Holderied and von Helversen 2003; Stilz and Schnitzler 2012) and, together with the low frequencies, the levels account for large maximum detection distances. The greater noctule bat, *Nyctalus lasiopterus*, can detect insects with target strengths between −40 and −60 dB up to distances of 11.3 m and 3.7 m, respectively. In all open space foragers, detection distances for prey exceeding 20 m are very unlikely even under the most favorable echolocation conditions (Stilz and Schnitzler 2012; Denzinger and Schnitzler 2013). Open space aerial foragers are mainly found in the families Rhinopomatidae, Emballonuridae, and Molossidae, Vespertilionidae.

6.5.1.2 Edge Space Aerial Foragers

Bats that fly in edge space use echoes from the background to determine their own position in relation to the background and to adjust their flight path and flight maneuvers to avoid collisions. They collect information necessary for biotope recognition and search for prey near background targets. While foraging, edge space aerial foragers often emit mixed signals consisting of a short, broadband frequency-modulated component followed or preceded by a longer and more shallowly modulated narrow bandwidth segment, called a quasi-constant-frequency (QCF) component (Figure 6.4). Bats using pure FM signals, such as many *Myotis* species, often produce a more shallowly modulated segment in the middle of their search signals, which has the same function as the QCF component. The frequency of the narrowband component is species specific, ranges mostly between 30 and 80 kHz in the relevant harmonic, and is suited for target detection at intermediate distances. The broadband, steeply modulated component is suited for more precise localization and classification of background targets and is most likely used to control flight maneuvers near the background, for biotope recognition, and navigation.

Signals have an intermediate duration of about 3–10 ms and are emitted every wingbeat or in groups of two per wingbeat in bats flying close to the background. The source levels range between 101 and 107 dB SPL at 1 m (Holderied and von Helversen 2003; Surlykke and Kalko 2008). Edge space aerial foragers are mainly found among the Emballonuridae, Molossidae, Mormoopidae, Vespertilionidae, Miniopteridae, and in at least one species of Phyllostomidae.

6.5.1.3 Edge Space Trawling Foragers

Trawling foragers fly at low height above calm water surfaces and forage for insects drifting on or flying very close above the water surface, or they forage for entirely aquatic organisms such as fish or small crustaceans. Submersed water-dwelling prey, such as fish, are detected directly when jumping out of the water or through water drops that are generated when they break through the water surface (Schnitzler et al. 1994). When trawling bats fly in the vicinity of the shore, they encounter similar echolocation scenes as edge space aerial foragers. If they fly far off the background, such as in the middle of a lake, the echolocation scene may even be similar to open spaces but with the difference that the emitted signal is always followed by the surface echo. If the water surface is covered with debris or if the water is turbulent, prey echoes are buried in clutter echoes and bats have difficulty finding the prey.

Edge space trawling foragers are found in Vespertilionidae (mainly in *Myotis* species), in Noctilionidae, and in the long-legged bat, *Macrophyllum macrophyllum* (Phyllostomidae). Signal structure not only depends on the echolocation task but is also related to phylogeny. *Myotis* species emit a mixed search phase signal with a steeply modulated component in the beginning and at the end and a shallowly modulated component in between. The species-specific peak frequency (frequency with the highest SPL) is between 30 and 60 kHz and the signals have an intermediate duration of 3–7 ms (Figure 6.4). *Macrophyllum macrophyllum* (Phyllostomidae), in contrast, emits multi-harmonic signals of 2–4 ms duration with the main energy in the second and third harmonic at frequencies above 50 kHz (Weinbeer and Kalko 2007; Brinkløv et al. 2010). The greater bulldog bat, *Noctilio leporinus*, and the lesser bulldog bat, *Noctilio albiventris* (Noctilionidae), produce a combination of pure CF signals and mixed signals consisting of a CF component followed by a FM component. The CF component is species specific and signal durations range from 6 ms, when flying close to the water surface, to 21 ms in *Noctilio albiventris*, when flying in high search flight (Schnitzler et al. 1994; Kalko et al. 1998).

The source levels recorded in the field vary between species. *N. leporinus* and *N. albiventris* produce source levels up to 116 dB SPL (at 1 m), which is louder than the source level of open space bats. In the Daubenton's bat, *Myotis daubentonii*, the mean source level is about 100 dB SPL at 1 m (Surlykke et al. 2009). *Macrophyllum macrophyllum* reach source levels of 91 dB SPL at 1 m in an open situation and 85 dB SPL at 1 m in edge settings (Brinkløv et al. 2010).

Edge space trawling foragers show morphological adaptations to this foraging mode with specializations of the hind legs and of the interfemoral membrane between the legs. In piscivorous species the claws of the hind legs are very sharp to facilitate grasping slippery fish, and the fur is often oily to facilitate taking off from the water surface after occasional accidents.

6.5.1.4 Narrow Space Flutter-Detecting Foragers

Bats belonging to the guild of narrow space flutter-detecting foragers have evolved a very specialized echolocation system that allows the evaluation of flutter information in the returning echoes (reviewed in Schnitzler and Denzinger 2011). These bats emit echolocation calls consisting of a long CF component followed by, and sometimes also preceded by, a brief frequency-modulated component (Figure 6.4). The long CF-FM signals account for a very high duty cycle, hence flutter-detecting foragers have also been called high duty cycle bats (Fenton 1995). CFs in flutter-detecting foragers are species specific and range from 28 kHz in the Bourret's horseshoe bat, *Rhinolophus paradoxolophus*, to more than 150 kHz in some small Old World leaf-nosed bats (Hipposideridae), such as the ashy leaf-nosed bat, *Hipposideros cineraceus*.

In flight, the bats lower the emission frequency to compensate for the Doppler shift generated by their own flight speed, thus keeping the echo frequency constant within the specialized frequency range of the auditory fovea. When the long CF-FM signal reflects off a fluttering insect, the CF component of the echo is modulated in frequency and amplitude in the rhythm of the wingbeat and can easily be discriminated from unmodulated echoes of background targets. The modulated echoes are analyzed in the auditory fovea with many sharply tuned neurons specialized for the evaluation of flutter information, encoding size, type and aspect angle of the prey (von der Emde and Menne 1989; von der Emde and Schnitzler 1990; Roverud et al. 1991). In addition to flutter information, the CF component contains flow field information bats might use to follow landscape contours during commuting flights (Müller and Schnitzler 1999; Schnitzler et al. 2003). The short FM component is suited for localizing both insects and background targets.

All flutter-detecting foragers maintain a CF component in their signals, even in the shortest calls of the terminal group in the approach. Flutter detection has evolved in the Old World Hipposideridae and Rhinolophidae and in the Neotropics independently in one species of the Mormoopidae family, the mustached bat (*Pteronotus parnellii*), which actually represents several cryptic species (Clare et al. 2013). Narrow space flutter-detecting foragers correspond to high duty cycle bats according to Fenton (1995), whereas the bats of all other guilds belong to the group of low duty cycle bats.

6.5.1.5 Narrow Space Active-Gleaning Foragers

Bats that are able to detect the prey sitting quietly on a substrate by using echolocation alone are assigned to the guild of narrow space active-gleaning foragers. So far only a single insectivorous species from the phyllostomid family has been identified as belonging to this guild. The common big-eared bat, *Micronycteris microtis*, forages for stationary prey items that sit silently and motionless on large leaves. By ensonifying the leaves obliquely from above, the bat encounters a micro-trawling echolocation scene that resembles that of edge space trawling bats but on a micro time scale (Figure 6.2). This species emits very low amplitude, ultra-short (0.2 ms) broadband high-frequency calls (Figure 6.4). The signals are emitted in groups (Denzinger and Schnitzler 2004, 2013; Geipel et al. 2013). Most likely there are other active gleaners, e.g., in the phyllostomid family, that have not yet been identified.

6.5.1.6 Narrow Space Passive-Gleaning Foragers

Bats that glean animals such as insects, other arthropods, and small vertebrates from substrates, where they encounter echolocation scenes that do not allow the separation of prey and clutter echoes, must rely on prey-generated cues to find and localize their prey (Fenton 1990; Goerlitz et al. 2008; Page and Ryan 2008). In most cases, the prey-generated cues are sounds, but under favorable conditions, vision may also play a role in prey detection (Bell 1985; Eklöf and Jones 2003). These bats are assigned to the guild of narrow space passive-gleaning foragers. The prey-generated cues must allow these bats to approach the prey site with sufficient accuracy to catch the prey using tactile and olfactory cues (Kolb 1958). Echolocation is used only to guide the approach flight to the prey site. In the laboratory, some passive-gleaning foragers are able to make the transition to active gleaning under favorable conditions, e.g., if the prey item is offered on a flat surface (Marimuthu et al. 1995; Schmidt et al. 2000; Flick 2008). All narrow space, passive-gleaning foragers operate with short broadband signals with low source levels (Figure 6.4) (Holderied et al. 2011) suited for spatial orientation and biotope recognition. Narrow space passive-gleaning foragers are found in Phyllostomidae, Megadermatidae, Nycteridae, and Vespertilionidae.

6.5.1.7 Narrow Space Passive-/Active-Gleaning Foragers

The guild of narrow space, passive-/active-gleaning foragers comprises all bats that are phytophagous and have to find food that is part of the background. In contrast to prey animals that do their best to avoid being found, plants advertise the nature and position of their fruits or flowers by species-specific odor bouquets, exposed positions in relation to the background, and/or specific echo properties. In frugivorous and nectarivorous bats, the primary cue for the attraction of the bats over long distances is odor (von Helversen et al. 2000; Kalko and Ayasse 2009; Hodgkison et al. 2013). Many bat-pollinated flowers have an intense and, sometimes for humans, unpleasant odor (von Helversen 1993; Tschapka and Dressler 2002). Such odor plumes present a gradient that guides the bats into close vicinity of the food item. However, the odor plumes are probably too diffuse for approach and precise localization. For that purpose, bats switch from passive mode (with low spatial accuracy) to a precise echolocation-guided approach in active mode.

Plants may facilitate acoustic detection and precise localization of fruits or flowers by presenting them in exposed positions and by having reflective properties that produce conspicuous echoes. Some bat-pollinated plants present specific, dish-shaped petals or other leaves with distinct sound reflective properties that produce spatially invariant echoes with a characteristic spectral and amplitude pattern over a wide range of sound incidence angles (von Helversen and von Helversen 1999; von Helversen et al. 2003; Simon et al. 2011). Thereby, these echoes, which are largely invariant of the bat's approach angle, contrast from the more variable background echoes.

Another common property of bat-pollinated flowers is cauliflory: flowers are borne on the main trunk or branches of the plant, which also improves the separation between target and background echoes. Flagelliflory, the presentation of flowers on long peduncles, distances the reproductive parts from the background clutter and also facilitates an echolocation-guided approach (Dobat and Peikert-Holle 1985). So far, all behavioral studies with phytophagous bats have shown that they use a combination of initially passive and rather imprecise localization of a food source by odor with a subsequent precise active localization of the food with echolocation. It is conceivable that the degree of overlap between these two modes and their relative importance for the foraging process may differ between species. Passive-/active-gleaning foragers emit short, multi-harmonic broadband signals often operating in the high-frequency range of the third and fourth harmonic (Figure 6.4). The signals have low source levels (Brinkløv et al. 2011), which reduce the clutter echoes from background. They are well-suited for spatial orientation and biotope recognition but also to guide bats to fruits and flowers. Narrow space, passive/active foragers are exclusively known in the family Phyllostomidae.

6.5.2 Approach Behavior

The echolocation behavior during approach to a food item depends primarily on the foraging mode (Figure 6.4). All bats that home in on moving aerial prey shorten pulse duration and pulse interval and increase bandwidth with decreasing distance to prey. The signals are often arranged in groups and the approach sequences end with distinct terminal groups consisting of buzz I and buzz II. In buzz I, pulse interval is further reduced; in buzz II, signals have a minimal and constant pulse interval of approximately 5–6 ms and in some species also a lower signal frequency. Gleaning bats use echolocation for control of landing on either the prey or the prey site. The signals are also arranged in groups but the approach sequences lack a distinct terminal group with many signals at a high repetition rate, such as the buzzes of aerial-hawking bats. The number of signals within a group depends on whether the prey is approached in active or passive mode. Narrow space active-gleaning foragers and narrow space passive-/active-gleaning foragers localize the stationary prey item by echolocation and emit more signals per group than narrow space passive-gleaning foragers, which only home in on the food site but do not localize the food item by echolocation (Figure 6.4) (Kalko and Schnitzler 1998; Schnitzler et al. 2003; Denzinger and Schnitzler 2013).

6.5.3 Assigning Bat Species to Guilds

Most bats have a preferred habitat type and foraging mode, reflected in clear-cut adaptations of their sensory and motor systems. Provided there is sufficient information, this makes it easy to assign them to a specific guild. Especially easy is the assignment of bats to the guild of narrow space flutter-detecting foragers. These bats use long CF-FM signals to detect fluttering insects flying near vegetation or even sitting on it. They maintain a long CF component in their calls in all behavioral situations, even in the shortest signals during the approach, and they have specific adaptations such as Doppler shift compensation, an auditory fovea, and high duty cycle echolocation (for a review see Schnitzler and Denzinger 2011). Some species are more variable in the habitat type and foraging mode they use. In particular, edge space aerial foragers often switch back and forth between edge and open space, and some species that glean prey insects from surfaces are also able to catch flying insects in the aerial-hawking mode.

Bats switching from their preferred habitat type to another also change to the foraging and echolocation behavior that is appropriate for the new habitat type and foraging mode. In these species the assignment to guilds can be rather difficult. Fenton (1990) pointed out that species adapted to foraging in closed habitats also have access to edge and open habitats, whereas those adapted to open habitats only have limited access to more narrow habitats. Limitations of the motor and sensory systems of strictly open space foragers make it difficult to fly in more narrow spaces and to find prey in the presence of clutter echoes. Narrow space gleaning foragers are able to fly in edge space and forage there in the aerial mode, and edge space aerial foragers often search for prey in open space. The access to more open habitats is possible, but the reverse is not.

In general, bats that are able to switch from one habitat to another are best adapted in echolocation and wing morphology to the more preferred habitat type. Therefore Denzinger and Schnitzler (2013) proposed that species should be assigned to a specific guild based on their dominant foraging behavior for which their echolocation and wing morphology are best adapted. The foraging and echolocation behaviors of the members of a specific guild are so similar that behavioral patterns of well-studied species have a high predictive value for other species of the same guild.

6.6 Niche Differentiation

6.6.1 Niche Dimensions and Niche Spaces

Bat species from different guilds differ distinctly in the environmental resources they exploit and in the way they forage for prey. Thus, species belonging to different guilds usually do not compete for food. An exception may occur if narrow space flutter-detecting bats (Hipposideridae and Rhinolophidae) also hunt in edge space for similar prey as edge space aerial foragers, such as species of the genera *Murina* and *Kerivula*. Among species of the same guild, the exploited resources, foraging strategies, and echolocation behaviors are far more similar than in species from different guilds. Therefore, guild members have to partition the available resources by niche differentiation to avoid competition (Simberloff and Dayan 1991). To understand the structure, dynamics, and functionality of local bat assemblages and guilds, it is essential to know the fine grain resource partitioning and niche differentiation between species.

The foraging niche of a species is determined by its resources and the environmental conditions (Hutchinson 1959). Niches are therefore characterized by a combination of temporal, spatial, food-specific, and environmental niche dimensions. Each species has a unique combination of niche dimensions and is adapted to its niche in a unique way that is reflected in its morphology, physiology, echolocation, and foraging behavior.

Based on similarities in the combination of niche dimensions, three distinctly different main types of niche spaces can be discriminated in bats, which are reflected in different types of foraging behaviors. One type comprises the niches of all open and edge space aerial-hawking and trawling foragers. Another type contains the niches of narrow space flutter-detecting foragers. The third type comprises the niches of all narrow space gleaning foragers that take stationary food—either animals or plants—from vegetation or the ground.

6.6.2 Niche Space of Aerial-Hawking and Trawling Bats

The niche space of aerial-hawking and trawling bats includes the niches of all bats that forage in open and edge space for airborne prey or for prey drifting on water surfaces. The position of the prey in relation to the background is the most important niche dimension and had the greatest impact on the evolution of species-specific adaptations, especially in the echolocation and flight systems. Other important niche dimensions are set by the prey type and depend on its size, movement pattern, or other prey-specific information encoded in the prey echoes. Combinations of these dimensions offer bats a large variety of niches and promoted the evolutionary radiation of bats into many species.

In aerial-hawking foragers, niche partitioning is mainly indicated by differences in the echolocation behavior (such as patterning, frequency, harmonic structure, bandwidth, duration, pulse interval, and SPL of the echolocation signals) and in the bats' maneuverability and agility, which depend on parameters that characterize the flight abilities such as aspect ratio and wing loading. Low frequencies increase the maximum detection distance for prey but reduce the target strength due to Raleigh scattering at frequencies below 30 kHz. Calls with low frequency often have long signal durations that affect the detection ranges for prey. According to the size filtering hypothesis of Schnitzler and Kalko (1998), the detection of weak echoes from small insects at close range is hampered in long signals as they produce wide signal overlap zones and the weak echoes of nearby small insects are masked. Long signals are adapted for the long-range detection of larger insects, whereas shorter signals with higher frequencies are suited for smaller prey at shorter distances. The higher emission SPLs of the long low-frequency signals of open space aerial foragers also increase the maximum detection range, and, consequently, the search volume. Bandwidth determines how close to the background a species is able to catch prey. Comparative studies on the prey detection performance in front of a clutter-producing background in *Myotis* species (Siemers and Schnitzler 2004) and in paleotropical species of the vespertilionid subfamilies Kerivoulinae and Murininae (Schmieder et al. 2012) revealed that the minimal distance between prey and background at which the bats just could detect a prey item decreased with increasing bandwidth, thus indicating that bandwidth accounts for sensory-based niche partitioning.

In aerial-hawking foragers, echolocation delivers only limited information on the nature and quality of prey. The pulse-echo pairs that encode the position of prey also contain some information on the sizes and movement patterns of insects but little information on type and quality of the prey. Bats can probably evaluate the movement pattern of prey and prefer moving prey over immobile items (Barclay and Brigham 1994). They certainly can obtain some information about prey dimensions; however, there is little evidence that echolocation is used to select between specific favorable prey and other prey (reviewed in Jones and Rydell 2003). There is evidence that bats learn quickly to discriminate between targets in specific situations, e.g., within a short time some species are able to discriminate a thrown pebble from a flying insect or to select a mealworm out of a cluster of three small discs (Webster and Durlach 1963). This ability, perfected under natural conditions, may explain food selection. Some food specialists, like *Barbastella barbastellus*, that use a highly specialized stealth echolocation system (Goerlitz et al. 2010; Seibert et al. 2015), feed nearly exclusively on moths and discriminate those from other insects. In summary, in aerial-hawking bats, differences in echolocation behavior and flight morphology are the main factors that determine resource partitioning and reflect niche differentiation.

6.6.3 Niche Space of Flutter-Detecting Bats

The niche space of flutter-detecting bats is available to all rhinolophid and hipposiderid bats of the Old World and one mormoopid bat of the New World. These bats mainly differ in size and in the frequency of the long CF component of their echolocation signals. Their radiation into many different niches is reflected in a large number of species. In general, size and frequency are inversely correlated, and in species of similar sizes, the CF frequencies of the hipposiderids are generally above those of rhinolophids (Thong 2011). Wing morphology of hipposiderids and rhinolophids is rather similar (Norberg and Rayner 1987). The frequency of the long CF component determines the maximum operational range and therefore the search volume of the bats. The search volume, the ability to classify prey according to the species-specific modulation pattern in the echoes, and the size-dependent maneuverability and ability to handle prey seem to be the principal factors affecting niche differentiation. In addition, different foraging areas, small differences in wing morphology, and searching for fluttering insects from perches or on the wing are further adaptations that account for resource partitioning (Dietz et al. 2006, 2007).

6.6.4 *Niche Space of Gleaning Bats*

The niche space of narrow space gleaning bats comprises the niches of all bats that forage for food items that are either positioned on surfaces (animals) or are part of the background (fruits and flowers with nectar). These food items are all stationary and differ in niche dimensions, such as position relative to the background, size, reflection properties, and food-specific passive cues, such as sounds, odors, and even temperature in the case of vampire bats. Combinations of these niche dimensions offer a great variety of niches, which have promoted the radiation into many species that cover a wide range of body mass. All bats that glean their food from surfaces have to perform similar motor and echolocation tasks that are reflected in similar adaptations in morphology and echolocation behavior. However, the wide variety of animal and plant diets has led to species-specific adaptations that indicate niche differentiation. Thus, differences in the sensory modalities that are used to find the food and differences in morphology and physiology needed to handle the food are important factors for resource partitioning.

6.6.4.1 Animalivorous Gleaning Bats

In animalivorous gleaning bats, the mere type of food is not relevant for a subdivision in different feeding types. This is supported by Giannini and Kalko (2005), who found that in animal-eating leaf-nosed bats (Phyllostomidae: Phyllostominae) the percentage of insectivory decreases with increasing body size and is gradually replaced by carnivory (i.e., preying on vertebrates) in association with increasing mass and limited dental modifications. They concluded that carnivory is a size-dependent extreme of animalivory rather than a qualitatively distinct feeding habit. Carnivorous bats are bigger and only modestly modified versions of smaller insect gleaners. In passive gleaners, prey-generated acoustic cues, such as rustling sounds of moving prey or courtship signals (e.g., from katydids and frogs), deliver the sensory cues for prey selection (Russo et al. 2007; Jones et al. 2011). Active gleaners such as *Micronycteris microtis*, and maybe some other not yet identified species, use echolocation to find silent prey at specific locations (Denzinger and Schnitzler 2004, 2013) and can discriminate prey from dummy targets (Geipel et al. 2013). Prey-specific locations and echolocation cues thus define the foraging niches in these bats.

6.6.4.2 Phytophagous Gleaning Bats

Phytophagous gleaning bats are assigned to the guild of narrow space passive-/active-gleaning foragers. They all have the big advantage that prey selection is much easier than in animalivorous gleaners, since bat plants advertise the nature and position of their fruits or flowers by species-specific odor bouquets, positions of food in relation to the background, and specific echo properties of the food item (von Helversen et al. 2000; Kalko and Ayasse 2009). The diversity of the available food plants and their distribution determines species richness and the vertical stratification of phytophagous bats (Bernard 2001; Kalko and Handley 2001; Rex et al. 2008). Discrete dietary groups are mostly compatible with clades of the phylogenetic tree of phyllostomid bats (Giannini and Kalko 2004; Datzmann et al. 2010). Frugivorous phyllostomid bats partition fruits based largely on fruit and body size (e.g., Kalko et al. 1996a, b). However, some species also show particular adaptations for feeding on extremely hard fruits (Dumont et al. 2009) or even consume the seeds of certain figs (Nogueira and Peracchi 2003; Wagner et al. 2015). Insects, nectar, and leaves are used opportunistically by some frugivores as additional food types but without obvious morphological or behavioral adaptations that might permit/merit a finer-grained guild classification (Kunz and Diaz 1995; Tschapka and Dressler 2002).

Within the nectarivorous passive-/active-gleaning, there is a broad overlap in the use of floral resources and no partitioning based on flower size is obvious. Body size and wing proportions may together form the basis for the observed relations between specific flower preferences and energy density of flowers, i.e., the quantity and quality of nectar resources within a particular area (Tschapka 2004). Additionally, there are specific differences in the dependence on nectar. While morphologically more specialized, longer jawed species (e.g., *Musonycteris harrisoni*, *Hylonycteris underwoodi*) seem to rely on a year-round nectar supply (Tschapka 2004; Tschapka et al. 2008); shorter-jawed species, such as *Glossophaga* spp., may switch seasonally to a more frugivorous diet (Tschapka 2004, 2005); or insects may form a significant part of their diet year-round (Sperr et al. 2011).

In summary, all narrow space gleaning bats have to perform similar foraging tasks so that differences in echolocation behavior and flight morphology are less useful as indicators for resource partitioning and niche differentiation. However, adaptations for dietary specialization based on sensory modality for prey selection and the wide range of body sizes are good indicators for niche differentiation.

6.7 Conclusion

The assignment of bats to seven guilds according to the preferred habitat type and foraging mode and the identification of three types of niche spaces according to similarities in the combinations of niche dimensions may seem rather arbitrary. Such an assignment can be difficult, especially in behaviorally flexible bats that can switch between habitats and foraging modes, that vary in their preferred food, and that forage for animals as well as fruits and/or flowers. Without doubt there is still much to be learned about feeding habits, sensory capabilities, and behavior, particularly of the rarer bat species. Nevertheless, this approach is pursued because the assignment of a species to a guild and to a niche space indicates, rather precisely, its ecological position within the local ecosystems.

Using the guild concept, species are identified that live under similar ecological conditions and share similar sensory and motor adaptations. Members of different guilds do not directly compete for food, and within guilds the echolocation and foraging behaviors are so similar that the behavioral patterns of well-studied species have a high predictive value for other species of the same guild. The attribution of the many different niches of bats into three main types of niche spaces indicates that the selective pressures for the evolution of adaptations in morphology, physiology, echolocation, and foraging behaviors are highly differentiated. In the niche space of aerial-hawking and trawling bats, niche differentiation is dominated by the position of the preferred prey to the background and is reflected mainly in differences in body size, echolocation behavior, and flight morphology. Within the niche space of flutter-detecting bats, niche differentiation is mainly indicated by body size and corresponding differences in frequency of the CF components of the echolocation signals. In the niche space of gleaning bats, niche differentiation is not reflected by differences in the echolocation system but by large differences in body size and by adaptations for dietary specialization in sensory and feeding systems.

Acknowledgements This paper summarizes the work of many years of field work that were made possible and enjoyable by the help of many students and volunteers. We also want to thank Ingrid Kaipf for technical assistance with the figures. This work was supported by the Deutsche Forschungsgemeinschaft (Schn 138/29-1).

References

Aldridge, H. D. J. N., & Rautenbach, I. L. (1987). Morphology, echolocation and resource partitioning in insectivorous bats. *Journal of Animal Ecology*, 56, 763–778.

Barclay, R. M. R., & Brigham, R. M. (1994). Constraints on optimal foraging: A field test of prey discrimination by echolocating insectivorous bat. *Animal Behaviour*, 48, 1013–1021.

Bell, G. P. (1985). The sensory basis of prey location by the California leaf-nosed bat *Macrotus californicus* (Chiroptera: Phyllostomidae). *Behavioral Ecology and Sociobiology*, 16, 343–347.

Bernard, E. (2001). Vertical stratification of bat communities in primary forests of Central Amazon, Brazil. *Journal of Tropical Ecology*, 17, 115–126.

Blood, B. R., & Clark, M. K. (1998). *Myotis vivesi. Mammalian Species*, 588, 1–5.
Brinkløv, S., Kalko E. K. V., & Surlykke, A. (2010). Dynamic adjustment of biosonar intensity to habitat clutter in the bat *Macrophyllum macrophyllum* (Phyllostomidae). *Behavioral Ecology and Sociobiology*, 64, 1867–1874.
Brinkløv, S., Jakobsen, L., Ratcliffe, J. M., Kalko E. K. V., & Surlykke, A. (2011). Echolocation call intensity and directionality in flying short-tailed fruit bats, *Carollia perspicillata* (Phyllostomidae). *Journal of the Acoustical Society of America*, 129, 427–435.
Clare, E. L., Adams, A. M., Maya-Simoes, A. Z., Eger, J. L., Hebert, P. D. N., & Fenton, M.B. (2013). Diversification and reproductive isolation: Cryptic species in the only New World high-duty cycle bat, *Pteronotus parnellii. BMC Biology*, doi: 10.1186/1471-2148-13-26
Datzmann, T., von Helversen, O., & Mayer, F. (2010). Evolution of nectarivory in phyllostomid bats (Phyllostomidae Gray, 1825, Chiroptera: Mammalia). *BMC Evolutionary Biology*, 10, 165, http://www.biomedcentral.com/1471-2148/10/165
Denzinger, A., & Schnitzler, H.-U. (2004). Perceptual tasks in echolocating bats. In U. J. Ilg, H. H. Bülthoff, & H. A. Mallot (Eds.), *Dynamic perception* (pp. 33–38). Berlin: Akademische Verlagsgesellschaft.
Denzinger A., & Schnitzler, H.-U. (2013). Bat guilds, a concept to classify the highly diverse foraging and echolocation behaviors of microchiropteran bats. *Frontiers in Physiology*, doi: 10.3389/fphys2013.00164
Denzinger, A., Kalko E. K. V., & Jones, G. (2004). Ecological and evolutionary aspects of echolocation in bats. In J. A. Thomas, C. F. Moss, & M. Vater (Eds.), *Echolocation in bats and dolphins* (pp. 311–326). Chicago: The University of Chicago Press.
Dietz, C., Dietz, I., & Siemers, M. B. (2006). Wing measurement variations in the five European horseshoe bat species (Chiroptera: Rhinolophidae). *Journal of Mammalogy*, 87, 1241–1251.
Dietz, C., von Helversen, O., & Nill, D. (2007). *Handbuch der Fledermäuse Europas und Nordwestafrikas*. Stuttgart: Franckh Kosmos Verlag.
Dobat, K., & Peikert-Holle, T. (1985). *Blüten und Fledermäuse*. Frankfurt am Main: Verlag Waldemar Kramer.
Dumont E. R., Herrel, A., Medellín, R. A., Vargas-Contreras, J. A., & Santana, S. E. (2009). Built to bite: Cranial design and function in the wrinkle-faced bat. *Journal of Zoology*, 279, 329–337.
Eklöf, J., & Jones, G. (2003). Use of vision in prey detection by brown long-eared bats, *Plecotus auritus. Animal Behaviour*, 66, 949–953.
Fenton, M. B. (1990). The foraging behaviour and ecology of animal-eating bats. *Canadian Journal of Zoology*, 68, 411–422.
Fenton, M. B. (1995). Natural history and biosonar signals. In A. N. Popper & R. R. Fay (Eds), *Hearing by Bats* (pp. 37–86). New York: Springer.
Flick, J. (2008). Der Wechsel vom passiven zum aktiven Substratfang beim Großen Mausohr (*Myotis myotis*, Borkhausen 1797). Diploma thesis. Tübingen: Universität Tübingen.
Gardner, A. L. (1977). Feeding habits. In R. J. Baker, J. K. Jones, Jr., & D. C. Carter (Eds.), *Biology of bats of the New World family Phyllostomatidae*. Part II (pp. 293–350). Lubbock, TX: The Museum, Texas Tech University.
Geipel, I., Jung, K., & Kalko, E. K. V. (2013). Perception of silent and motionless prey on vegetation by echolocation in the gleaning bat *Micronycteris microtis. Proceedings of the Royal Society of London B: Biological Sciences*, doi: 10.1098/rspb.2012.2830
Giannini, N. P., & Kalko, E. K. V. (2004). Trophic structure in a large assemblage of phyllostomid bats in Panama. *OIKOS*, 105, 209–220.
Giannini, N. P., & Kalko, E. K. V. (2005). The guild structure of animalivorous leaf-nosed bats of Barro Colorado Island, Panama, revisited. *Acta Chiropterologica*, 7, 131–146.
Goerlitz, H. R., Greif, S., & Siemers, B. M. (2008). Cues for acoustic detection of prey: Insect rustling sounds and the influence of walking substrates. *The Journal of Experimental Biology*, 211, 2799–2806.
Goerlitz, H. R., ter Hofstede, H. M., Zeale, M. R. K., Jones, G., & Holderied, M. W. (2010). An aerial hawking bat uses stealth echolocation to counter moth hearing. *Current Biology*, 20, 1568–1572. doi: 10.1016/j.cub.2010.07.046

Greif, S., & Siemers, B. M. (2010). Innate recognition of water bodies in echolocating bats. *Nature Communications* 1, 107. doi: 10.1038/ncomms1110

Grunwald J. E., Schornich, S., & Wiegrebe, L. (2004). Classification of natural textures in echolocation. *Proceedings of the National Academy of Sciences of the USA*, 101, 5670–5674.

Hessel, K., & Schmidt, U. (1994). Multimodal orientation in *Carollia perspicillata* (Phyllostomidae). *Folia Zoologica*, 43(4), 339–346.

Hill, J. E., & Smith, J. D. (1984). *Bats: A natural history*. London: British Museum (Natural History).

Hodgkison, R., Ayasse, M., Häberlein, C., Schulz, S., Zubaid, A., Mustapha, W. A. W., Kunz, T. H., & Kalko, E. K. V. (2013). Fruit bats and bat fruits: The evolution of fruit scent in relation to the foraging behaviour of bats in the New and Old World tropics. *Functional Ecology*, 27, 1075–1084.

Holderied, M. W., & von Helversen, O. (2003). Echolocation range and wingbeat period match in aerial-hawking bats. *Proceedings of the Royal Society of London B: Biological Sciences*, 270, 2293–2299.

Holderied, M., Korine, C., & Moritz, T. (2011). Hemprich's long-eared bat (*Otonycteris hemprichii*) as a predator of scorpions: Whispering echolocation, passive gleaning and prey selection. *Journal of Comparative Physiology A: Neuroethology, Sensory, Neural, and Behavioral Physiology*, 197, 425–433.

Holland, R. A., Thorup, K., Vonhof, M., Cochran, W., & Wikelski, M. (2006). Navigation: Bat orientation using Earth's magnetic field. *Nature*, 444, 702.

Hutchinson, G. E. (1959). Homage to Santa Rosalia of why are there so many kinds of animals? *The American Naturalist*, 93, 145–159.

Jensen, M. E., & Miller, L. A. (1999). Echolocation signals of the bat *Eptesicus serotinus* recorded using a vertical microphone array: Effect of flight altitude on searching signals. *Behavioral Ecology and Sociobiology*, 47, 60–69.

Jones, G., & Rydell, J. (2003). Attack and defense: Interactions between echolocating bats and their insect prey. In T. H. Kunz & M. B. Fenton (Eds.), *Bat ecology* (pp. 301–345). Chicago: The University of Chicago Press.

Jones, G., & Teeling, E. C. (2006). The evolution of echolocation in bats. *Trends in Ecology and Evolution*, 21, 149–156.

Jones, P. L., Page, R. A., Hartbauer, M., & Siemers, B. M. (2011). Behavioral evidence for eavesdropping on prey song in two Palearctic sibling bat species. *Behavioral Ecology & Sociobiology*, 65(2), 333–340.

Kalko, E. K. V. (1997). Diversity in tropical bats. In H. Ulrich (Ed.), *Tropical biodiversity and systematics* (pp. 13–43). Bonn: Zoologisches Forschungsinstitut und Museum König.

Kalko, E. K. V. (1998). Organization and diversity of tropical bat communities through space and time. *Zoology*, 101, 281–297.

Kalko, E. K. V., & Schnitzler, H.-U. (1989). The echolocation and hunting behavior of Daubenton's bat, *Myotis daubentoni*. *Behavioral Ecology and Sociobiology*, 24, 225–238.

Kalko, E. K. V., & Schnitzler, H.-U. (1993). Plasticity in echolocation signals of European pipistrelle bats in search flight: Implications for habitat use and prey detection. *Behavioural Ecology and Sociobiology*, 33, 415–428.

Kalko, E. K. V., & Schnitzler, H.-U. (1998). How echolocating bats approach and acquire food. In T. H. Kunz & P. A. Racey (Eds.), *Bat biology and conservation* (pp. 197–204). Washington DC: Smithsonian Institution Press.

Kalko, E. K. V., & Handley, C. O. (2001). Neotropical bats in the canopy: Diversity, community structure and implications for conservation. *Plant Ecology*, 153, 319–333.

Kalko, E. K. V., & Ayasse, M. (2009). Study and analysis of odor involved in the behavioral ecology of bats. In T. H. Kunz & S. Parsons (Eds.), *Ecological and behavioral methods for the study of bats* (pp. 491–499). Baltimore: The Johns Hopkins University Press.

Kalko, E. K. V., Handley, C. O., & Handley, D. (1996a). Organization, diversity, and long-term dynamics of a neotropical bat community. In M. Cody & J. Smallwood (Eds.), *Long-term studies in vertebrate communities* (pp. 503–553). Los Angeles: Academic Press.

Kalko, E. K. V., Herre, E. A., & Handley, C. O. (1996b). Relation of fig fruit characteristics to fruit-eating bats in the New and Old World tropics. *Journal of Biogeography*, 23, 565–576.
Kalko, E. K. V., Schnitzler, H.-U., Kaipf, I., & Grinnell, A. D. (1998). Echolocation and foraging behavior of the lesser bulldog bat, *Noctilio albiventris*: Preadaptations for piscivory? *Behavioral Ecology and Sociobiology*, 42, 305–319.
Kolb, A. (1958). Nahrung und Nahrungsaufnahme bei Fledermäusen. *Zeitschrift für Säugetierkunde*, 23, 84–95.
Krausman, P. R. (1999). Some basic principles of habitat use. In K. L. Launchbaugh, K. D. Sanders & J. C. Mosley (Eds.), *Grazing behavior of livestock and wildlife* (pp. 85–90). Idaho: Idaho Forest, Wildlife & Range Experimental Station Bulletin, No. 70, University of Idaho.
Kulzer, E. (1956). Flughunde erzeugen Orientierungslaute durch Zungenschlag. *Naturwissenschaften*, 43, 117–118.
Kunz, T. H., & Diaz, C. A. (1995). Folivory in fruit-eating bats, with new evidence from *Artibeus jamaicensis* (Chiroptera: Phyllostomidae). *Biotropica*, 27(1), 106–120.
Laska, M. (1990). Olfactory sensitivity to food odor components in the short-tailed fruit bat, *Carollia perspicillata* (Phyllostomatidae, Chiroptera). *Journal of Comparative Physiology, A: Neuroethology, Sensory, Neural, and Behavioral Physiology*, 66, 395–399.
McGuire, L. P., Guglielmo, C. G., Mackenzie, S. A., & Taylor, P. D. (2012). Migratory stopover in the long-distance migrant silver-haired bat, *Lasionycteris noctivagans*. *Journal of Animal Ecology*, 81(2), 377–385.
McNab, B. K. (1971). The structure of tropical bat faunas. *Ecology* 52, 353–358.
Marimuthu, G., Habersetzer, J., & Leipert, D. (1995). Active gleaning from the water surface by the Indian false vampire bat, *Megaderma lyra*. *Ethology*, 99, 61–74.
Müller, R., & Schnitzler, H.-U. (1999). Acoustic flow perception in CF-bats: Properties of the available cues. *Journal of Acoustical Society of America*, 105, 2958–2966.
Neuweiler, G. (1989). Foraging ecology and audition in echolocating bats. *Trends in Ecology and Evolution*, 4, 160–166.
Nogueira, M. R., & Peracchi, A. L. (2003). Fig-seed predation by two species of *Chiroderma*: Discovery of a new feeding strategy in bats. *Journal of Mammalogy*, 84, 225–233.
Norberg, U. M. (1994). Wing design, flight performance and habitat use in bats. In P. C. Wainwright & S. M. Reilly (Eds.), *Ecological morphology: Integrative organismal biology* (pp. 205–239). Chicago: University of Chicago Press.
Norberg, U. M., & Rayner, J. M. V. (1987). Ecological morphology and flight in bats (Mammalia: Chiroptera): Wing adaptation, flight performance, foraging strategy and echolocation. *Philosophical Transactions of the Royal Society B: Biological Sciences*, 316, 335–427.
Page, R. A., & Ryan, M. J. (2005). Flexibility in assessment of prey cues: Frog-eating bats and frog calls. *Proceedings of the Royal Society of London B: Biological Sciences*, 272, 841–847.
Page, R. A., & Ryan, M. J. (2008). The effect of signal complexity on localization performance in bats that localize frog calls. *Animal Behaviour*, 76, 761–769.
Rex, K., Kelm, D. H., Wiesner, K., Kunz, T. H., & Voigt, C. C. (2008). Species richness and structure of three Neotropical bat assemblages. *Biological Journal of the Linnean Society*, 94, 617–629.
Rojas, D., Vale, A., Ferrero, V., & Navarro, L. (2012). The role of frugivory in the diversification of bats in the Neotropics. *Journal of Biogeography*, 39(11), 1948–1960.
Rojas-Martinez, A., Valiente-Banuet, A., Arizmendi, M. d. C., Alcantara-Eguren, A., & Arita, H. T. (1999). Seasonal distribution of the long-nosed bat (*Leptonycteris curasoae*) in North America: Does a generalized migration pattern really exist? *Journal of Biogeography*, 26(5), 1065–1077.
Root, R. B. (1967). The niche exploitation pattern of the blue-gray gnatcatcher. *Ecological Monographs*, 37, 317–350.
Roverud, R. C., Nitsche, V., & Neuweiler, G. (1991). Discrimination of wing beat motion by bats, correlated with echolocation sound pattern. *Journal of Comparative Physiology A: Neuroethology, Sensory, Neural, and Behavioral Physiology*, 168, 259–263.
Russo, D., Jones, G., & Arlettaz, R. (2007). Echolocation and passive listening by foraging mouse-eared bats *Myotis myotis* and *M. blythii*. *Journal of Experimental Biology*, 210(1), 166–176.

Rydell, J. (1993). Variation in the sonar of an aerial-hawking bat (*Eptesicus nilssonii*). *Ethology*, 93, 275–284.
Schaub, A., & Schnitzler, H.-U. (2007). Echolocation behavior of the bat *Vespertilio murinus* reveals the border between the habitat types "edge" and "open space". *Behavioral Ecology* and *Sociobiology*, 61, 513–523.
Schmidt, S., Hanke, S., & Pillat, J. (2000). The role of echolocation in the hunting of terrestrial prey – new evidence for an underestimated strategy in the gleaning bat, *Megaderma lyra*. *Journal of Comparative Physiology A: Neuroethology, Sensory, Neural, and Behavioral Physiology*, 186, 975–988.
Schmidt, U., Joermann, G., & Rother, G. (1988). Acoustical vs. visual orientation in neotropical bats. In P. E. Nachtigall & P. W. B. Moore (Eds.), *Animal sonar processes and performance* (pp. 589–594). New York: Plenum Press.
Schmieder, D. A., Kingston, T., Hashim, R., & Siemers, B. M. (2012). Sensory constraints on prey detection performance in an ensemble of vespertilionid understorey rain forest bats. *Functional Ecology*, 26, 1043–1053.
Schnitzler, H.-U. (1968). Die Ultraschall-Ortungslaute der Hufeisen-Fledermäuse (Chiroptera-Rhinolophidae) in verschiedenen Orientierungssituationen. *Zeitschrift für Vergleichende Physiologie*, 57, 376–408.
Schnitzler, H.-U., & Kalko, E. K. V. (1998). How echolocating bats search and find food. In T. H. Kunz & P. A. Racey (Eds.), *Bat biology and conservation* (pp. 183–196). Washington and London: Smithsonian Institution Press.
Schnitzler, H.-U., & Kalko, E. K. V. (2001). Echolocation by insect-eating bats. *BioScience*, 51, 557–569.
Schnitzler, H.-U., & Denzinger, A. (2011). Auditory fovea and Doppler shift compensation: Adaptations for flutter detection in echolocating bats using CF-FM signals. *Journal of Comparative Physiology A: Neuroethology, Sensory, Neural, and Behavioral Physiology*, 197, 541–559.
Schnitzler, H.-U., Kalko, E. K. V., Kaipf, I., & Grinnell, A. D. (1994). Fishing and echolocation behavior of the greater bulldog bat, *Noctilio leporinus*, in the field. *Behavioral Ecology and Sociobiology*, 35, 327–345.
Schnitzler, H.-U., Moss, C. F., & Denzinger, A. (2003). From spatial orientation to food acquisition in echolocating bats. *Trends in Ecology and Evolution*, 18, 386–394.
Schnitzler, H.-U., Kalko, E. K. V., & Denzinger, A. (2004). Evolution of echolocation and foraging behavior in bats. In J. A. Thomas, C. F. Moss, & M. Vater (Eds.), *Echolocation in bats and dolphins* (pp. 331–339). Chicago: The University of Chicago Press.
Seibert, A.-M., Koblitz, J. C., Denzinger, A., & Schnitzler, H.-U. (2015). Bidirectional echolocation in the bat *Barbastella barbastellus*: Different signals of low source level are emitted upward through the nose and downward through the mouth. *PLoS ONE*, 10(9): e0135590. doi: 10.1371/journal.pone.0135590
Siemers, B. M., & Schnitzler, H.-U. (2004). Echolocation signals reflect niche differentiation in five sympatric bat species. *Nature*, 429, 657–661.
Siemers, B. M., Stilz, P., & Schnitzler, H.-U. (2001) The acoustic advantage of hunting at low heights above water: Behavioural experiments on the European "trawling" bats *Myotis capaccinii*, *M. dasycneme* and *M. daubentonii*. *The Journal of Experimental Biology*, 204, 3843–3854.
Siemers, B. M., Baur, E., & Schnitzler, H.-U. (2005). Acoustic mirror effect increases prey detection distance in trawling bats. *Naturwissenschaften*, 92, 272–276.
Simberloff, D., & Dayan, T. (1991). The guild concept and the structure of ecological communities. *Annual Review of Ecology and Systematics*, 22(1), 115–143.
Simon, R., Holderied, M. W., Koch, C. U., & von Helversen, O. (2011). Floral acoustics: Conspicuous echoes of a dish-shaped leaf attract bat pollinators. *Science*, 333, 631–633.
Sperr, E. B., Caballero-Martinez, L. A., Medellin, R. A., & Tschapka, M. (2011). Seasonal changes in species composition, resource use and reproductive patterns within a guild of nectar-feeding bats in a west Mexican dry forest. *Journal of Tropical Ecology*, 27, 133–145.
Stilz, W.-P., & Schnitzler H.-U. (2012). Estimation of the acoustic range of bat echolocation for extended targets. *Journal of the Acoustical Society of America*, 132, 1765–1775.

Surlykke, A., & Kalko, E. K. V. (2008). Echolocating bats cry out loud to detect their prey. *PLoS ONE*, 3(4), e2036. doi: 10.1371/0002036

Surlykke, A., Pedersen, S. B., & Jacobsen, L. (2009). Echolocating bats emit a highly directional sonar sound beam in the field. *Proceedings of Royal Society of London B: Biological Sciences*, 276, 853–860.

Thies, W., Kalko, E. K. V., & Schnitzler, H.-U. (1998). The roles of echolocation and olfaction in two Neotropical fruit-eating bats, *Carollia perspicillata* and *C. castanea*, feeding on *Piper. Behavioral Ecology and Sociobiology*, 42, 397–409.

Thong, V. D. (2011). Systematics and echolocation of rhinolophoid bats (Mammalia: Chiroptera) in Vietnam. PhD dissertation. Tuebingen: University of Tuebingen.

Tschapka, M. (2004). Energy density patterns of nectar resources permit coexistence within a guild of Neotropical flower-visiting bats. *Journal of Zoology*, 263, 7–21.

Tschapka, M. (2005). Reproduction of the nectarivore bat *Glossophaga commissarisi* in the Costa Rican rainforest. *Biotropica*, 37, 408–414.

Tschapka, M., & Dressler, S. (2002). Chiropterophily: On bat-flowers and flower-bats. *Curtis' Botanical Magazine*, 19, 114–125.

Tschapka, M., Sperr, E. B., Caballero-Martínez, L. A., & Medellín, R. A. (2008). Diet and cranial morphology of *Musonycteris harrisoni*, a highly specialized nectar-feeding bat in Western Mexico. *Journal of Mammalogy*, doi: 10.1644/07-MAMM-A-038.1

von der Emde, G., & Menne, D. (1989). Discrimination of insect wingbeat-frequencies by the bat *Rhinolophus ferrumequinum. Journal of Comparative Physiology A: Neuroethology, Sensory, Neural, and Behavioral Physiology*, 164, 663–671.

von der Emde, G., & Schnitzler, H.-U. (1990). Classification of insects by echolocating greater horseshoe bats. *Journal of Comparative Physiology A: Neuroethology, Sensory, Neural, and Behavioral Physiology*, 167, 423–430.

von Helversen, O. (1993). Adaptation of flowers to pollination by Glossophagine bats. In W. Barthlott, C. M. Naumann, K. Schmidt-Loske, & K. Schuhmann (Eds.), *Animal-plant interaction in tropical environments*. Bonn: Museum Koenig.

von Helversen, D., & von Helversen, O. (1999). Acoustic guide in bat-pollinated flower. *Nature*, 398, 759–760.

von Helversen, O., Winkler, L., & Bestmann, H. J. (2000). Sulphur-containing "perfumes" attract flower-visiting bats. *Journal of Comparative Physiology A: Neuroethology, Sensory, Neural, and Behavioral Physiology*, 186, 143–153.

von Helversen, D., Holderied, M. W., & von Helversen, O. (2003). Echoes of bat-pollinated bell-shaped flowers: Conspicuous for nectar-feeding bats? *The Journal of Experimental Biology*, 206, 1025–1034.

Wagner, I., Ganzhorn, J. U., Kalko, E. K. V., & Tschapka M. (2015). Cheating on the mutualistic contract: nutritional gain through seed predation in the frugivorous bat *Chiroderma villosum* (Phyllostomidae). *The Journal of Experimental Biology*, 218, 1016–1021.

Webster, F. A., & Durlach, N. I. (1963). *Echolocation systems of a bat*. MIT Lincoln Lab. Rep. No. 41-G-3. Lexington, MA.

Weinbeer, M., & Kalko, E. K. V. (2007). Ecological niche and phylogeny: The highly complex echolocation behavior of the trawling long-legged bat, *Macrophyllum macrophyllum. Behavioral Ecology* and *Sociobiology*, 61, 1337–1348.

Yovel, Y., Stilz, P., Franz, M. O., Boonman, A., & Schnitzler H.-U. (2009). What a plant sounds like: The statistics of vegetation echoes as received by echolocating bats. *PLoS Computational Biology*, 5(7), e1000429. doi: 10.1371/journal.pcbi.1000429

Yovel, Y., Franz, M. O., Stilz, P., & Schnitzler H.-U. (2011). Complex echo classification by echolocating bats: A review. *Journal of Comparative Physiology A: Neuroethology, Sensory, Neural, and Behavioral Physiology*, 197, 475–490.

Chapter 7
Neural Coding of Signal Duration and Complex Acoustic Objects

Paul A. Faure and Uwe Firzlaff

Keywords Acoustic communication • Auditory cortex • Auditory objects • Complex echoes • Echolocation • Excitation • Inferior colliculus • Inhibition • Object normalization • Perception • Psychophysics • Species-specific vocalizations

Abbreviations

2-AFC	Two-alternative forced choice
AC	Auditory cortex
AIp	Posterior primary auditory cortex
BD	Best duration
BEF	Best excitatory frequency
CF	Constant frequency
CNS	Central nervous system
dB	Decibels
DSCF	Doppler-shifted constant frequency
DTN	Duration-tuned neuron
EPSP	Excitatory post-synaptic potential
FI	Fisher information
FM	Frequency modulated
FSL	First-spike latency
GABA	γ-Aminobutyric acid
HRTF	Head-related transfer function
IC	Inferior colliculus

P.A. Faure (✉)
Department of Psychology, Neuroscience and Behaviour,
McMaster University, Hamilton, ON, Canada L8S 4K1
e-mail: paul4@mcmaster.ca

U. Firzlaff
Lehrstuhl für Zoologie, Technische Universität München,
Liesel-Beckmann.-Str. 4, Freising-Weihenstephan 85354, Germany
e-mail: uwe.firzlaff@wzw.tum.de

M.B. Fenton et al. (eds.), *Bat Bioacoustics*, Springer Handbook of Auditory Research 54, DOI 10.1007/978-1-4939-3527-7_7

IPSP	Inhibitory post-synaptic potential
IR	Impulse response
LFP	Local field potential
OFF_E	Offset-evoked excitation
ON_E	Onset-evoked excitation
PSTH	Peristimulus time histogram
q-CF	Quasi-constant frequency
ROC	Receiver operating characteristic
SAM	Sinusoidal amplitude modulation
SPL	Sound pressure level
SSI	Stimulus specific information

7.1 Introduction

This chapter focuses on two aspects of hearing by bats: neural tuning to sound duration, a somewhat unappreciated form of auditory temporal processing, and neural coding of complex acoustic features. The first section examines the importance of signal duration and reviews the basic electrophysiological response properties of duration-tuned neurons in the bat's central auditory system. It focuses on neural mechanisms of duration tuning in the inferior colliculus as evidenced by in vivo single-unit extracellular recording, intracellular recording, application of neuropharmacological agents in combination with single-unit recording, and computational simulations. It concludes by speculating on possible functions of duration selectivity to hearing and echolocation by bats. The second section examines the neural representation of complex objects and species-specific vocalizations in the bat auditory cortex. It highlights recent findings on the coding of spectrotemporal features in complex echoes and relates these neural data to behavioral object recognition tasks in bats. This chapter concludes by discussing the processing of conspecific vocalizations in the auditory cortex and amygdala of bats.

7.2 Neural Coding of Signal Duration in the Central Auditory System

7.2.1 *Signal Duration Is Important for Echolocation*

Temporal features of sound convey information vital for behavior as varied as speech recognition by humans and echolocation by bats. Examples of temporal features include the duration of a sound, the time interval between sounds (i.e., signal period = 1/pulse repetition rate), the rate and direction of frequency modulation (FM), the rate and depth of amplitude modulation (AM), and the sequence of elements within a complex acoustic signal. To capture flying insects, bats listen for reflected echoes of their vocalizations and perform neural computations to obtain information about target size, position, velocity, and range. The type of echolocation sounds emitted by a bat relates to the type of foraging behavior it employs and the stage of prey capture (Casseday and Covey 1995). The time between sound production and echo reception varies with target distance, and most bats precisely adjust their call duration to avoid perceptual problems associated with temporal overlap between outgoing calls and the returning echoes. For echolocation to work, the central nervous system (CNS) must represent auditory signal duration. Knowing how the bat's brain encodes and decodes sound duration represents a fundamental advancement in mammalian hearing because it increases foundational knowledge on the normal functioning auditory system.

Echolocating bats precisely control signal duration. The stylized spectrogram in Figure 7.1 illustrates the changes that occur in the sequence of sounds emitted by a big brown bat (*Eptesicus fuscus*) during an attack on a flying insect. The pulse repetition rate systematically increases and the signal duration decreases across the search, approach, and terminal phases of prey capture. In the search phase, the bat probes the air by emitting loud (high-energy), relatively long-duration, narrowband quasi-constant-frequency (q-CF) sounds that may or may not contain harmonic elements (see also Fenton, Grinnell, and Gould, Chapter 1). Upon detecting an insect, the bat decreases the duration and increases both the FM bandwidth and pulse repetition rate, presumably to increase the amount of information acquired during the approach phase. In the terminal phase of an attack, the bat emits a series of very short duration sounds (e.g., 1–5 ms) at an extremely fast repetition rate (e.g., 50 to >100 Hz; Elemans et al. 2011) that culminates in a terminal feeding buzz when the bat attempts prey capture.

Echolocation works best when the bat's loud outgoing calls do not interfere with its ability to hear the later and fainter echoes returning from nearby objects and prey. Bats that use low duty cycle echolocation (duty cycle = ratio of signal duration to signal period), such as big brown bats, avoid forward masking effects by separating pulse and echo in time by controlling the onset and duration of their signals. In contrast, pulse-echo overlap is an important aspect of signal perception in high duty cycle bats that emit long duration CF sounds (Fenton et al. 2012). High duty cycle bats extract acoustic information by encoding the Doppler shifts in reflected echoes from animated targets such as fluttering insects (see Hirya, Mora, and Riquimaroux, Chapter 9). The amazing sensory capabilities of low and high duty cycle echolocating bats raise a number of important questions. How do bats precisely control their biosonar pulse duration? Do bats have neurons with responses selective for signal duration? If so, do the responses of these cells contain sufficient information for the CNS to encode and decode signal duration? Bats and other mammals have specialized central auditory neurons with responses highly "tuned" to specific signal durations. Not surprisingly, these cells are called duration-tuned neurons (DTNs).

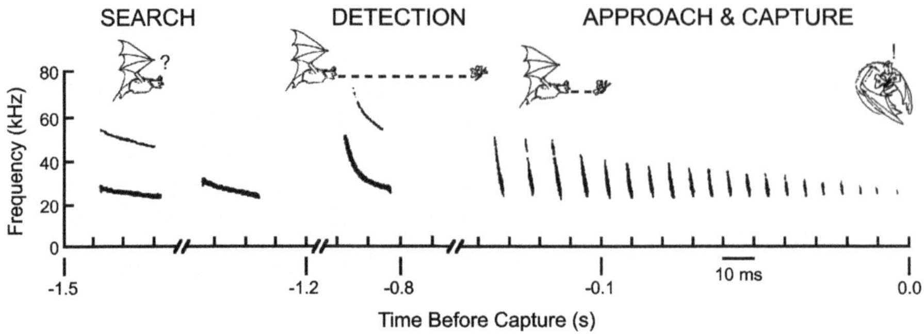

Fig. 7.1 Schematic illustrating the sequence of sounds emitted by a foraging big brown bat (*Eptesicus fuscus*) during various stages of aerial insect pursuit. Bats like *E. fuscus* that employ low duty cycle echolocation exhibit fine control over the duration of their biosonar vocalizations, mainly to avoid pulse-echo overlap. During the search phase of echolocation the bat emits long-duration, quasi-constant-frequency (i.e., shallow FM) calls at a low repetition rate (5–20 Hz). Once an insect has been detected, the bat increases its signal bandwidth and shortens its call duration. While transitioning from the approach to the capture phase of echolocation, the bat continues to emit very short duration sounds (ca. 1–5 ms) and increases its pulse repetition rate from 20–50 Hz to 50–100+ Hz, respectively. The rapid increase in pulse emissions during this final stage is also called the terminal feeding buzz. Signal frequency is shown on the *ordinate*; time to capture is shown on the *abscissa* (note time breaks in the *abscissa*). Adapted from Casseday and Covey (1995), with the sequence of echolocation sounds taken from Simmons (1987) (Line drawings by Ellen Covey)

7.3 Duration-Tuned Neurons in Bats

Echolocating bats are naturally suited for auditory electrophysiological studies because of the importance of hearing for orientation, prey detection, and social interactions (see Ratcliffe, Page, and Surlykke, Chapter 4; Moss, Simmons, and Wiegrebe, Chapter 10). Each bat species emits characteristic sounds adjusted in amplitude, frequency, and duration to meet the perceptual demands (i.e., acoustical constraints) imposed by different foraging environments and sensory tasks (e.g., Faure and Barclay 1994; Siemers and Schnitzler 2004; Denzinger, Kalko, Tschapka, Grinnell, and Schnitzler, Chapter 6). Perhaps the main reason why bats are exceptionally interesting for studies on hearing and the neural control of acoustically evoked behavior is because the ecological and evolutionary context of echolocation is well understood. Knowing the significance of bat vocalizations to natural behavior simplifies the effort to discover anatomical and/or neural adaptations of hearing (e.g., Veselka et al. 2010). Their ability to rapidly process complex acoustic signals makes echolocating bats ideal for studying temporal processing phenomena such as neural tuning to signal duration and duration discrimination behavior.

The auditory midbrain or inferior colliculus (IC) of an echolocating bat is hypertrophied and extends to the dorsal surface of the brain, making it readily accessible for in vivo physiological recording and experimentation (Figure 7.2). Because the components and anatomical patterns of connectivity within the bats' auditory system are fundamentally mammalian (Casseday et al. 2002), temporal processing mechanisms that can be readily discovered in bats are likely to be of general relevance and importance to all mammals, including humans.

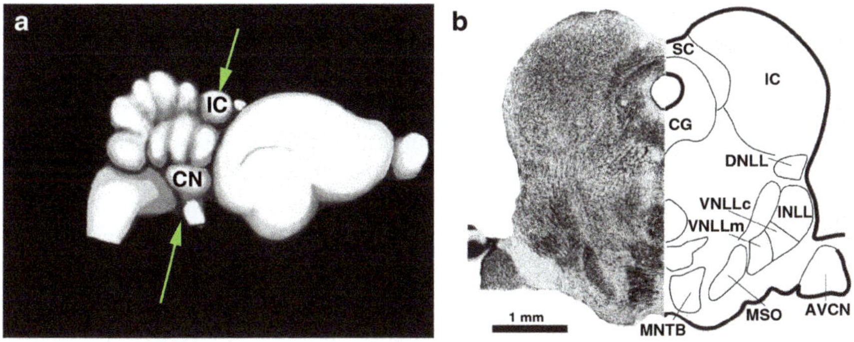

Fig. 7.2 Brain morphology and major central auditory nuclei of the big brown bat (*Eptesicus fuscus*). (**a**) Drawing of the brain of *E. fuscus* with the skull removed (rostral *right*; caudal *left*). Primary auditory afferents from the cochlea project via the vestibulocochlear nerve (cranial nerve VIII) to the CNS and enter to the first central auditory processing station, the cochlear nucleus (CN). In echolocating bats, the auditory midbrain or inferior colliculus (IC) is hypertrophied and extends to the dorsal surface of the brain. In the image, the IC is sandwiched between the cerebellum (*left*) and the superior colliculus (*right*). *Arrows* indicate the plane of section shown in panel (**b**). (**b**) Coronal (frontal) view showing a Nissl-stained tissue section (*left*) and line drawing (*right*) of some of the major central auditory nuclei in the brain of *E. fuscus* (note: not all auditory nuclei are shown in this plane of section). *Abbreviations*: *AVCN* anteroventral cochlear nucleus, *CG* central grey, *DNLL* dorsal nucleus of the lateral lemniscus, *IC* inferior colliculus, *INLL* intermediate nucleus of the laternal lemniscus, *MNTB* medial nucleus of the trapezoid body, *MSO* medial superior olive, *VNLLc* columnar subdivision of the ventral nucleus of the lateral lemniscus, *VNLLm* medial subdivision of the ventral nucleus of the lateral lemniscus, *SC* superior colliculus [Drawing in panel (**a**) by Ellen Covey; panel (**b**) adapted from Huffman and Covey 1995]

Duration tuning is an emergent electrophysiological property created de novo within the bat's auditory midbrain. Thus far, DTNs have not been reported from the auditory periphery (i.e., primary cochlear afferents are not duration tuned) nor from the lower brain stem auditory nuclei (i.e., cochlear nucleus, medial nucleus of the trapezoid body, olivary complex, and nuclei of the lateral lemniscus). First discovered from the torus semicircularis of frogs (Potter 1965; Narins and Capranica 1980) and subsequently from the IC of bats (Jen and Schlegel 1982), DTNs have now been reported from the auditory midbrain and cortex of a variety of mammals. That DTNs exist across multiple vertebrate species and in different sensory modalities strongly suggests that duration selectivity is a general processing feature of sensory systems (Faure et al. 2003; see Sayegh et al. 2011 for review). It is important to emphasize that the ability to echolocate is not a prerequisite for the evolution of duration selectivity within the mammalian CNS; however, this does not preclude DTNs from serving a highly specialized role in the neural basis of echolocation and hearing by bats.

7.3.1 *Types of Duration-Tuned Neurons*

A neural representation of signal duration is achieved within the central auditory system of echolocating bats by the spiking responses of DTNs with different temporal profiles and tuning to different signal durations (or ranges of durations). A cell's best duration (BD) is defined as the stimulus duration evoking the maximum spike count (or spike rate). Usually, the BD and duration-tuning profile are determined by stimulating the neuron with constant-amplitude, variable-duration, pure-tone pulses presented at the cell's characteristic frequency (i.e., the frequency with the lowest threshold on the cell's frequency tuning curve) or best excitatory frequency (BEF). A cell's cut-off duration(s) is (are) defined as the signal duration(s) where the spiking response falls to ≤50 % of the maximum (see Jen and Zhou 1999; Faure et al. 2003). The temporal tuning profiles of DTNs fall into one of five duration-filter classes based on the shape of the duration-tuning curve (Figure 7.3). The names of these classes are analogous to the shapes of different frequency filters used in resonant electrical circuits even though this analogy has not been especially useful in elucidating the synaptic mechanisms underlying or the functions of DTNs to normal hearing.

1. Short-pass DTNs respond maximally to sounds presented at or below the cell's BD, with a ≥50 % reduction in spiking to signals longer than the cut-off duration (Figure 7.3a).
2. Band-pass DTNs also respond maximally at BD but have a ≥50 % reduction in spiking at stimulus durations both shorter and longer than BD. Band-pass DTNs have two cut-off durations: one below and one above the BD (Figure 7.3b).
3. Long-pass DTNs lack a BD but have a ≥50 % reduction in spiking at stimulus durations shorter than the cut-off duration (Figure 7.3c). By definition, long-pass cells do not have a BD because, in theory, they respond to any stimulus that exceeds some minimum duration.
4. Band-reject DTNs have complex duration tuning curves with a temporal selectivity opposite to that of band-pass DTNs. Strictly speaking, band-reject DTNs have at least one BD plus another response peak with a ≥50 % reduction in spiking for the band-reject durations in-between. Although comparatively uncommon, band-reject DTNs have been recorded from the IC of the big brown bat (Pinheiro et al. 1991); Pallas' mastiff bat, *Molossus molossus*, (Mora and Kössl 2004); and the least horseshoe bat, *Rhinolophus pusillus*, a high duty cycle species that emits long-duration CF-FM signals and employs Doppler-shift compensation (Luo et al. 2008). Plausible biological mechanisms that reproduce the spiking responses of band-reject DTNs have been described (Mora and Kössl 2004; Aubie et al. 2009).
5. Multi-peak DTNs have complex duration tuning curves, like band-reject DTNs. By definition, multi-peaked DTNs have multiple response peaks, including a BD with local response minima in-between. Multi-peaked DTNs have been reported from the IC and AC of big brown bats (Pinheiro et al. 1991; Hou et al. 1992) and the IC of least horseshoe bats (Luo et al. 2008).

Although DTNs share response properties across different bats, the range of neuronal BDs and temporal tuning bandwidths vary substantially among species with different echolocation strategies and auditory constraints. Short-pass DTNs are typically tuned to short signal durations and have narrower duration-tuning curves than band-pass DTNs tuned to similar durations and recorded from the same species (e.g., Fremouw et al. 2005; Wu and Jen 2008a). Offset-responding DTNs with short BDs are frequently recorded from the IC of low duty cycle FM bats, whereas DTNs

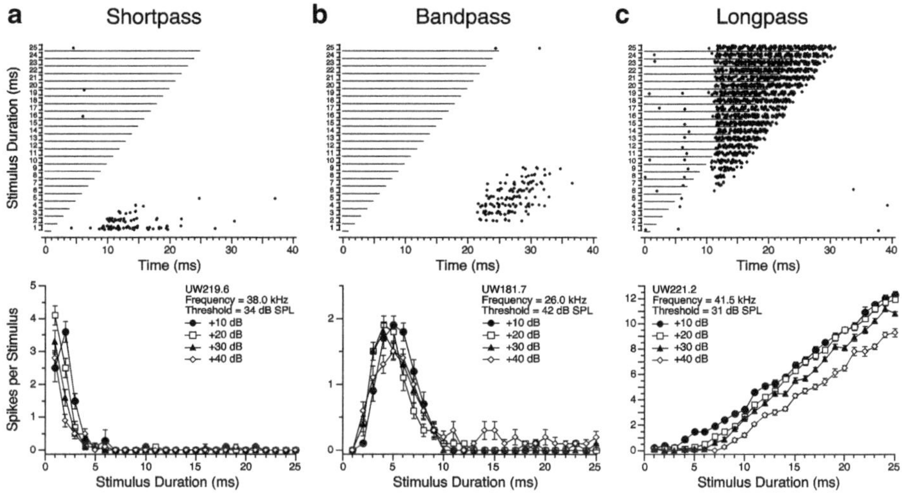

Fig. 7.3 The three most common temporal tuning types of duration-tuned neurons (DTNs) in bats. (**a**) Short-pass DTN. (**b**) Band-pass DTN. (**c**) Long-pass DTN. *Top row*: Post-stimulus time dot raster displays illustrating the timing of spikes in response to best excitatory frequency (BEF) tone pulses that were randomly varied in stimulus duration and presented at +30 decibels (dB) above the best duration (BD) threshold. *Bottom row*: Mean (±SE) spikes per stimulus as a function of stimulus duration at three sound pressure levels above threshold. Note that the duration selectivity of each cell is relatively stable with changes in stimulus level. (**a**, **c**) 15 trials/stimulus; (**b**) 20 trials/stimulus (Reproduced from Faure et al. 2003)

with onset-evoked responses tuned to longer BDs appear to be more common in the IC of high duty cycle CF-FM bats (e.g., Luo et al. 2008; Macías et al. 2011). Interestingly, the number of IC cells that exhibited duration selectivity increased and their amplitude tolerance increased when least horseshoe bats were stimulated with complex CF-FM sounds representative of natural biosonar pulses instead of pure tones (Luo et al. 2008). In general, duration tuning curves in echolocating bats are more sharply tuned than in non-echolocating vertebrates (see Table 1 in Sayegh et al. 2011), suggesting that the ability to echolocate has naturally selected for DTNs to have narrower temporal tuning profiles, presumably to facilitate rapid processing of pulses and echoes (e.g., Condon et al. 1996).

Computationally, short-pass, band-pass, and band-reject DTNs are interesting because their response selectivities cannot be explained by the integration of stimulus energy, requiring special intrinsic properties and/or the temporal interaction of excitatory and inhibitory synaptic inputs (see Section 7.4). Although the responses of long-pass DTNs appear similar to primary auditory afferents, most vertebrate auditory neurons decrease their first-spike latency (FSL) when presented with higher amplitude signals. In contrast, the FSL and minimum duration required to evoke spiking in a long-pass DTN does not shorten with increasing stimulus amplitude (Brand et al. 2000; Faure et al. 2003).

7.3.2 Duration-Tuned Neuron Response Properties

Key electrophysiological and topographical properties of DTNs are known mainly from recordings in the IC of big brown bats, the species with the most comprehensive data on duration tuning. When the acoustic spiking thresholds of a large population of DTNs are plotted as a function of their BEF and shown with the behavioral audiogram of big brown bats, it is clear that cells with the lowest thresholds mirror the bat's best behavioral sensitivity, and the range of DTN thresholds covers the dynamic range of hearing (Morrison et al. 2014).

Like other types of central auditory neurons, DTNs in the IC of big brown bats are tonotopically organized and have BEFs that systematically increase with recording electrode depth (Pinheiro et al. 1991; Wu and Jen 2008a). Acoustic thresholds of DTNs in big brown bats are also topographically organized; cells with the lowest thresholds are found at shallower depths in the IC (Morrison et al. 2014). Jen and Wu (2006) reported a positive correlation between BD and BEF, implying a topographic map of neuronal BDs within the IC of big brown bats; however, this finding has not been replicated. While it is easy to imagine how a topographic map of duration tuning might prove useful in neural computations of echolocation, it remains an open question whether a spatial map of duration selectivity exists within the central auditory system.

Spontaneous firing rates of DTNs are typically low (Aubie et al. 2014). The phasic spiking responses of most short-pass and band-pass DTNs in the IC of big brown bats occur after stimulus offset (Wu and Jen 2008a) and have FSLs that increase with the tone burst duration, making them offset-responding cells (Casseday et al. 1994, 2000; Ehrlich et al. 1997). Faure et al. (2003) reported that the FSLs of short-pass and band-pass DTNs in the IC of big brown bats were always longer than their BD and that different cells tuned to the same BD could have a wide range of FSLs. Moreover, cells responding with two or more action potentials had spike burst durations (last-spike latency minus first-spike latency) that were typically greater than the cell's BD. Interestingly, many short-pass DTNs in the IC of the pallid bat (*Antrozous pallidus*) (Fuzessery and Hall 1999) and Pallas' mastiff bat (Mora and Kössl 2004), and in the AC of little brown myotis (Galazyuk and Feng 1997) had onset-evoked responses with more-or-less constant FSLs. In contrast, long-pass DTNs typically exhibit tonic (i.e., primary-like responses) with sustained spiking during the ongoing portion of the stimulus (e.g., Brand et al. 2000; Faure et al. 2003). The FSL and burst duration are important physiological parameters because they provide clues about the underlying mechanisms responsible for creating DTNs and may be relevant as inputs for other temporal processing mechanisms in higher central auditory pathways (e.g., delay-tuned neurons).

Although they vary across species, the temporal specificities of DTN responses are largely tolerant to changes in SPL. Spike counts and FSLs of midbrain DTNs in big brown bats (Zhou and Jen 2001; Fremouw et al. 2005) and Parnell's mustached bat (Macías et al. 2011) are amplitude tolerant over a large dynamic range, in some cases up to 50 dB. Figure 7.4 demonstrates the amplitude tolerance of DTNs recorded from the IC of big brown bats. In other bats, amplitude tolerance of duration tuning is more variable. For example, DTNs recorded from the IC of little brown myotis

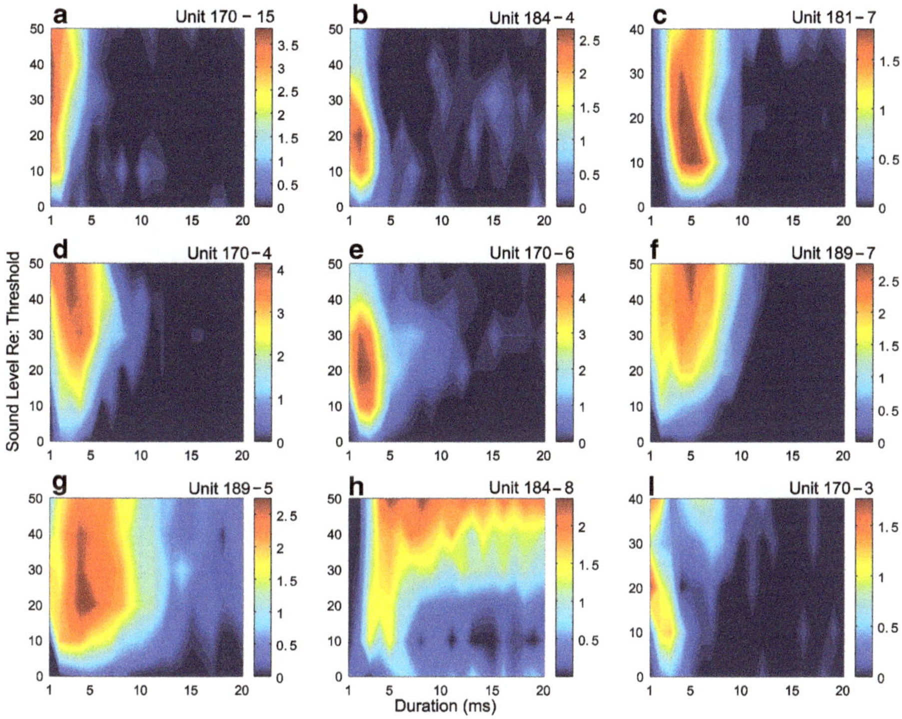

Fig. 7.4 Amplitude tolerance of duration-tuned neurons in the inferior colliculus of *E. fuscus*. Nine examples of typical temporal response areas showing duration tuning as a function of SPL above neural threshold. The number of spikes per stimulus is indicated by the color scale varying from dark blue (*lowest*) to dark red (*highest*), with a scale bar beside each panel (Reproduced from Fremouw et al. 2005)

were level tolerant, whereas those recorded from the AC were level dependent (Galazyuk and Feng 1997), suggesting that mechanisms of duration selectivity may differ between the midbrain and cortex. In contrast, of the few DTNs recorded from the AC of pallid bats that responded to pure tones, none maintained neural tuning for stimulus duration at multiple sound intensities (Razak and Fuzessery 2006).

Neural inhibition is essential for establishing the duration-tuned response because iontophoretic application of antagonists of inhibitory neurotransmitters—either bicuculline, a $GABA_A$ antagonist (Fuzessery and Hall 1999; Jen and Feng 1999; Jen and Wu 2005), and/or strychnine, a glycine antagonist (Casseday et al. 2000; Yin et al. 2008)—abolishes or broadens duration tuning. Intracellular recordings (Covey et al. 1996; Leary et al. 2008) and single-unit extracellular recordings combined with paired-tone stimulation (Faure et al. 2003; Covey and Faure 2005) have confirmed that the onset of inhibition acting on DTNs usually precedes the onset of excitation. In general, inhibition sharpens the temporal selectivity of DTNs (Aubie et al. 2009) and can also enhance echo duration and frequency selectivity (Jen and Wu 2008; Wu and Jen 2008a, b). The effective time course of the inhibition acting on DTNs lasts as long as, or longer than, the duration of the stimulus evoking it (Faure et al. 2003). Moreover, the effects of inhibition can accumulate and influence the ability of DTNs to respond to repeated stimuli (Wang et al. 2008, 2010; Sayegh et al. 2012).

Experiments using monaural stimulation of the ear contralateral to the IC being recorded have shown that monaural, contralateral auditory pathways contain all of the circuitry necessary to create DTNs. Sayegh et al. (2014) measured the relative

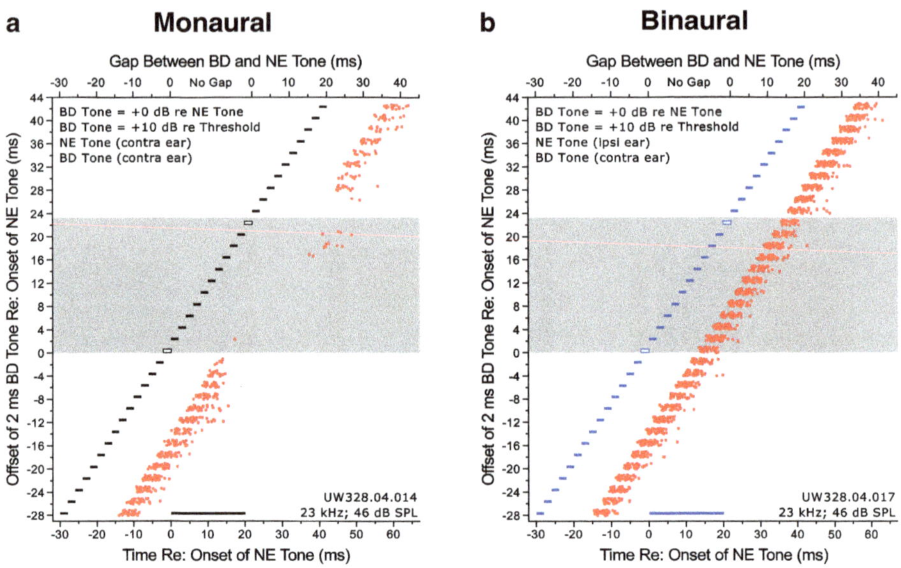

Fig. 7.5 Monaural and binaural inhibition acting on duration-tuned neurons (DTNs) in the inferior colliculus of *E. fuscus*. Dot raster displays of the spiking responses (*red dots*) of a short-pass DTN during monaural and binaural paired-tone stimulation. (**a**) Monaural stimulation. Cell stimulated with a pair of pure tones (*black bars*): a roving tone presented at the cell's best duration (BD tone = 2 ms) and a stationary tone presented at a non-excitatory duration (NE tone = 20 ms). The onset time of the BD tone was randomized; the onset time of the NE tone was fixed. The BD and NE tones were electronically mixed and broadcast to the contralateral ear. When the two tones temporally overlapped they summed to form a single, composite tone with an amplitude pedestal. Note how the cell's response became suppressed before, during, and after the presentation of the NE tone, thus demonstrating the leading, sustained, and persistent monaural inhibition that is responsible for creating the duration-tuned response. Also note that the monaural inhibition evoked by the NE tone had a shorter latency than the first excitatory spikes evoked by the BD tone. (**b**) Binaural stimulation. The BD and NE tones (*blue bars*) were split, with the BD tone presented to the contralateral ear and the NE tone presented to the ipsilateral ear. Note how responses from the cell are no longer suppressed by the presentation of the NE tone, indicating that the ipsilateral ear was not inhibitory for this cell and demonstrating that the inhibition that is responsible for creating the duration-tuned response is purely monaural in nature. The NE tone is shown with a long solid bar above the *x*-axis; the BD tone is shown as a series of vertically stacked short solid bars. BD tone bars with a white fill indicate intervals when the BD tone was contiguous with, but did not overlap, the NE tone. The *gray box* indicates the range of stimulus onset times over which the BD tone was contiguous with or overlapped the NE tone. All tones presented at the DTN's BEF. BD and NE tone frequency = 23 kHz; BD and NE tone amplitude = 31 dB SPL; BD tone threshold = 21 dB SPL; 15 trials per stimulus (Data graphed from Sayegh et al. 2014)

contributions (i.e., strength and time course) of the neural inhibitions acting on midbrain DTNs recruited by the separate monaural and binaural auditory pathways in big brown bats, and found that about half of the cells received only monaural (i.e., contralateral) inhibition (Figure 7.5). When ipsilateral inhibition was observed, it was weaker in strength and longer in latency compared to the contralateral inhibition evoked monaurally in the same cells. Thus, neural inhibition responsible for creating the temporally selective responses of midbrain DTNs originates from monaural central auditory pathways.

Two recent studies suggest that temporal and spectral tuning interact in bats (Wu and Jen 2008a, b). Electrical engineering theory dictates that resonant filter circuits with broad frequency tuning have less temporal distortion (i.e., ringing) than narrow frequency filters. Consistent with this expectation, DTNs with broad frequency tuning had sharper temporal tuning curves than DTNs with narrow frequency tuning (Morrison et al. 2014). In the IC of big brown bats, DTNs with sharp frequency and broad duration tuning were located in the dorsal IC, whereas cells with wide spectral and narrow temporal tuning were located in the ventral IC (Morrison et al. 2014).

7.4 Conceptual and Computational Mechanisms of Duration Tuning

The mammalian auditory brain stem consists of multiple parallel and serial pathways that converge at the IC. The IC also receives descending modulatory input from cortical and subcortical nuclei and transmits information to motor control systems of the cerebellum and superior colliculus. Thus, the IC is an integrative center for information processing (Casseday et al. 2002). Because neural latencies to the IC are longer in some auditory pathways than in others (Haplea et al. 1994), it is possible to create neural circuits that compare sounds through the coincidence of synaptic inputs. Some forms of temporal specificity are created by complex sequences of neural excitation and inhibition (Casseday et al. 1994; Fuzessery and Hall 1999) that may converge at the same time or at different times depending on the nature of the stimulus (Covey et al. 1996).

The IC is the first auditory nucleus where DTNs are found. Within the IC, duration selectivity is created through the convergence and temporal interaction of excitatory and inhibitory synaptic inputs offset in time (e.g., Ehrlich et al. 1997; Fuzessery and Hall 1999; Faure et al. 2003). Neuropharmacological experiments and intracellular recordings have shown that duration tuning is not relayed from the lower brain stem but is a response property created de novo in the IC (Covey et al. 1996). The mechanisms responsible for creating DTNs in the auditory thalamus and cortex remain unknown.

Various conceptual mechanisms have been proposed to explain the temporal selectivity of DTNs (Casseday et al. 1994; Fuzessery and Hall 1999; Hooper et al. 2002; Leary et al. 2008). In theory, any central auditory neuron could be duration tuned if the relative timing and/or strength of its sound-evoked inputs varied as a function of stimulus duration. Evidence from extracellular and intracellular recordings, along with computational modeling studies, have proposed two basic mechanisms thought to underlie duration-tuned circuits. Although details of these mechanisms differ among animal species and between classes of DTNs, all may be classified as either coincidence detection or anti-coincidence mechanisms. These models predict that a cell's BD, duration-tuning profile, and temporal tuning bandwidth are controlled, in part, by the amount of time that inhibition precedes excitation.

7.4.1 Coincidence Detection Mechanism

Figure 7.6 summaries two of many circuit-based mechanisms that use the timing of excitatory and inhibitory synaptic inputs to recreate short-pass and band-pass electrophysiological responses in the IC of the bat (Aubie et al. 2009, 2012; Sayegh et al. 2011).

The coincidence detection mechanism relies on the simultaneous occurrence (i.e., temporal coincidence) of at least two excitatory (depolarizing) events (Narins and Capranica 1980): one linked to stimulus onset (ON_E), the other linked to stimulus offset (OFF_E). While neither is sufficient to evoke spiking on its own, when the two coincide and sum (or their effects sufficiently overlap), the resultant excitation may be suprathreshold and evoke spiking (Figure 7.6a, b). The coincidence detection mechanism incorporates three inputs to a DTN: (1) a transient, onset-evoked, excitatory post-synaptic potential (EPSP); (2) a sustained, onset-evoked, inhibitory post-synaptic potential (IPSP) whose latency is as short or shorter than the EPSP latency; and (3) a transient, offset-evoked EPSP that can be explicitly modeled as a dedicated excitatory input or as post-inhibitory rebound from the sustained inhibition. When excitations (1) and (3) coincide, the DTN fires action potentials. But the DTN fails to spike when the signal duration is too short so that the OFF_E event occurs before the ON_E event or when it is too long so that the sustained inhibition overlaps and overrides the ON_E excitation. Because the latency of the OFF_E varies directly with stimulus duration, the responses of the cell become temporally selective. When the temporal window of excitation is symmetrical, the coincidence detection mechanism generates a band-pass duration-tuned response (Figure 7.6a, b). This mechanism can also generate a short-pass duration-tuned response when the peak excitatory coincidence occurs at short stimulus durations (Aubie et al. 2009). Offset-responding DTNs are commonly seen in recordings from the IC of big brown bats (Faure et al. 2003), pallid bats (Fuzessery and Hall 1999), and Pallas' mastiff bats (Mora and Kössl 2004). The coincidence detection mechanism naturally results in the FSL tracking stimulus offset (i.e., spiking occurs at a constant latency re stimulus offset) because action potentials can only be evoked during the arrival of the OFF_E event.

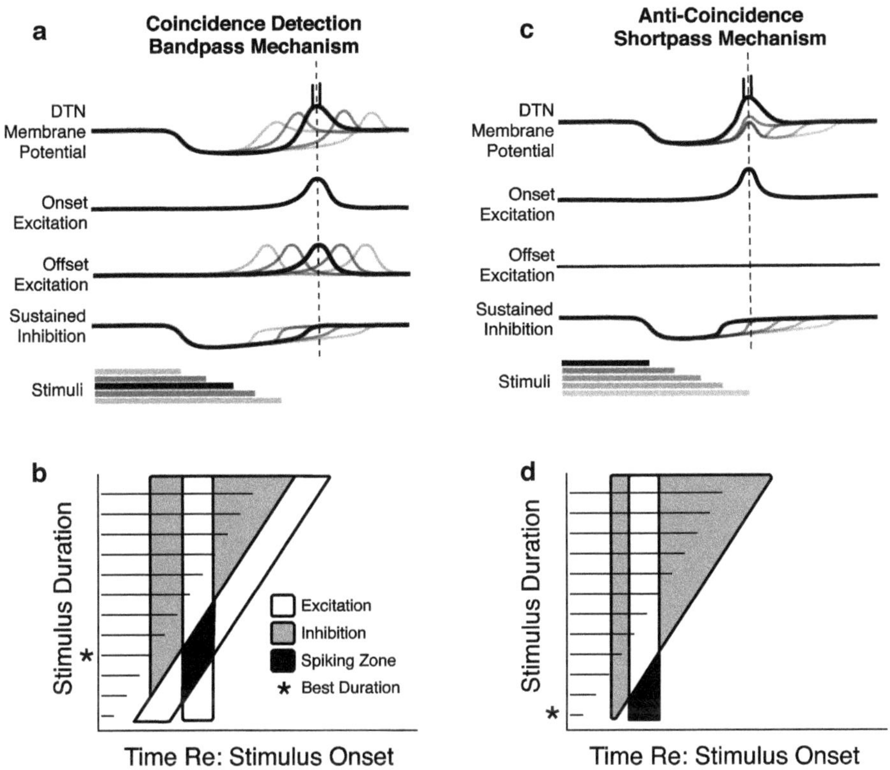

Fig. 7.6 Two representations of the coincidence detection and anti-coincidence mechanisms of duration tuning. (**a**, **c**) The top trace represents the membrane potential of the model DTN in response to a best duration (BD) stimulus (*bold trace*) and to stimuli not at BD (*shades of gray*, with lighter grays corresponding to responses evoked by stimuli further from BD). The bottom three traces represent the magnitude and timing of synaptic inputs to the DTN at BD (*bold*) and not at BD (*shades of gray*). Inputs are: an onset-evoked transient excitation (second trace), an offset-evoked transient excitation (third trace), and an onset-evoked sustained inhibition (fourth trace) that grows with stimulus duration. Stimulus duration is represented by horizontal bars below the traces. *Black bars* represent the BD stimulus; *gray bars* represent stimuli not at BD, with lighter grays corresponding to shorter or longer duration signals. Excitatory input latencies in the coincidence detection mechanism were set so that the maximum temporal coincidence (*vertical dashed line*) occurred at an intermediate stimulus duration resulting in a band-pass DTN. For the anti-coincidence mechanism, the vertical dashed line is the first point when there is no temporal coincidence between the excitatory and inhibitory inputs. (**b**, **d**) Schematic dot raster displays illustrating how temporal interaction of excitatory and inhibitory inputs result in offset spiking responses in a band-pass and short-pass DTN. The latency and time course of excitatory (*white*) and inhibitory (*gray*) inputs, and the zone of high spiking probability by the DTN (*black*), are illustrated for a range of stimulus durations. Maximum spiking (*) occurs in response to the BD stimulus. Stimulus durations are shown as *horizontal black lines*. The axes are intentionally unlabeled to highlight the timescale invariance of the models (Reproduced from Aubie et al. 2012)

7.4.2 Anti-Coincidence Mechanism

Not all DTNs have offset-evoked responses. Many short-pass DTNs in the IC of big brown bats (Ehrlich et al. 1997), pallid bats (Fuzessery and Hall 1999), and Parnell's mustached bats (Macías et al. 2011), some band-pass DTNs in the IC of least horseshoe bats (Luo et al. 2008) and Parnell's mustached bats, and almost all long-pass DTNs in bats have constant FSLs relative to stimulus onset and appear to lack an OFF_E input (e.g., Faure et al. 2003). To account for this, Fuzessery and Hall (1999) proposed the anti-coincidence mechanism of duration tuning (Figure 7.6c, d). This mechanism requires at least two events: a transient, onset- or offset-evoked, suprathreshold excitation (EPSP); and a sustained, onset-evoked, inhibition (IPSP). The anti-coincidence mechanism predicts that DTNs will fire only when the EPSP fails to coincide with the IPSP. At short stimulus durations, spikes are evoked when the EPSP latency is longer than the offset of the IPSP. At longer durations, the sustained IPSP overlaps with the EPSP rendering it subthreshold. The anti-coincidence mechanism naturally generates a short-pass duration-tuned response with either constant or increasing FSLs at the shortest stimulus durations. Modified versions of the anti-coincidence mechanism can also generate band-pass duration-tuning responses when the EPSP input is weak or absent at short durations because of low stimulus energy (e.g., Sayegh et al. 2011).

An anti-coincidence mechanism can also generate a long-pass duration-tuned response with paradoxical latency shift (Faure et al. 2003). Long-pass tuning requires two components: a strong, onset-evoked, transient IPSP and a suprathreshold, onset-evoked EPSP that is sustained for the duration of the stimulus. Thus, just like short-pass and band-pass cells, long-pass DTNs receive an early inhibition and a delayed excitation; however, unlike short-pass and band-pass DTNs, the inhibition is transient and the excitation is sustained. Therefore, the leading inhibition that creates short-pass and band-pass duration tuning can be reconfigured to form a long-pass DTN that requires a minimum stimulus duration before it can respond. This minimum duration depends on the amount of time that inhibition leads excitation. Thus, by simply reversing the pattern of synaptic inputs, it is possible to create cells with different types of duration tuning.

In addition to single cell in vivo electrophysiological recordings, formal mathematical models have been used to evaluate the biological plausibility of the proposed mechanisms of duration tuning. These models mirror known neuroanatomical network connections from the auditory brain stem through to the midbrain and can replicate the in vivo responses of DTNs in bats and other vertebrates (Aubie et al. 2009, 2012). When validated, such models will permit researchers to explore how a duration-tuned network would respond to novel stimulation or a simulated hearing deficit, with the ultimate goal being to predict behavioral performance.

7.4.3 Multiple Mechanisms of Duration Tuning

The coincidence and anti-coincidence mechanisms predict that a cell's BD, FSL, temporal bandwidth, and duration tuning class are controlled, in part, by the amount of time that inhibition precedes excitation. This detail has factored into the design of experiments intended to elucidate the mechanisms underlying the creation of DTNs.

Recall that the above models can be combined in interesting ways to produce different classes of duration tuning (Sayegh et al. 2011). Although band-pass tuning naturally results from network connections that employ a coincidence detection mechanism, an anti-coincidence network can also generate a band-pass response. Similarly, the anti-coincidence mechanism illustrates how to create a short-pass DTN, yet band-pass DTNs can be created using essentially the same mechanism. When the latencies of the subthreshold ON_E and OFF_E events are delayed relative to stimulus onset, different duration-tuning classes emerge from the coincidence detection mechanism. Band-reject tuning is possible when the OFF_E event is subthreshold at short stimulus durations (thus requiring coincidence with an ON_E event to evoke spiking), but is suprathreshold at longer durations with higher stimulus energy (thus requiring anti-coincidence with the sustained inhibition). Band-reject and/or multi-peaked duration tuning could also arise from the coincidence of an OFF_E event with multiple ON_E events (e.g., Dear et al. 1993), each with its own FSL (see Mora and Kössl 2004).

7.5 Duration Tuning and Echolocation

While the contribution(s) of DTNs to hearing and echolocation by bats is (are) still unknown, they must play some functional role. Because the BDs and range of temporal selectivity closely mirror the range of echolocation call durations (see Table 1 in Sayegh et al. 2011), this suggests that one function of DTNs in bats is to encode echo duration changes. For example, a disproportionate number of DTNs in the IC of big brown bats (Pinheiro et al. 1991; Faure et al. 2003) and Pallas' mastiff bats (Mora and Kössl 2004) have BEFs that fall within the spectral bandwidth of the fundamental FM element used for echolocation. There is evidence suggesting that the amplitude tolerance of DTNs varies among bat species that emit different types of echolocation and employ different signaling strategies. For example, in low duty cycle bats, like big brown bats, the spike counts and FSLs of DTNs at BD are tolerant to large changes in signal amplitude (Zhou and Jen 2001; Fremouw et al. 2005), whereas DTNs in high duty cycle bats, like Pallas' mastiff bat*s* and least horseshoe bats, show more variation in amplitude tolerance (Mora and Kössl 2004; Luo et al. 2008). If DTNs operated in neural circuits that detected the delay between pulses and echoes, amplitude tolerance would ensure that the spike count and FSL of a DTN remained stable when responding to both loud outgoing vocalizations and weaker returning echoes.

Neurons with similar BDs but different FSLs could act as delay lines for cells in higher auditory centers. For example, if the spiking response of a delay-tuned neuron from the auditory thalamus or cortex depended on the coincidence of inputs from DTNs in the IC, then this circuit would detect the delay between specific durations and frequencies contained in the bat's pulses and echoes (Faure et al. 2003; Covey and Faure 2005). By incorporating DTNs tuned to similar (different) frequencies and SPLs, more (less) specific and complex responses could emerge.

The ability to integrate across inter-stimulus intervals is important because echolocating bats naturally experience variation in pulse-echo timing. Despite having amplitude tolerance, some DTNs sharpen their temporal selectivity when they are repeatedly stimulated with pairs of BD tones that mimic pulses and echoes (see review by Wu and Jen 2010). This suggests that DTNs could serve as auditory filters, responding only when stimulated by specific signal durations, amplitudes, and frequencies. In bats, DTNs tuned to short BDs have shorter recovery cycles for stimulation with BD tones mimicking pulse-echo pairs than cells tuned to longer BDs (Wang et al. 2008, 2010; Sayegh et al. 2012). This type of response could have been naturally selected if DTNs play a role in hunting and target pursuit.

Electrophysiological recordings from the IC of big brown bats suggest that DTNs may play a role in the perception of biosonar sounds shorter than a neuron's BD. Consider, for example, a band-pass DTN with a BD of 5 ms (e.g., Figure 7.3b) stimulated with a pair of 2 ms BEF tones. Stimulation with such sounds would normally evoke relatively weak spiking, but the cell's response would strengthen as the inter-stimulus interval (i.e., gap) decreased so that the combined duration of tone1 + gap + tone2 approximated the BD of the cell. In this situation, when the interval between the onset of an outgoing echolocation call (Signal 1) and the offset of its returning echo (Signal 2) was close to BD and when the pulse-echo interval was shorter than the cell's recovery time (defined as the minimum interval for the response to Signal 2 to be ≥50 % of the response to Signal 1), the tone pair might be "integrated and perceived" by the DTN as a single, unified stimulus. A pleasing feature of this mechanism is that it naturally predicts that DTNs should respond after signal offset (and many do). It also predicts that cells tuned to longer BDs will integrate over longer pulse-echo intervals, thus rendering them useful for detecting distant targets compared to cells with shorter BDs. This mechanism requires the inter-stimulus interval to be shorter than the cell's recovery time so that it does not respond to each independent sound. Such a mechanism would seem useful for detecting pulse-echo intervals shorter than 8–30 ms, the preferred range of delays encoded by most delay-tuned neurons in the auditory midbrain (Dear and Suga 1995).

So far the emphasis has been on DTNs serving as temporal filters, but these cells are also tuned in frequency and have V-shaped, U-shaped, and O-shaped tuning curves just like other auditory neurons (Sutter 2000). Like temporal tuning, spectral tuning enhances neural selectivity. A DTN fires action potentials only when it "hears" a sound of the correct frequency (i.e., spectral bandwidth), duration (i.e., temporal bandwidth), and amplitude (i.e., SPL). Therefore, midbrain DTNs could serve as spectro-temporal filters for auditory processing

(Morrison et al. 2014). Populations of DTNs with different BEFs, frequency response areas, BDs, and temporal tuning curves could act as inputs to the neural equivalent of a spectrogram, providing the brain with specific information about the acoustic environment. Such filters would be useful for hearing in general and echolocation by bats.

7.6 Neural Coding of Complex Acoustic Features in the Bat Auditory System

The formation of perceptual objects from physical stimuli is an essential task that echolocating bats tackle by analyzing echoes of their sonar emissions. In doing so, they must cope with all of the problems common to object recognition, independent of the sensory domain: Object size is variable, objects can be viewed from different perspectives, and/or objects can move or change their structure over time. Bats also gain information from passive hearing, including the analysis of prey-generated sounds, which are especially important for substrate-gleaning bats (see Ratcliffe, Page, and Surlykke, Chapter 4), and conspecific vocalizations, as many bats possess rich vocal repertoires and have complex social interactions (see Gillam and Fenton, Chapter 5).

7.6.1 Perception of Complex Auditory Objects

The acoustic image of an object can be characterized by its impulse response (IR). The IR is the sum of the reflections from an object ensonified by an impulse of infinitely short duration and infinitely high amplitude (i.e., a Dirac impulse). When a bat's call reflects off an object, its acoustic properties are imprinted on the signal and object-specific echo properties are created. In technical terms, the perceived echo is the convolution of the bat's sonar emission with the object's IR.

The short integration times of cochlear filters (e.g., 200 µs in the Indian false vampire bat, *Megaderma lyra*; Weissenbacher and Wiegrebe 2003) induce some constraints on echo-acoustic object analysis. Small objects with narrowly spaced reflecting surfaces can only be analyzed in the frequency domain by spectral interference patterns because cochlear filter integration times are too long for a temporal analysis of the echo. In contrast, large objects with widely spaced reflecting surfaces can be analyzed in the time domain because the temporal separation of echoes is longer than the integration times of cochlear filters.

Psychophysical experiments have shown that bats can detect and perceive object structure based on spectral interference patterns (Schmidt 1988). The ecological relevance of an object's spectral interference pattern was demonstrated in nectar-feeding bats that detect neotropical flowers with echolocation (von Helversen and von Helversen 1999). Different developmental stages of the same flower, but also flowers or leaves from different plants, gave rise to distinguishable and directional echo spectral interference patterns (von Helversen et al. 2003; Simon et al. 2011). For example, nectar-feeding Pallas's long-tongued bats (*Glossophaga soricina*) characterized hollow spheres of a given size by their unique spectral echo pattern and generalized object shape independent of object size. Although the spectral composition of echoes reflected from hollow spheres changed with sphere size, the spectral interference pattern remained constant for a given size and was independent of the angle of incidence of ensonification (von Helversen 2004).

Larger complex objects, like bushes or trees, typically consist of many separate surfaces. Consequently, echoes reflected from such objects consist of the sum of many single reflections, and they will have no systematic spectral interference pattern. Moreover, movement of parts of these objects will impose an unstable time structure on the reflected echoes. Therefore, large complex acoustic objects are best described by statistical measures, such as the degree of fluctuation in the envelope of an objects' IR. Indeed, IR roughness was shown to be a good predictor for correctly associating the IR to the corresponding object (Müller and Kuc 2000). This was confirmed in a free-field study demonstrating the high predictive power of IR roughness for the classification of natural objects (Stilz 2004). In a psychophysical playback experiment with phantom objects, pale spear-nosed bats (*Phyllostomus discolor*) were trained to discriminate echoes generated with a smooth or a rough IR. Once bats had learned the task, they spontaneously classified unknown IRs according to IR roughness (Grunwald et al. 2004). In addition to IR roughness, the regular (periodic) anatomical structure of plants (i.e., the species-specific distances between leaves and branches) contributes to the temporal envelope statistics of echoes, and bats reliably use this information for object classification (Yovel et al. 2011).

7.6.2 Passive Hearing and Communication Sounds

Passive hearing is important for substrate-gleaning bats that find food by listening for prey-generated sounds (Schnitzler et al. 2003), and bats use spectral and temporal cues to detect rustling sounds in psychophysical (e.g., Hübner and Wiegrebe 2003; Goerlitz et al. 2008) and prey detection tasks (e.g., Fuzessery et al. 1993). For gleaning pallid bats, separate parallel thalamocortical pathways for passive and active sound localization have been proposed (Razak et al. 2007). Information from both pathways appears to converge because neurons in the AC are bi-modally tuned to the low- and high-frequency ranges used for passive hearing and active echolocation, respectively (Razak et al. 1999).

Passive hearing is also important in the analysis of intraspecific communication vocalizations (Kanwal et al. 1994; Wright et al. 2013). Bats have a rich vocal repertoire and their social calls have a complex spectral and temporal structure. The detection and discrimination thresholds of AM and FM stimuli have been investigated in psychophysical experiments (e.g., Esser and Lud 1997). Bats also have a remarkable ability for vocal learning (see Gillam and Fenton, Chapter 5). In the last two decades the investigation of the neural circuitry underlying the processing of bat communication sounds, especially with respect to semantic meaning and emotional content, has opened a broad field of research.

The following sections focus on the neural representation of spectral and temporal features of complex objects in echolocating bats, with an emphasis on cortical mechanisms and relating electrophysiological findings to the behavioral performance of bats in psychophysical tasks. The chapter concludes with a section on neural processing of conspecific vocalizations in the AC and the amygdala of bats. More recently, some studies have investigated the processing of bat vocalizations by IC neurons (see Pollak, Chapter 8).

7.7 Neural Coding of Complex Echo-Acoustic Objects

7.7.1 Neural Coding of Object Spatial Extent

In the visual system, the spatial extent of an object (i.e., its visual aperture) is explicitly encoded by the extent of the image on the retina; however, in the auditory system, spatial information must be computed. For echolocating bats, the situation is even more complicated because object width is encoded not only by echo amplitude or intensity but also by the naturally co-varying spread of angles of incidence of echoes impinging upon the bat's ears (hereafter called its "sonar aperture"). Because sonar aperture also increases with object width, bats might use echo intensity and/or sonar aperture to estimate object width.

This problem was addressed in electrophysiological experiments with pale spear-nosed bats where virtual objects were presented to bats via earphones (Heinrich et al. 2011). The technique of virtual stimulus presentation makes use of the bat's HRTF, which describes the frequency-specific ratio of SPLs near the tympanum to that measured in the free field (Young et al. 1996). To generate object echoes with a specific width, Heinrich et al. (2011) convolved a typical pale spear-nosed bat echolocation call with HRTFs measured from the same species (Firzlaff and Schuller 2003), corresponding to several adjacent horizontal positions in space, and summed the resulting echoes (Figure 7.7a–e). Echoes were presented in different intensity steps over a range of 24 dB while single and multi-unit extracellular responses were recorded from anesthetized bats. In the AC and IC of pale spear-nosed bats, a subset of neurons responded strongest to object echoes with a specific sonar aperture, independent of echo intensity. Response-strength contour plots from these neurons (Figure 7.7f–i) showed robust changes along the horizontal (object width) axis but varied little along the vertical (intensity) axis. The observed selectivity depended on binaural inputs, as monaural stimulation decreased the number of sonar aperture selective cells.

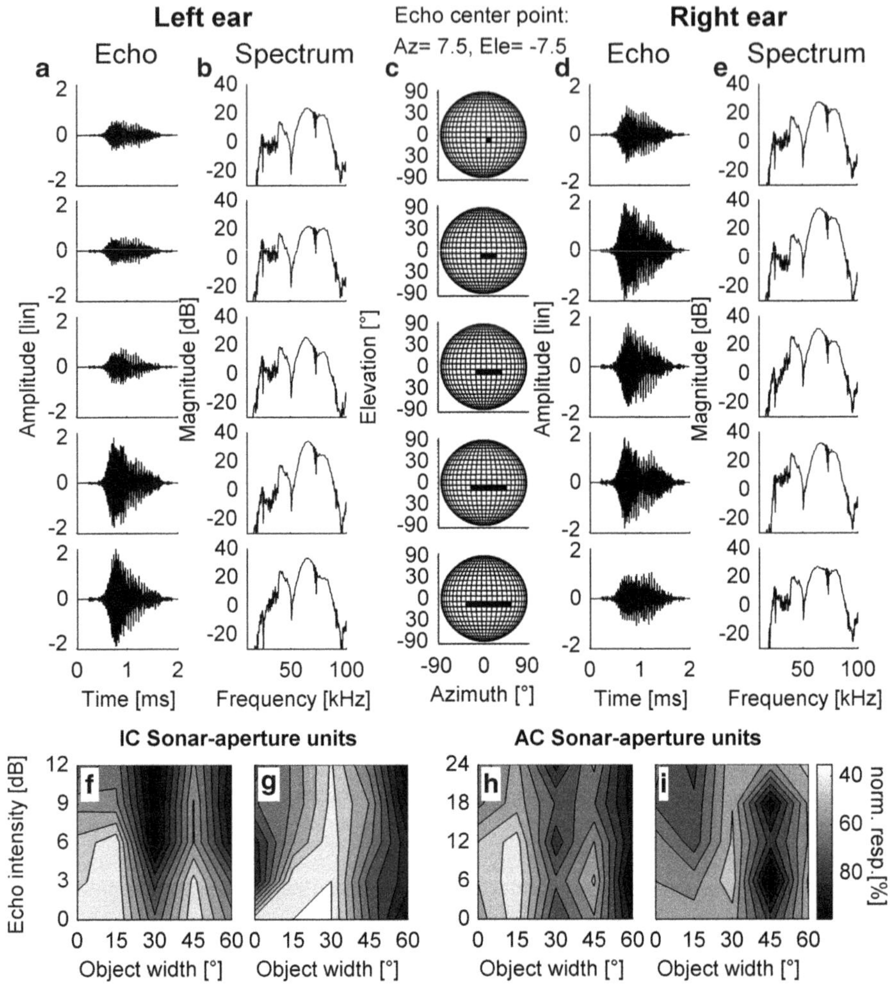

Fig. 7.7 Spatially extended echoes presented at the bat's eardrums via earphones. Temporal (**a**, **d**) and spectral (**b**, **e**) characteristics of the virtual object stimulus for increasing width (from *top* to *bottom*). (**c**) Virtual objects were centered at 7.5° in azimuth and –7.5° in elevation and extended symmetrically. (**f**, **g**, **h**, **i**) Sonar-aperture units in the IC (**f**, **g**) and the AC (**h**, **i**). Note the vertical orientation patterns of response contours. These units encode a certain object width independent of echo intensity. Normalized response strength (%) is coded on a gray scale bar. Contours drawn in steps of 10 % decrease from the maximal response strength (Adapted from Heinrich et al. 2011)

In the same study, pale spear-nosed bats were trained in virtual-object playback experiments to discriminate reflections of their own echolocation calls differing in intensity or sonar aperture (or both). Discrimination for objects with physically correct co-variation of sonar aperture and echo intensity ("object width") did not differ from performance when only sonar aperture was varied. Thus, bats detected changes in object width in the absence of intensity cues. To compare psychophysical and electrophysiological performance, a sensitivity index (d′) was calculated either from the psychometric functions or the response-strength differences in sonar aperture selective units. Responses of AC sonar aperture units reflected psychophysical performance sufficiently well to explain the bat's behavioral performance. This study also demonstrated that bats perceive and behaviorally exploit the sonar aperture of an ensonified object (Heinrich et al. 2011).

7.7.2 *Object Normalization*

Natural variation in objects (e.g., size) imposes special demands on recognition. For reliable recognition, the auditory system must segregate information about object size from object structure to achieve a size-invariant (i.e., normalized) representation about an object. In passively hearing animals, including humans, the size of a sound source is well perceived and of behavioral importance. For example, information about body size is well preserved in human speech. The human auditory system can segregate size information from speech content and compensate for the effect of speaker size on speech perception. Although a vowel sound may differ dramatically in its spectral content when pronounced by an adult and a child, it is still perceived as the same vowel (Smith et al. 2005). In non-human animals such as dogs, lions, deer, and primates, elongation of the vocal tract often serves to exaggerate the perceived size of the sender (see Fitch 2000 for a review). As echolocating bats rely on object identification to find flowers and fruit (von Helversen and von Helversen 2003; Kalko and Condon 1998), object normalization seems crucial for dealing with inevitable variations in object size.

Electrophysiological and psychophysical approaches have been used to search for a size-invariant, echo-acoustic object representation in pale spear-nosed bats (Firzlaff et al. 2007). Impulse responses of two different virtual objects were generated, each consisting of 12 single echo reflections with random temporal intervals and amplitudes (Figure 7.8a). The IRs were then scaled in the time (object depth) and amplitude domains (object surface area).

Electrophysiological experiments were used to test the ability of neurons in the AC of pale spear-nosed bat to encode normalized representations. Stimuli were scaled IRs convolved with a pale spear-nosed bat echolocation call. Object depth and surface area co-varied in a meaningful manner only along the diagonal axis, thus representing properly scaled versions of the object (Figure 7.8b). Acoustically, object surface area was encoded by echo SPL and object depth was encoded by echo duration.

The results showed that 13 % of neurons in the AC of pale spear-nosed bat encoded object depth (Figure 7.8c), whereas 57 % encoded object surface area (Figure 7.8d). Interestingly, 13 % of neurons responded best to scaled versions of the objects (Figure 7.8e). Responses of these cells were not solely dependent on echo amplitude or duration but on a meaningful combination of both parameters. These units had sophisticated response properties because they responded equally

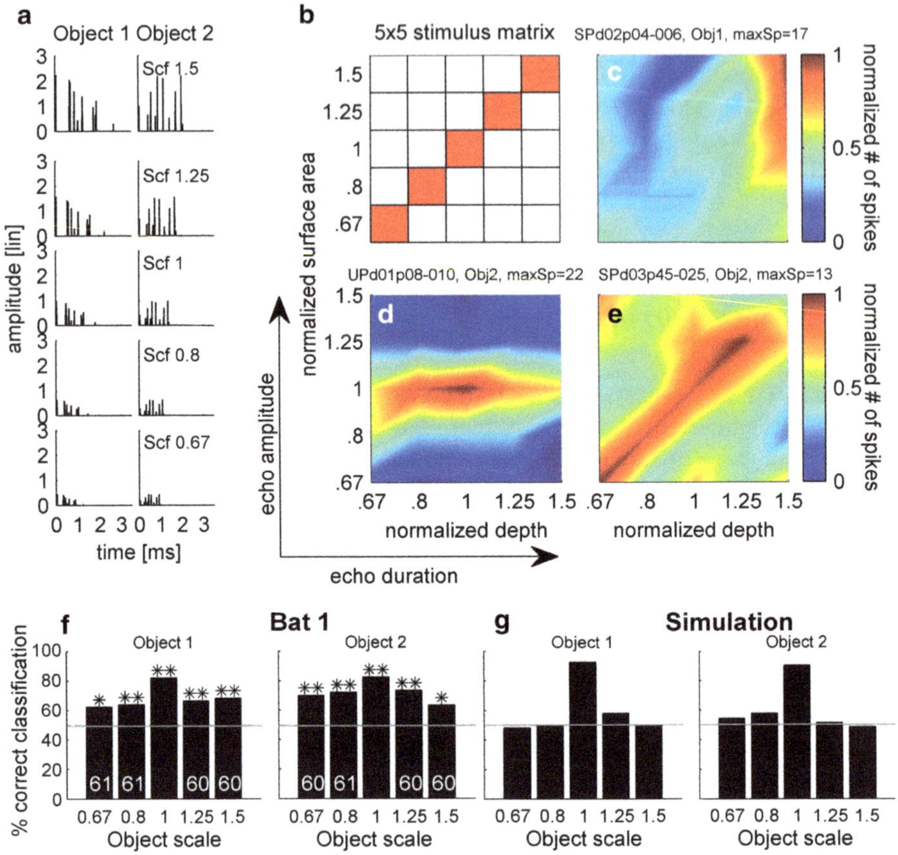

Fig. 7.8 Object normalization in bats. (**a**) Impulse responses of echoes of object 1 and object 2 (*left* and *right* column, respectively) used for the psychophysical and electrophysiological experiments investigating size-invariant echo-acoustic object representation in the pale spear-nosed bat. (**b**, **c**, **d**, **e**) Electrophysiology: (**b**) 5×5 stimulus matrix of the IRs shown in (**a**) scaled in terms of delay and amplitude of the reflections and convolved with an echolocation call. Object surface area and object depth vary along the vertical and the horizontal dimensions, respectively. *Red squares* mark the properly scaled versions of the objects shown in (**a**). The physical parameters that change in the *vertical* and *horizontal* dimension are amplitude and echo duration. (**c**) Normalized responses of a cortical unit that responded best to a particular object depth ("depth" category). These units were largely insensitive to changes of object surface area. (**d**) Responses of units that encoded object surface area ("surface" category). These units were largely insensitive to changes in object depth. (**e**) Examples of cortical units that encoded scaled versions. The unit responds best to stimuli roughly along the *diagonal axis* of the stimulus matrix shown in (**b**). (**f**, **g**) Example psychophysical classification performance of one bat, and simulated performance of a spectrotemporal pattern recognizer. Correct classification of scaled objects marked either with a single asterisk ($p<0.05$) or two asterisks ($p<0.01$). The number of test trials for each condition is superimposed on each bar. Although the spectrotemporal pattern recognizer cannot correctly classify the scaled objects, the performance of bat 1 is in the majority of test conditions significantly better (Adapted from Firzlaff et al. 2007)

well no matter whether the object was small (i.e., faint, short-duration echo) or large (i.e., loud, long-duration echo). Thus, their firing rates reflected normalized object features.

In psychophysical experiments, pale spear-nosed bats were trained with food rewards in a 2-alternative forced-choice (2-AFC) task to discriminate different virtual objects (standard versions with scaling factor = 1). Echolocation calls were recorded, convolved with the IRs, and then played back to the bats in real time. After bats learned to discriminate the two objects, scaled versions of both were randomly interspersed with a probability of 25 %. The bats correctly classified scaled versions of the objects to the trained standard (Figure 7.8f) even though a spectrotemporal pattern recognizer (based on a computational model of the auditory periphery of pale spear-nosed bats) could not reliably perform the classification (Figure 7.8g).

Altogether, these results support the hypothesis that an object-oriented (rather than a stimulus-parameter–oriented) representation of object echoes is achieved at the level of the bat's AC (Firzlaff et al. 2007). They also suggest the existence of a dedicated central (cortical) processing mechanism for object normalization to explain the bats' compensation for size-induced echo variations. Just like the visual system, the bat's sonar system meets the important requirement of having an effective object-recognition system that allows for identification of objects independent of their size.

For objects of small size, the spectral interference pattern of the reflected echoes encodes information about object structure. Neurophysiological studies in bats have shown that the short temporal delay of overlapping sounds, and thus the spectral interference pattern, is encoded by the response strength of neurons in the IC and AC (Sanderson and Simmons 2000, 2002). For some naturally occurring objects, such as flowers, the spectral interference pattern and echo amplitude regularly change with object size. For reliable, size-invariant object recognition, bats should be able to compensate for both of these changes.

The neural representation of objects scaled in the frequency domain was investigated in pale spear-nosed bat using sonar calls convolved with the IR of a two-front object (Firzlaff and Schuller 2007). The IR of a two-front object can be imagined as the IR recorded from a disc with several holes drilled to the same depth, with the surface area of the disc equal to the total surface areas of the bottom of the drilled holes. Thus, reflections arising from the disc surface and the bottom of the holes will be of equal amplitude. However, due to different time delays, the magnitude spectrum of the resultant echo will contain an interference pattern characteristic of the hole depth. Echoes of the virtual two-front object were scaled in amplitude (disk area) and delay (hole depth), and convolved with a pale spear-nosed bat sonar call. Stimuli were presented via earphones and extracellular responses were recorded from the AC of anesthetized bats.

The results showed that 20 % of AC neurons encoded a specific echo spectral structure independent of echo amplitude. A small population of cells (3 %) responded best to scaled versions of the objects, when spectral structure and echo amplitude co-varied in a meaningful manner. Therefore, some neurons in the AC of bats might serve for the recognition of object-specific spectral echo patterns independent of object size (i.e., echo amplitude). Responses of these cells could not be predicted from a simple linear model based on their excitatory frequency response areas because the cells seemed to integrate spectral information in a nonlinear manner.

When interpreting the above results, it is important to remember that echolocation is an active sensing system requiring bats to compare received echoes with their preceding calls. This is also true for object recognition. The responses of AC neurons to a special spectral envelope pattern in big brown bats were influenced by the temporal sequence of pulse–echo pairs as the preceding pulse influenced neural activity in response to a two-glint stimulus (Sanderson and Simmons 2002). Thus, some AC neurons could be encoding both echo delay (object distance) and spectral pattern (object structure).

7.7.3 Neural Coding of Stochastic Echoes

Some natural objects, such as trees, have a complex structure with a large number of reflecting surfaces. This complicates the recognition process because the objects are not stable over time (e.g., leaves and branches of the tree move in the wind), and a bat is not likely to approach them from the same position; hence the reflected echoes received by the bat will always differ and have no systematic spectral pattern. Nevertheless, echolocating bats are able to identify and navigate around such complex objects, and in the case of frugivorous bats, they can recognize and classify the types of trees that provide food.

Different trees can be classified based on a statistical description of the degree of fluctuation in the envelopes of their IRs (Müller and Kuc 2000). A broad-leafed tree with relatively few but large reflecting surfaces will produce a rough IR, whereas the small leaves of a conifer will produce many low-amplitude reflections resulting in a smooth IR. The roughness of an IR can be quantified by its fourth moment (M4), calculated as the IR waveform raised to the power of four, divided

Fig. 7.9 (continued) sharing the IR roughness given on the *abscissa*. *Vertical bars* represent the range of the 25 and 75 % percentiles. Significant response differences (Kruskal–Wallis test, $p<0.05$) indicated by *horizontal lines* with *asterisks*. (**c**) Results of the psychophysical experiments. *Inset*: Psychometric function for IR-roughness discrimination. *Vertical dotted line*: Reference IR roughness (1.8). *Dashed lines*: Discrimination threshold (75 % correct value). *Horizontal bars* show average results for the four values of reference roughness. The *left end* of each bar shows the reference roughness, and the *right end* shows the discrimination threshold. Error bars: Across-animal SE. (**d**) Comparison of the psychophysical and neurophysiological roughness discrimination performance. *Crosses*: Psychophysical performance. *Dotted line*: Sigmoidal fit to the psychophysical data. *Open circles*: Performance of an ideal observer (using a ROC analysis) basing its decisions on the pooled responses of roughness-sensitive cortical units (Adapted from Firzlaff et al. 2006)

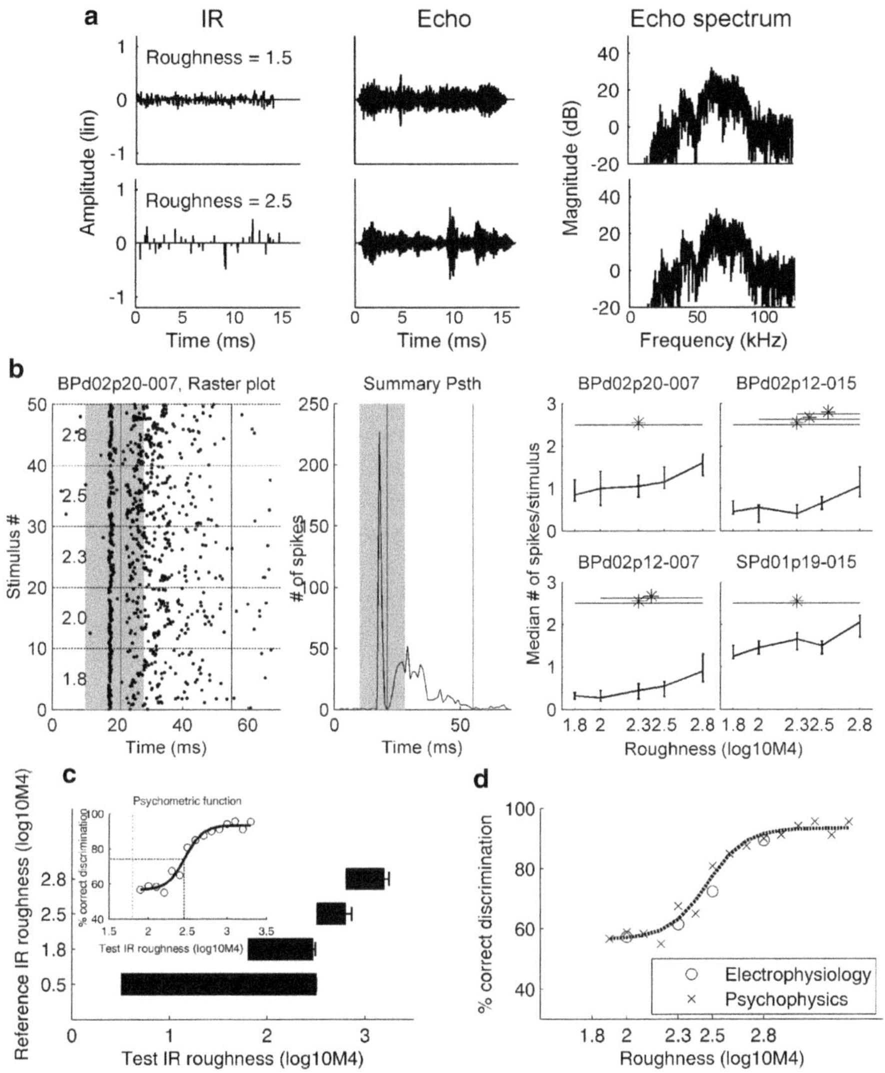

Fig. 7.9 Examples from the psychophysical and neurophysiological experiments in pale spear-nosed bats investigating neural processing and perception of stochastic echoes. (**a**) Examples are given for an IR roughness of 1.5 (*top row*, quantified as the base 10 logarithm of the waveform fourth moment) and 2.5 (*bottom row*). *First column*: IRs used to generate the echoes in the psychophysical playback experiment. *Second column*: Echoes generated from the IRs in the first column with a standard pale spear-nosed bat echolocation call. These echoes were used as stimuli for the neurophysiological experiments. *Third column*: Magnitude spectra of the echoes. Note that the magnitude spectra and the overall SPL are independent of roughness. (**b**) Electrophysiological results. *Left*: raster plot of a roughness-sensitive unit. *Middle*: PSTH. *Solid vertical lines* in the PSTH and raster plot indicate the analysis window duration (onset response was excluded). The time of stimulus presentation indicated by the *gray area* in the PSTH and raster plot. *Right*: Response strength as a function of IR roughness for four roughness-sensitive units in the AC of *P. discolor*. Response strength is shown as the median number of spikes per stimulus across ten different echoes

by the squared waveform raised to the power of two. In the following, IR roughness values are given as the base 10 logarithm of M4 (i.e., $\log_{10}$M4). Because the IR of an object is imprinted on the echoes received by a bat, echo roughness can be used for object classification. For example, in a psychophysical study, pale spear-nosed bats were able to classify unknown IRs according to their roughness (Grunwald et al. 2004).

This behavioral classification study inspired electrophysiological experiments in which the neural coding of echo roughness was investigated with extracellular recordings in the AC of anesthetized pale spear-nosed bats (Firzlaff et al. 2006). Stimuli consisted of a typical echolocation call of pale spear-nosed bats convolved with a stochastic IR (Figure 7.9a). Fifty IRs from five roughness groups (1.8, 2.0, 2.3, 2.5, and 2.8 $\log_{10}$M4) were randomly presented to bats over earphones. All IRs had the same root-mean-square amplitude and magnitude spectrum. The results showed that 15 % of AC neurons encoded echo roughness. In each of the four example cells shown in Figure 7.9b, response strength increased with increasing echo roughness.

The electrophysiological experiments were complemented by psychophysical playback experiments using the same IRs to measure echo roughness discrimination thresholds. Bats were trained to discriminate IR roughness in a 2-AFC paradigm, and the resulting psychometric function showed a fixed behavioral threshold (Figure 7.9c). A roughness of at least ~2.5 was required by pale spear-nosed bats to reliably discriminate an IR with this roughness from a smoother one. Above this threshold, roughness discrimination performance improved substantially.

To compare the psychophysical performance of roughness discrimination against the responses of AC neurons to roughness stimulation, a receiver operating characteristic (ROC) analysis was used to obtain a neurometric function with the same axis as the behavioral function. The neurometric function reflects the probability that an ideal observer could discriminate IR roughness based on spiking information of AC units. The analysis revealed good agreement between psychophysical and ideal-observer performance based on the subgroup of roughness-sensitive units tested (Figure 7.9d). Overall, the results indicated that psychophysical sensitivity to IR roughness as an ecologically meaningful parameter was quantitatively encoded by the responses of AC neurons in pale spear-nosed bats.

7.8 Neural Coding of Species-Specific Vocalizations

Acoustic signals are crucial for conspecific communication in both vertebrates and invertebrates. In mammals, communication calls can be especially complex (e.g., Kanwal et al. 1994). For a description of acoustic communication in bats, see Gillam and Fenton (Chapter 5). For a description of neural coding principles of communication calls in different mammals, see Wang (2000) and Suta et al. (2008).

A special situation exists in bats because their auditory system processes sounds used for both communication and echolocation (i.e., self-communication). This raises two important questions: Are communication and echolocation signals processed separately within the central auditory system, and do special neural mechanisms exist for extracting both types of information? Answers to these questions have been pursued in several studies with Parnell's mustached bats as a model organism because its auditory cortex is one of the best studied in bats. Numerous reports have identified neurons with facilitative responses to combinations of acoustic elements from the bat's CF-FM biosonar pulses and echoes (e.g., the fundamental and a higher harmonic) that specifically encode target range (see Wenstrup and Portfors 2011 for a review). Several areas of non-primary AC contain an orderly spatial map of tuning for target delay (range) along the cortical surface (see O'Neill 1995 for a review). But most important for studies on the neural processing of communication sounds is the rich repertoire of social calls emitted by Parnell's mustached bat. These vocalizations consist of one or more of 33 discrete syllable types emitted singly and/or in combination (Kanwal et al. 1994). Typically, the syllables consist of multiple elements of harmonically related CF and FM sounds or noise bursts.

7.8.1 Non-Primary Auditory Cortex

Extracellular recordings were used to investigate the coding of communication sounds in the FM-FM area of the AC in awake Parnell's mustached bats (Ohlemiller et al. 1996). The stimuli consisted of 14 natural syllables and 6 synthetic frequency variants of each syllable. The variants were generated by systematically varying the first harmonic. After screening cells with all stimuli at different SPLs, the most effective syllable variant was used to determine a neuron's overall preference for syllable types. Additionally, syllable types evoking the strongest responses were spectrally and temporally decomposed to determine the acoustic preferences of each neuron.

The results showed that the spiking responses of cortical FM-FM neurons were facilitated by stimulation with natural biosonar calls and pairs of syllables. The response facilitation involved acoustic elements separated in both frequency and time; however, the time course of inter-syllable facilitation and biosonar pulse-echo facilitation differed. Whereas some cells were tuned to inter-syllable intervals ranging from 20 to 60 ms, the range of delay tuning for pulse-echo stimuli in the same cells was restricted to intervals <10 ms.

A more detailed follow-up study extended the investigation to the processing of syntax in naturalistic communication calls in the cortical FM-FM area of Parnell's mustached bats (Esser et al. 1997). "Syntax" describes the system of rules for predicting the sequencing of communication sounds, which is an important feature of human speech. Awake bats were presented with ten natural composites of different syllables while extracellular recordings were made from the AC. Syllable compos-

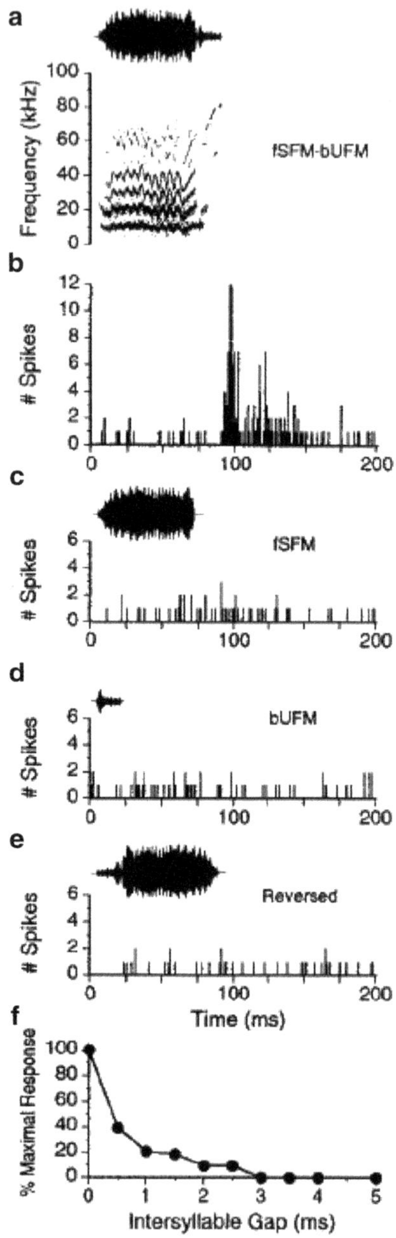

Fig. 7.10 Responses of neurons in the cortical FM-FM area of Parnell's mustached bat to communication calls. (**a**) Oscillogram (*top*) and spectrogram (*bottom*) of composite communication call fSFM-bUFM, the best composite for the unit from the cortical FM–FM area of the mustached bat shown in (**b–d**). (**b**) PSTH shows the unit's robust response to the original composite fSFM-bUFM communication call. The time axis in (**b**) also applies to (**a**). (**c**, **d**) The unit almost failed to respond to both the first syllable (fSFM) and the second syllable (bUFM) of the communication call when presented alone. (**e**) No response to playback of the reversed composite. (**f**) Unit's response decreased dramatically and eventually ceased when a silent period ≥0.5 ms was inserted between the two components of the composite fSFM-bUFM (Reproduced from Esser et al. 1997; ©National Academy of Sciences, USA)

ites were presented in their natural order, in time-reversed order, and with additional silent periods inserted between the syllables. Individual syllables were also presented alone. To study the interactions between spectral signal components, filtered versions of composites and single syllables were presented. As shown in a previous study (Ohlemiller et al. 1996), a large proportion of FM-FM neurons responded selectively and showed response facilitation when they were stimulated with syllable composites but not single syllables. The temporal structure of composite calls was especially important because introducing additional silence between syllables, reversing the syllable order, or playing the entire syllable composite in reversed order resulted in a reduction of response strength in AC neurons compared to naturalistic composites (Figure 7.10). Moreover, the AC neurons also showed spectral combination sensitivity (Esser et al. 1997).

These findings suggest that in bats two functionally different but equally complex tasks share a common neural substrate and can be accommodated within single AC neurons (Ohlemiller et al. 1996; Esser et al. 1997). As neuronal responses were strongly affected by time domain manipulations to the natural composite structure, syntax processing seems to occur at least at the level of the non-primary AC.

Interestingly, neural recordings from the FM-FM area of the bat AC were in accordance with the processing of communication signals in other mammals, including primates. For example, non-linear temporal and spectral integration appear to underlie the combination sensitivity of AC neurons involved in social communication in both Parnell's mustached bats and rhesus monkeys (see Kanwal and Rauschecker 2007 for a review).

7.8.2 Primary Auditory Cortex

Another extracellular recording study in awake Parnell's mustached bats investigated the role of FM sweeps and communication calls in eliciting response facilitation from neurons in the so-called Doppler-shifted CF processing (DSCF) area of the primary AC (Washington and Kanwal 2008). Stimuli consisted of pure tones, linear FM sweeps, and different types of simple syllabic social calls. The results showed that DSCF neurons responded well to pure tones in the bandwidth of the CF2 component of the biosonar call. Cells also responded well and sometimes showed response facilitation to linear FM sweeps traversing this range. Temporal information was crucial because the responses of DSCF neurons were often reduced when the FM sweep was time reversed. The authors suggested that the ability of DSCF neurons to respond to species-specific vocalizations likely emerges from their tuning to linear FM sweeps.

In contrast to the mechanism that creates combination sensitivity in the DSCF area of the primary AC, the posterior primary auditory cortex (AIp) of Parnell's mustached bats uses a different mechanism for coding communication calls (Medvedev and Kanwal 2004, 2008). This mechanism is based on temporal changes in the firing patterns of single units and ensembles of neurons in AIp. These changes carry sufficient information to encode different social calls uniquely.

The AIp contains a tonotopic representation of sound frequencies below ~50 kHz, covering the spectral range of communication calls emitted by Parnell's mustached bats (Kanwal et al. 1994). Importantly, the AIp does not have combination-sensitive neurons (O'Neill 1995). Medvedev and Kanwal (2004) recorded event-related local field potentials (LFPs), single-unit, and small multi-unit cluster activity in the AIp in response to seven variants of 14 simple syllables of communication calls in awake Parnell's mustached bats. The results revealed that AIp neurons typically showed correlated temporal response patterns in their peristimulus time histograms (PSTHs) and LFP waveforms. Although neurons showed weak preferences for single call types, the PSTH patterns and LFP waveforms differed between call types. To establish a relationship between the unique temporal structure of the LFP evoked by a call and the acoustic structure of that call, multidimensional scaling was performed on the averaged call-evoked LFPs and PSTHs for all 14 calls. The analysis revealed calls were better segregated in a two-dimensional space based on the LFP compared to the PSTH data. In the two-dimensional "LFP space," one of the dimensions correlated with the predominant and fundamental frequency of a communication call, while the other correlated with its spectral fine structure. In other words, the harmonic structure of calls determined the temporal structure of LFPs.

In contrast to the combination-sensitive call selectivity found in the FM-FM area of the non-primary AC, communication calls in the AIp area seem to be processed by an independent mechanism. Given that single AIp neurons fire infrequently and probabilistically in coincidence with specific LFP components, the call specificity of the LFP waveform may be based on the network's response to the unique spectro-temporal structure of a call. Therefore, communication calls appear to be represented in the AIp by a population code rather than by the responses of specialized neurons.

A follow-up experiment analyzed call-evoked gamma-band activity in the AIp of Parnell's mustached bat (Medvedev and Kanwal 2008). This was to address the question of whether LFPs recorded in response to different simple syllabic call types in awake bats contained stimulus-specific oscillatory activity in the gamma band between 20 and 100 Hz. Gamma-band oscillations are thought to be important for perceptual grouping in different sensory domains (e.g., Kaiser and Lutzenberger 2005) and are assumed to play a role in the perception of human speech (Crone et al. 2001; Sinai et al. 2005). The results of Medvedev and Kanwal (2008) showed that the majority of communication calls triggered gamma-band activity. Moreover, different call types could be classified into at least three groups based on the power spectrum of the evoked gamma-band activity, suggesting that gamma-band activity patterns might provide a basis for the perceptual classification of different call types. Gamma-band activity might play an important role in coordinating the simultaneous activation of local populations of neurons. If true, then this provides further support for the notion that the communication calls of Parnell's mustached bat are represented in the AIp by a population code.

7.8.3 Hemispheric Asymmetries and Sex-Specific Differences

Functional hemispheric asymmetries and sex-specific differences in the processing of biosonar and communication signals were investigated with extracellular recording in the DSCF area of Parnell's mustached bats, where neurons respond to both social vocalizations and biosonar signals. Hemispheric differences in the representation and dynamics of neural responses to communication calls versus pulse-echo tone pairs used to mimic echolocation were reported by Kanwal (2012). Neurons in the right hemisphere DSCF area responded more strongly to pulse-echo tone pairs than those in the left DSCF area; however, the bias was evident only in males. Peak response magnitudes for communication calls lacked a hemispheric bias in the DSCF area in either sex. Generally, call-to-tone response ratios were significantly higher for neurons in the left versus the right hemisphere. It was concluded that DSCF neurons located in the right hemisphere were better adapted for processing biosonar signals. In contrast, DSCF neurons in the left hemisphere exhibited an enriched representation of diverse call types and seemed to be specialized for intra-specific communication.

Washington and Kanwal (2012) used synthetic linear FM sweeps to investigate the neural mechanism(s) underlying hemispheric differences in the DSCF area of Parnell's mustached bats. Based on peak response magnitudes, left hemispheric DSCF area neurons were not selective for FM sweep rates and/or bandwidths; hence, they were suited for processing the broad variety of FMs present in bat communication sounds. In males, right hemispheric DSCF area neurons preferred slow-rate and narrowband FM. Peak response latencies to preferred FM stimuli were significantly longer for right compared to left hemisphere neurons in both sexes. These findings are in accordance with results from humans showing a high temporal precision in the left AC enabling the processing of rapid formant transitions or short FM sweeps (e.g., Jamison et al. 2006).

These studies support a sex-specific, spectrotemporal hemispheric asymmetry in the representation of tones and FM sweeps in bats: the DSCF area of the left primary AC of male Parnell's mustached bats showed a bias for representing sounds differing in FM sweep rate and bandwidth. It has been hypothesized that such asymmetries underlie the lateralized processing of communication sounds and are common to species as divergent as bats and humans.

7.8.4 Coding of Emotional Content of Communication Calls in the Amygdala

If the main functions of the AC are to analyze and identify communication calls based on their acoustic features (e.g., harmonic structure, temporal pattern), then to complete this task the AC could rely on, at least in part, pre-existing mechanisms like combination-sensitive neurons that are also used for processing biosonar sounds. But the social salience of communication sounds (i.e., their emotional content with respect to intraspecific social interactions) might not be encoded in the AC. A more likely candidate for extracting the biological saliency and mediating emotional responses evoked by communication sounds is the amygdala (e.g., Cardinal et al. 2002).

The coding properties of single neurons recorded from the basolateral amygdala of bats were investigated in three recent studies. Peterson and Wenstrup (2012)

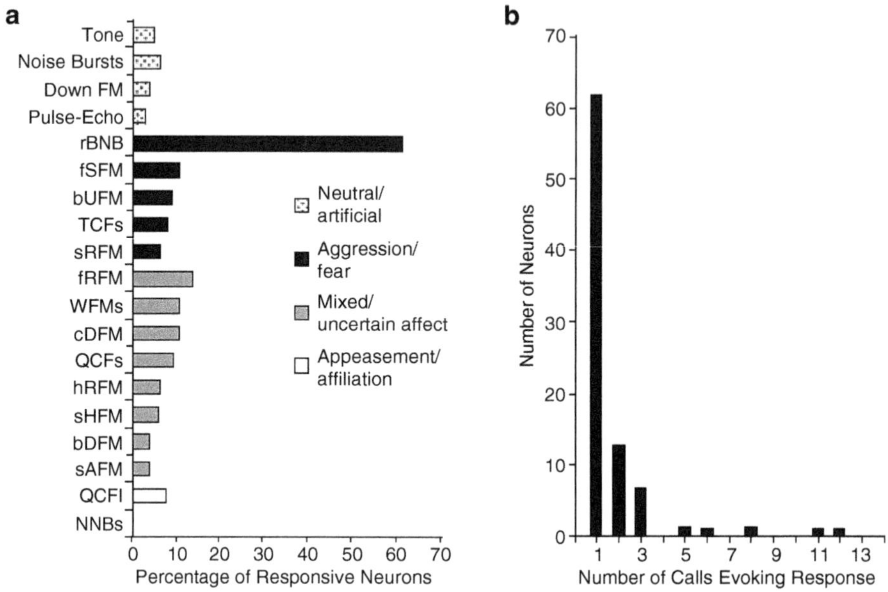

Fig. 7.11 Amygdalar neurons respond preferentially and selectively to social vocalizations. (**a**) Percentage of 87 neurons that discharged in response to different acoustic stimuli, from tones to individual syllables; see Peterson and Wenstrup (2012) for names of mustached bat syllables. (**b**) Most acoustically responsive amygdalar neurons responded to only one syllable type (Reproduced from Peterson and Wenstrup 2012)

presented social vocalizations and synthetic sounds (e.g., pure tones and FM sweeps) to awake Parnell's mustached bats while recording extracellular responses from the basolateral amygdala and found that single units were most likely to respond to social vocalizations and show selectivity for particular communication sounds (Figure 7.11a, b). Interestingly, calls associated with negative contexts (e.g., aggression or fear) evoked the largest proportion of responses, and playbacks of time-reversed calls and other temporal or spectral manipulations caused a decrease in response strength. A second important finding was that neurons often displayed a conspicuous and persistent firing pattern, with spike discharges extending beyond the duration of the stimulus.

This previous study corroborated most, but not all, of the findings from studies in the amygdala of big brown bats (Gadziola et al. 2012) and Parnell's mustached bats (Naumann and Kanwal 2011). Neurons in the amygdala of big brown bats also showed differential responses to different social vocalizations. Most of the information about a vocalization was encoded in the response duration and/or by persistent firing, with the most pronounced response differences occurring between aggressive and appeasement vocalizations (Gadziola et al. 2012). This was also demonstrated in Parnell's mustached bats where the spike rates of amygdalar neurons increased for call types associated with a negative affect, but they were suppressed by appeasement calls (Naumann and Kanwal 2011).

A striking difference in the studies on Parnell's mustached bats was the lack of response specificity for natural stimuli reported by Naumann and Kanwal (2011) in which 85 % of amygdalar neurons responded to pure tones compared to only 5 % in the study by Peterson and Wenstrup (2012). This discrepancy could be related to methodological differences because stimulus repetition rates, response threshold criteria, and loudspeaker placement differed between the studies.

In summary, the available evidence suggests that the social salience of communication sounds is encoded by neurons of the amygdala in different bat species. The amygdala seems to use persistent firing as a special coding strategy for discriminating between vocalizations. Function(s) of persistent firing has been discussed in the context of working memory formation in several neural systems (see Major and Tank 2004 for a review) and also seems to be crucial for the consolidation of emotional memories (e.g., Pelletier et al. 2005).

The second part of this chapter highlighted studies exploring the neuronal representation of complex objects and species-specific vocalizations in bats. Neurons in the AC encode several object-specific properties that are accessible through the analysis of echo-acoustic information; however, these features are not encoded on the basis of single acoustic parameters but rather on an object-oriented higher level. For example, object size is normalized and encoded as a combination of echo amplitude and duration or by the pattern of spectral interference. Stochastic features of the temporal echo envelope are extracted and encoded by neuron response rates. This also holds true for the spatial extent of an object, which in bats is represented independent of echo amplitude. Most important, the electrophysiological results were corroborated by the performance of bats in psychophysical classification and discrimination tasks. Thus, it can be assumed that the perception of echo-acoustic object information might directly be based on object-oriented neuronal representations in the AC.

It is important to emphasize that the electrophysiological studies described in this section were conducted on bats not actively vocalizing during the experiments. Echolocation is an active imaging process (e.g., Schnitzler et al. 2003; Ulanovsky and Moss 2008), and a comparison between the outgoing call and returning echo is necessary to extract spectrotemporal object features. Thus, studies on the neural coding of complex echo-acoustic objects should be extended to actively vocalizing and behaving bats.

Future studies on the neural coding of echo-acoustic objects should incorporate more realistic stimulus paradigms. For example, 3-dimensional information gained from ensonifying objects in only one direction is generally sparse. Single echo spectra from certain incident angles can be similar for objects of different sizes (von Helversen 2004), so bats are assumed to use the temporal sequence of echo spectra as a cue for object recognition. The ability to integrate sequences of acoustic information has been demonstrated psychophysically in bats (e.g., Genzel and Wiegrebe 2008) and other animals (e.g., Ralston and Herman 1995). Combining the techniques of virtual acoustic stimulation with electrophysiological recording

(Heinrich et al. 2011; Hoffmann et al. 2013) may help to incorporate spatial aspects into the analysis of neural mechanisms of echo-acoustic object coding in bats.

Studies of neural processing of bat vocalizations indicate that different coding strategies exist in the AC and amygdala. For example, combination-sensitive neurons are a major mechanism of biosonar call selectivity in the non-primary AC, whereas communication calls are represented in the AIp by a population code. Gamma-band activity might play an important role in coordinating the simultaneous activation of neural populations and may be a mechanism of perceptual classification of different call types. In the amygdala of bats, neurons use persistent firing as a strategy to encode the salience of communication signals. As for studying the processing of acoustic objects, electrophysiological recordings from freely behaving bats may yield insights into how communication sounds are coded by the brain. Especially important for the coding of the emotional content of a call is the behavioral context in which the bat acts. New techniques have made possible wireless single-unit recordings from freely behaving animals. These techniques have already been used to study the processing of vocalizations in primates (Roy and Wang 2012) and will undoubtedly contribute to our understanding of neural mechanisms underlying the analysis of species-specific vocalizations in bats in the future.

7.9 Summary

This chapter examined how the bat's brain keeps track of time by encoding signal duration and how it perceives information about complex acoustic objects and species-specific vocalizations. The first part showed that neurons from the bat auditory midbrain and cortex have spiking responses highly selective for signal duration, followed by a review of the neural mechanisms underlying the creation of duration selectivity in the IC and speculation on the possible functions of DTNs to hearing and echolocation by bats. The second part showed that neurons from the bat AC encode object-specific properties not simply on the basis of single acoustic parameters but on an object-oriented higher level, with the performance of bats in psychophysical tasks corroborating the neurophysiological findings. Thus, the perception of complex echo-acoustic object information might directly be based on an object-oriented cortical representation. The chapter concluded by showing that multiple coding strategies for vocal communication exist in different areas of the AC and amygdala of bats.

Acknowledgments We thank our undergraduate and graduate students, postdoctoral fellows, research collaborators, and colleagues for their assistance, support, and friendship. Research in the McMaster Bat Lab is supported by Discovery Grants from the Natural Sciences and Engineering Research Council (NSERC) of Canada, an Open Operating Grant from the Institute of Neurosciences, Mental Health and Addiction at the Canadian Institutes of Health Research (CIHR), and by infrastructure grants from the Canada Foundation for Innovation and the Ontario

Innovation Trust. U. Firzlaff (Lehrstuhl für Zoologie, Technische Universität München) is supported is supported by grants from the Deutsche Forschungsgemeinschaft (DFG).

References

Aubie, B., Becker, S., & Faure, P. A. (2009). Computational models of millisecond level duration tuning in neural circuits. *The Journal of Neuroscience*, 29, 9255–9270.

Aubie, B., Sayegh, R., & Faure, P. A. (2012). Duration tuning across vertebrates. *The Journal of Neuroscience*, 32, 6373–6390.

Aubie, B., Sayegh, R., Fremouw, T., Covey, E., & Faure, P. A. (2014). Decoding stimulus duration from neural responses in the auditory midbrain. *Journal of Neurophysiology*, 112, 2432–2445.

Brand, A., Urban, A., & Grothe, B. (2000). Duration tuning in the mouse auditory midbrain. *Journal of Neurophysiology*, 84, 1790–1799.

Cardinal, R. N., Parkinson, J. A., Hall, J., & Everitt, B. J. (2002). Emotion and motivation: The role of the amygdala, ventral striatum, and prefrontal cortex. *Neuroscience and Biobehavioral Reviews*, 26, 321–352.

Casseday, J. H., & Covey, E. (1995). Mechanisms for analysis of auditory temporal patterns in the brainstem of echolocating bats. In E. Covey, H. L. Hawkins, & R. F. Port (Eds.), *Neural representations of temporal patterns* (pp. 25–51). New York: Plenum Press.

Casseday, J. H., Ehrlich, D., & Covey, E. (1994). Neural tuning for sound duration: Role of inhibitory mechanisms in the inferior colliculus. *Science*, 264, 847–850.

Casseday, J. H., Ehrlich, D., & Covey, E. (2000). Neural measurement of sound duration: Control by excitatory-inhibitory interactions in the inferior colliculus. *Journal of Neurophysiology*, 84, 1475–1487.

Casseday, J. H., Fremouw, T., & Covey, E. (2002). The inferior colliculus: A hub for the central auditory system. In D. Oertel, R. R. Fay, & A. N. Popper (Eds.), *Integrative functions in the mammalian auditory pathway* (pp. 238–318). New York: Springer-Verlag.

Condon, C.J., White, K. R., & Feng, A. S. (1996). Neurons with different temporal firing patterns in the inferior colliculus of the little brown bat differentially process sinusoidal amplitude-modulated signals. *Journal of Comparative Physiology A: Neuroethology, Sensory, Neural, and Behavioral Physiology*, 178, 147–157.

Covey, E., & Faure, P. A. (2005). Neural mechanisms for analyzing temporal patterns in echolocating bats. In D. Pressnitzer, A. de Cheveigné, S. McAdams, & L. Collet (Eds.), *Auditory signal processing: physiology, psychoacoustics, and models* (pp. 251–257). New York: Springer-Verlag.

Covey, E., Kauer, J. A., & Casseday, J. H. (1996). Whole-cell patch-clamp recording reveals subthreshold sound-evoked postsynaptic currents in the inferior colliculus of awake bats. *The Journal of Neuroscience*, 16, 3009–3018.

Crone, N. E., Boatman, D., Gordon, B., & Hao, L. (2001). Induced electrocorticographic gamma activity during auditory perception. *Clinical Neurophysiology*, 112, 565–582.

Dear, S. P., & Suga, N. (1995). Delay-tuned neurons in the midbrain of the big brown bat. *Journal of Neurophysiology*, 73, 1084–1100.

Dear, S. P., Simmons, J. A., & Fritz, J. (1993). A possible neuronal basis for representation of acoustic scenes in the auditory cortex of the big brown bat. *Nature*, 364, 620–623.

Ehrlich, D., Casseday, J. H., & Covey, E. (1997). Neural tuning to sound duration in the inferior colliculus of the big brown bat, *Eptesicus fuscus*. *Journal of Neurophysiology*, 77, 2360–2372.

Elemans, C. P., Mead, A. F., Jakobsen, L., & Ratcliffe, J. M. (2011). Superfast muscles set maximum call rate in echolocating bats. *Science*, 333, 1885–1888.

Esser, K. H., & Lud, B. (1997). Discrimination of sinusoidally frequency-modulated sound signals mimicking species-specific communication calls in the FM-bat *Phyllostomus discolor*. *Journal*

of Comparative Physiology A: Neuroethology, Sensory, Neural, and Behavioral Physiology, 180, 513–522.
Esser, K. H., Condon, C. J., Suga, N., & Kanwal, J. S. (1997). Syntax processing by auditory cortical neurons in the FM-FM area of the mustached bat *Pteronotus parnellii*. *Proceedings of the National Academy of Sciences of the USA*, 94, 14019–14024.
Faure, P. A., & Barclay, R. M. R. (1994). Substrate-gleaning versus aerial-hawking: plasticity in the foraging and echolocation behaviour of the long-eared bat, *Myotis evotis*. *Journal of Comparative Physiology A: Neuroethology, Sensory, Neural, and Behavioral Physiology*, 174, 651–660.
Faure, P. A., Fremouw, T., Casseday, J. H., & Covey, E. (2003). Temporal masking reveals properties of sound-evoked inhibition in duration-tuned neurons of the inferior colliculus. *The Journal of Neuroscience*, 23, 3052–3065.
Fenton, M. B., Faure, P. A., & Ratcliffe, J. M. (2012). Evolution of high duty cycle echolocation in bats. *Journal of Experimental Biology*, 215, 2935–2944.
Firzlaff, U., & Schuller, G. (2003). Spectral directionality of the external ear of the lesser spear-nosed bat, *Phyllostomus discolor*. *Hearing Research*, 185, 110–122.
Firzlaff, U., & Schuller, G. (2007). Cortical responses to object size-dependent spectral interference patterns in echolocating bats. *European Journal of Neuroscience*, 26, 2747–2755.
Firzlaff, U., Schornich, S., Hoffmann, S., Schuller, G., & Wiegrebe, L. (2006). A neural correlate of stochastic echo imaging. *The Journal of Neuroscience*, 26, 785–791.
Firzlaff, U., Schuchmann, M., Grunwald, J. E., Schuller, G., & Wiegrebe, L. (2007). Object-oriented echo perception and cortical representation in echolocating bats. *Public Library of Science Biology*, 5(5), e100, doi: 10.1371/journal.pbio.0050100
Fitch, W. T. (2000). The evolution of speech: A comparative review. *Trends in Cognitive Sciences*, 4, 258–267.
Fremouw, T., Faure, P. A., Casseday, J. H., & Covey, E. (2005). Duration selectivity of neurons in the inferior colliculus of the big brown bat: tolerance to change in sound level. *Journal of Neurophysiology*, 94, 1869–1878.
Fuzessery, Z. M., & Hall, J. C. (1999). Sound duration selectivity in the pallid bat inferior colliculus. *Hearing Research*, 137, 137–154.
Fuzessery, Z. M., Buttenhoff, P., Andrews, B., & Kennedy, J. M. (1993). Passive sound localization of prey by the pallid bat (*Antrozous p. pallidus*). *Journal of Comparative Physiology A: Neuroethology, Sensory, Neural, and Behavioral Physiology*, 171, 767–777.
Gadziola, M. A., Grimsley, J. M. S., Shanbhag, S. J., & Wenstrup, J. J (2012). A novel coding mechanism for social vocalizations in the lateral amygdala. *Journal of Neurophysiology*, 107, 1047–1057.
Galazyuk, A. V., & Feng, A. S. (1997). Encoding of sound duration by neurons in the auditory cortex of the little brown bat, *Myotis lucifugus*. *Journal of Comparative Physiology A: Neuroethology, Sensory, Neural, and Behavioral Physiology*, 180, 301–311.
Genzel, D., & Wiegrebe, L. (2008). Time-variant spectral peak and notch detection in echolocation-call sequences in bats. *Journal of Experimental Biology*, 211, 9–14.
Goerlitz, H. R., Hübner, M, & Wiegrebe, L. (2008). Comparing passive and active hearing: Spectral analysis of transient sounds in bats. *Journal of Experimental Biology*, 211, 1850–1858.
Grunwald, J.E., Schornich, S., & Wiegrebe, L. (2004). Classification of natural textures in echolocation. *Proceedings of the National Academy of Sciences of the USA*, 101, 5670–5674.
Haplea, S., Covey, E., & Casseday, J. H. (1994). Frequency tuning and response latencies at three levels in the brainstem of the echolocating bat, *Eptesicus fuscus*. *Journal of Comparative Physiology A: Neuroethology, Sensory, Neural, and Behavioral Physiology*, 174, 671–683.
Heinrich, M., Warmbold, A., Hoffmann, S., Firzlaff, U., & Wiegrebe, L. (2011). The sonar aperture and its neural representation in bats. *The Journal of Neuroscience*, 31, 15618–15627.
Hoffmann, S., Warmbold, A., Wiegrebe, L., & Firzlaff, U. (2013). Spatiotemporal contrast enhancement and feature extraction in the bat auditory midbrain and cortex. *Journal of Neurophysiology*, 110, 1257–1268.

Hooper, S. L., Buchman, E., & Hobbs, K. H. (2002). A computational role for slow conductances: Single-neuron models that measure duration. *Nature Neuroscience*, 5, 552–556.
Hou, T., Wu, M., & Jen, P. H.-S. (1992). Pulse repetition rate and duration affect the responses of bat auditory cortical neurons. *Chinese Journal of Physiology*, 35, 259–278.
Hübner, M., & Wiegrebe, L. (2003). The effect of temporal structure on rustling-sound detection in the gleaning bat, *Megaderma lyra. Journal of Comparative Physiology A: Neuroethology, Sensory, Neural, and Behavioral Physiology*, 189, 337–346.
Huffman, R. F., & Covey, E. (1995). Origin of the ascending projections to the nuclei of the lateral lemniscus in the big brown bat, *Eptesicus fuscus. The Journal of Comparative Neurology*, 357, 532–545.
Jamison, H. L., Watkins, K. E., Bishop, D. V. M., & Matthews, P. M. (2006). Hemispheric specialization for processing auditory nonspeech stimuli. *Cerebral Cortex*, 16, 1266–1275.
Jen, P. H.-S., & Schlegel, P. A. (1982). Auditory physiological properties of the neurons in the inferior colliculus of the big brown bat, *Eptesicus fuscus. Journal of Comparative Physiology A: Neuroethology, Sensory, Neural, and Behavioral Physiology*, 147, 351–363.
Jen, P. H.-S., & Feng, R. B. (1999). Bicuculline application affects discharge pattern and pulse-duration tuning characteristics of bat inferior collicular neurons. *Journal of Comparative Physiology A: Neuroethology, Sensory, Neural, and Behavioral Physiology*, 184, 185–194.
Jen, P. H.-S., & Zhou, X. M. (1999). Temporally patterned pulse trains affect duration tuning characteristics of bat inferior collicular neurons. *Journal of Comparative Physiology A: Neuroethology, Sensory, Neural, and Behavioral Physiology*, 185, 471–478.
Jen, P. H.-S., & Wu, C. H. (2005). The role of GABAergic inhibition in shaping the response size and duration selectivity of bat inferior collicular neurons to sound pulses in rapid sequences. *Hearing Research*, 202, 222–234.
Jen, P. H.-S., & Wu, C. H. (2006). Duration selectivity organization in the inferior colliculus of the big brown bat, *Eptesicus fuscus. Brain Research*, 1108, 76–87.
Jen, P. H.-S., & Wu, C. H. (2008). Echo duration selectivity of the bat varies with pulse-echo amplitude difference. *Neuroreport*, 19, 373–377.
Kaiser, J., & Lutzenberger, W. (2005). Human gamma-band activity: a window to cognitive processing. *Neuroreport*, 16, 207–211.
Kalko, E. K. V., & Condon, M. A. (1998). Echolocation, olfaction and fruit display: how bats find fruit of flagellichorous cucurbits. *Functional Ecology*, 12, 364–372.
Kanwal, J. S. (2012). Right-left asymmetry in the cortical processing of sounds for social communication vs. navigation in mustached bats. *European Journal of Neuroscience*, 35, 257–270.
Kanwal, J. S., & Rauschecker, J. P. (2007). Auditory cortex of bats and primates: Managing species-specific calls for social communication. *Frontiers in Bioscience*, 12, 4621–4640.
Kanwal, J. S., Matsumura, S., Ohlemiller, K., & Suga, N. (1994). Analysis of acoustic elements and syntax in communication sounds emitted by mustached bats. *Journal of the Acoustical Society of America*, 96, 1229–1254.
Leary, C. J., Edwards, C. J., & Rose, G. J. (2008). Midbrain auditory neurons integrate excitation and inhibition to generate duration selectivity: an *in vivo* whole-cell patch study in anurans. *The Journal of Neuroscience*, 28, 5481–5493.
Luo, F., Metzner, W., Wu, F. J., Zhang, S. Y., & Chen, Q. C. (2008). Duration-sensitive neurons in the inferior colliculus of horseshoe bats: Adaptations for using CF-FM echolocation pulses. *Journal of Neurophysiology*, 99, 284–296.
Macías, S., Mora, E. C., Hechavarría, J. C., & Kössl, M. (2011). Duration tuning in the inferior colliculus of the mustached bat. *Journal of Neurophysiology*, 106, 3119–3128.
Major, G., & Tank, D. (2004). Persistent neural activity: prevalence and mechanisms. *Current Opinion in Neurobiology*, 14, 675–684.
Medvedev, A. V., & Kanwal, J. S. (2004). Local field potentials and spiking activity in the primary auditory cortex in response to social calls. *Journal of Neurophysiology*, 92, 52–65.

Medvedev, A. V., & Kanwal, J. S. (2008). Communication call-evoked gamma-band activity in the auditory cortex of awake bats is modified by complex acoustic features. *Brain Research*, 1188, 76–86.

Mora, E. C., & Kössl, M. (2004). Ambiguities in sound duration selectivity by neurons in the inferior colliculus of the bat *Molossus molossus* from Cuba. *Journal of Neurophysiology*, 91, 2215–2226.

Morrison, J. A., Faranak, F., Fremouw, T., Sayegh, R., Covey, E., & Faure, P. A. (2014). Organization and trade-off of spectro-temporal tuning properties of duration-tuned neurons in the mammalian inferior colliculus. *Journal of Neurophysiology*, 111, 2047–2060.

Müller, R., & Kuc, R. (2000). Foliage echoes: A probe into the ecological acoustics of bat echolocation. *Journal of the Acoustical Society of America*, 108, 836–845.

Narins, P. M., & Capranica, R. R. (1980). Neural adaptations for processing the two-note call of the Puerto Rican treefrog, *Eleutherodactylus coqui. Brain, Behavior and Evolution*, 17, 48–66.

Naumann, R. T., & Kanwal, J. S. (2011). Basolateral amygdala responds robustly to social calls: Spiking characteristics of single unit activity. *Journal of Neurophysiology*, 105, 2389–2404.

O'Neill, W. E. (1995). The bat auditory cortex. In A. N. Popper & R. R. Fay (Eds.), *Hearing by bats* (pp. 416–480). New York: Springer.

Ohlemiller, K. K., Kanwal, J. S., & Suga, N. (1996). Facilitative responses to species-specific calls in cortical FM-FM neurons of the mustached bat. *Neuroreport*, 7, 1749–1755.

Pelletier, J. G., Likhtik, E., Filali, M., & Pare, D. (2005). Lasting increases in basolateral amygdala activity after emotional arousal: Implications for facilitated consolidation of emotional memories. *Learning and Memory*, 12, 96–102.

Peterson, D. C., & Wenstrup, J. J. (2012). Selectivity and persistent firing responses to social vocalizations in the basolateral amygdala. *Neuroscience*, 217, 154–171.

Pinheiro, A. D., Wu, M., & Jen, P. H.-S. (1991). Encoding repetition rate and duration in the inferior colliculus of the big brown bat, *Eptesicus fuscus. Journal of Comparative Physiology A: Neuroethology, Sensory, Neural, and Behavioral Physiology*, 169, 69–85.

Potter, H. D. (1965). Patterns of acoustically evoked discharges of neurons in the mesencephalon of the bullfrog. *Journal of Neurophysiology*, 28, 1155–1184.

Ralston, J. V., & Herman, L. M. (1995). Perception and generalization of frequency contours by a bottle-nosed dolphin (*Tursiops truncatus*). *Journal of Comparative Psychology*, 109, 268–277.

Razak, K. A., & Fuzessery, Z. M. (2006). Neural mechanisms underlying selectivity for the rate and direction of frequency-modulated sweeps in the auditory cortex of the pallid bat. *Journal of Neurophysiology*, 96, 1303–1319.

Razak, K. A., Fuzessery, Z. M., & Lohuis, T. D. (1999). Single cortical neurons serve both echolocation and passive sound localization. *Journal of Neurophysiology*, 81, 1438–1442.

Razak, K. A., Shen, W. M., Zumsteg, T., & Fuzessery, Z. M. (2007). Parallel thalamocortical pathways for echolocation and passive sound localization in a gleaning bat, *Antrozous pallidus. The Journal of Comparative Neurology*, 500, 322–338.

Roy, S., & Wang, X. Q. (2012). Wireless multi-channel single unit recording in freely moving and vocalizing primates. *Journal of Neuroscience Methods*, 203, 28–40.

Sanderson, M. I., & Simmons, J. A. (2000). Neural responses to overlapping FM sounds in the inferior colliculus of echolocating bats. *Journal of Neurophysiology*, 83, 1840–1855.

Sanderson, M. I., & Simmons, J. A. (2002). Selectivity for echo spectral interference and delay in the auditory cortex of the big brown bat *Eptesicus fuscus. Journal of Neurophysiology*, 87, 2823–2834.

Sayegh, R., Aubie, B., & Faure, P. A. (2011). Duration tuning in the auditory midbrain of echolocating and non-echolocating vertebrates. *Journal of Comparative Physiology A: Neuroethology, Sensory, Neural, and Behavioral Physiology*, 197, 571–583.

Sayegh, R., Aubie, B., Fazel-Pour, S., & Faure, P. A. (2012). Recovery cycle times of inferior colliculus neurons in the awake bat measured with spikes counts and latencies. *Frontiers in Neural Circuits*, doi: 10.3389/fncir.2012.00056

Sayegh, R., Casseday, J. H., Covey, E., & Faure, P. A. (2014). Monaural and binaural inhibition underlying duration-tuned neurons in the inferior colliculus. *The Journal of Neuroscience*, 34, 481–492.

Schmidt, S. (1988). Evidence for a spectral basis of texture-perception in bat sonar. *Nature*, 331, 617–619.

Schnitzler, H.-U., Moss, C. F., & Denzinger, A. (2003). From spatial orientation to food acquisition in echolocating bats. *Trends in Ecology and Evolution*, 18, 386–394.

Siemers, B. M., & Schnitzler, H.-U. (2004). Echolocation signals reflect niche differentiation in five species of sympatric congeneric bat species. *Nature*, 429, 657–661.

Simmons, J. A. (1987). Acoustic images of target range in the sonar of bats. *Naval Research Reviews*, 39, 11–26.

Simon, R., Holderied, M. W., Koch, C. U., & von Helversen, O. (2011). Floral acoustics: conspicuous echoes of a dish-shaped leaf attract bat pollinators. *Science*, 333, 631–633.

Sinai, A., Bowers, C. W., Crainiceanu, C. M., Boatman, D., Gordon, B., Lesser, R. P., Lenz, F. A., & Crone, N. E. (2005). Electrocorticographic high gamma activity versus electrical cortical stimulation mapping of naming. *Brain*, 128, 1556–1570.

Smith, D. R. R., Patterson, R. D., Turner, R., Kawahara, H., & Irino, T. (2005). The processing and perception of size information in speech sounds. *Journal of the Acoustical Society of America*, 117, 305–318.

Stilz, W. P. (2004). *Akustische Untersuchungen zur Echoortung bei Fledermäusen*. PhD dissertation, University of Tuebingen, Tuebingen.

Suta, D., Popelar, J., & Syka, J. (2008). Coding of communication calls in the subcortical and cortical structures of the auditory system. *Physiological Research*, 57, 149–159.

Sutter, M. L. (2000). Shapes and level tolerances of frequency tuning curves in primary auditory cortex: quantitative measures and population codes. *Journal of Neurophysiology*, 84, 1012–1025.

Ulanovsky, N., & Moss, C. F. (2008). What the bat's voice tells the bat's brain. *Proceedings of the National Academy of Sciences of the USA*, 105, 8491–8498.

Veselka, N., McErlain, D. D., Holdsworth, D. W., Eger, J. L., Chem, R. K., Mason, M. J., Brain, K. L., Faure, P. A., & Fenton, M. B. (2010). A bony connection signals laryngeal echolocation in bats. *Nature*, 463, 939–942.

von Helversen, D. (2004). Object classification by echolocation in nectar feeding bats: size-independent generalization of shape. *Journal of Comparative Physiology A: Neuroethology, Sensory, Neural, and Behavioral Physiology*, 190, 515–521.

von Helversen, D., & von Helversen, O. (1999). Acoustic guide in bat-pollinated flower. *Nature*, 398, 759–760.

von Helversen, D., & von Helversen, O. (2003). Object recognition by echolocation: a nectar-feeding bat exploiting the flowers of a rain forest vine. *Journal of Comparative Physiology A: Neuroethology, Sensory, Neural, and Behavioral Physiology*, 189, 327–336.

von Helversen, D., Holderied, M. W., & von Helversen, O. (2003). Echoes of bat-pollinated bell-shaped flowers: conspicuous for nectar-feeding bats? *Journal of Experimental Biology*, 206, 1025–1034.

Wang, X. Q. (2000). On cortical coding of vocal communication sounds in primates. *Proceedings of the National Academy of Sciences of the USA*, 97, 11843–11849.

Wang, X., Luo, F., Wu, F. J., Chen, Q.-C., & Jen, P. H.-S. (2008). The recovery cycle of bat duration-selective collicular neurons varies with hunting phase. *Neuroreport*, 19, 861–865.

Wang, X., Luo, F., Jen, P. H.-S., & Chen, Q.-C. (2010). Recovery cycle of neurons in the inferior colliculus of the FM bat determined with varied pulse-echo duration and amplitude. *Chinese Journal of Physiology*, 53, 119–129.

Washington, S. D., & Kanwal, J. S. (2008). DSCF neurons within the primary auditory cortex of the mustached bat process frequency modulations present within social calls. *Journal of Neurophysiology*, 100, 3285–3304.

Washington, S. D., & Kanwal, J. S. (2012). Sex-dependent hemispheric asymmetries for processing frequency-modulated sounds in the primary auditory cortex of the mustached bat. *Journal of Neurophysiology*, 108, 1548–1566.
Weissenbacher, P., & Wiegrebe, L. (2003). Classification of virtual objects in the echolocating bat, *Megaderma lyra. Behavioural Neuroscience*, 117, 833–839.
Wenstrup, J. J., & Portfors, C. V. (2011). Neural processing of target distance by echolocating bats: functional roles of the auditory midbrain. *Neuroscience & Biobehavioral Reviews*, 35, 2073–2083.
Wright, G. S., Chiu, C., Xian, W., Wilkinson, G. S., & Moss, C. F. (2013). Social calls of flying big brown bats (*Eptesicus fuscus*). *Frontiers in Physiology*, doi: 10.3389/fphys.2013.00214
Wu, C. H., & Jen, P. H.-S. (2008a). Echo frequency selectivity of duration-tuned inferior collicular neurons of the big brown bat, *Eptesicus fuscus*, determined with pulse-echo pairs. *Neuroscience*, 156, 1028–1038.
Wu, C. H., & Jen, P. H.-S. (2008b). Bat inferior colliculus neurons have the greatest frequency selectivity when determined with best-duration pulses. *Neuroscience Letters*, 438, 362–367.
Wu, C. H., & Jen, P. H.-S. (2010). A duration coding mechanism underlying bat echo recognition. *Adaptive Medicine*, 2, 71–77.
Yin, S., Chen, Z., Yu, D., Feng, Y., & Wang, J. (2008). Local inhibition shapes duration tuning in the inferior colliculus of guinea pigs. *Hearing Research*, 237, 32–48.
Young, E. D., Rice, J. J., & Tong, S. C. (1996). Effects of pinna position on head-related transfer functions in the cat. *Journal of the Acoustical Society of America*, 99, 3064–3076.
Yovel, Y., Franz, M. O., Stilz, P., & Schnitzler, H.-U. (2011). Complex echo classification by echolocating bats: a review. *Journal of Comparative Physiology A: Neuroethology, Sensory, Neural, and Behavioral Physiology*, 197, 475–490.
Zhou, X., & Jen, P. H.-S. (2001). The effect of sound intensity on duration-tuning characteristics of bat inferior collicular neurons. *Journal of Comparative Physiology A: Neuroethology, Sensory, Neural, and Behavioral Physiology*, 187, 63–73.

Chapter 8
The Neural Processing of Frequency Modulations in the Auditory System of Bats

George D. Pollak

Keywords Communication calls • Echolocation • Frequency modulation • Inferior colliculus • Inhibition

8.1 Introduction

Bats rely to an inordinate degree on sound for survival. Bats, of course, are best known for their ability to echolocate, a form of biosonar used for both orientation and detecting, identifying, and capturing prey. Not only do they rely on hearing for orientation and hunting through echolocation, but hearing is also critically important for social communication. Many bats live in large colonies where they engage in a myriad of social interactions that are accomplished almost entirely with sound since they live in dark environments where visual displays are of no use (Griffin 1958). Their communication signals can be quite elaborate (Bohn et al. 2009), and some species are capable of vocal learning (Boughman 1998; Knornschild et al. 2006). Indeed, the repertoire of signals bats use for vocal communication is rich and sophisticated (Kanwal 1999; Bohn et al. 2008).

This chapter is concerned with the neural processing of signals that bats employ in both their echolocation and communication calls. Particular attention is given to the processing of communication calls and the roles of inhibition in creating response selectivity that enables the auditory system to distinguish among the various signals bats receive. Vocal communication was presumably used by their ancestors before bats took to the night sky to exploit a food supply for which there was little competition. It is noteworthy in this regard that the primary acoustic feature of virtually all echolocation calls is a brief, frequency-modulated (FM) sweep (Simmons et al. 1975; Neuweiler 1990); strikingly similar FM sweeps are also acoustic features of many communication calls emitted by bats (Schwartz et al. 2007; Bohn et al. 2008). Thus, the view presented in this

G.D. Pollak (✉)
Department of Neuroscience and Center for Perceptual Systems, The University of Texas at Austin, Austin, TX 78712, USA
e-mail: gpollak@austin.utexas.edu

M.B. Fenton et al. (eds.), *Bat Bioacoustics*, Springer Handbook of Auditory Research 54, DOI 10.1007/978-1-4939-3527-7_8

chapter is that one of the primary challenges that drove the evolution of their auditory systems was to process communication signals, and adaptations required for echolocation were subsequently added or co-opted to enable the various species of bats to compete successfully for food resources in a wide range of different habitats.

8.1.1 Some General Comments on Echolocation

Echolocation, while exotic, is not as unique a perceptual ability as many believe. It evolved several times and is present in two species of birds, in cetaceans, in one species of megachiropteran bat, the Egyptian Rousette bat, *Rossetus aegyptiacus*, as well as in all microchropteran bats (Griffin 1958). The proposition that echolocation is not a unique perception that required new and special modes of processing is supported by the fact that humans can learn to echolocate with high precision (Thaler et al. 2011).

There have always been anecdotal stories about one or another blind person who displayed echolocation abilities that were so good that they appeared not to be blind at all (Griffin 1944, 1958). Recently, one person, Daniel Kish, has received particular attention. Kish has been sightless since he was a year old, yet he can mountain bike, navigate the wilderness alone, and recognize a building hundreds of feet away, all with echolocation. You can see his abilities to recognize buildings and other objects at http://www.youtube.com/watch?v=PGMpswJtCdI&feature=fvwrel. He echolocates by emitting clicks with his tongue, the same way birds and tomb bats do, and he can form remarkably precise images from the echoes he receives. The progress Kish has made and how he is teaching other blind people to use echolocation can be seen at http://www.youtube.com/watch?v=CRA-asTuP_Y and at http://www.youtube.com/watch?v=xATIyq3uZM4&feature=related.

Studies on echolocation in the blind suggest that the echoes they perceive evoke some activity in the visual portions of the cortex (Thaler et al. 2011). The importance of visual involvement is intriguing, but unclear, since these people did not report that they experienced "visual" images while echolocating, even though they had sight early in life. Nevertheless, they could form images of objects in the environment through sound, which shows that whatever circuits and mechanisms that enable bats to form images of objects in their environment by listening to echoes are also present in humans. Consistent with this notion is the evidence that the auditory systems of bats are similar to the auditory systems of all other mammals, with the same structures, wiring, and mechanisms for processing information that are possessed by other mammals (Pollak and Casseday 1986; Pollak et al. 1995; Winer et al. 1995). The distinguishing features of the auditory system of bats are not novel mechanisms. Rather, the distinguishing feature is that some mechanisms, which are common to all mammals, are far more pronounced in their auditory systems than in other mammals. The difference is quantity, not quality.

8.1.2 Themes of the Chapter

In the first section of the chapter, some of the communication signals bats emit are presented with the behaviors during which the signals are emitted to illustrate the richness and variety of their vocal repertoires. The subsequent sections then deal with the neural processing of communication calls, focusing on the processing of FM sweeps. Additionally, focus is on processing of FM sweeps in the inferior colliculus (IC), the midbrain auditory nucleus that receives projections from almost all lower auditory nuclei and then synthesizes that convergent information (Pollak and Casseday 1986; Casseday et al. 2002). The net result of those syntheses is that a variety of new response properties are either formed de novo in the IC or other response properties that have been formed in lower nuclei are sharpened or further modified in the IC.

The following sections on neural processing have three themes. The first theme is that the various computations employed by the IC endow its neurons with selectivities for features of FM sweeps, and those selectivities, in turn, are a principal feature for creating the response selectivities for echolocation as well as the various conspecific communication calls these animals hear in their daily lives. The second theme is that inhibition is the major sculptor in the auditory system, where inhibition acts to shape selective response properties out of excitatory inputs that are far less selective or even unselective. The third theme is that the IC population is heterogeneous, in that some cells form their selective response properties in a straightforward, linear manner, whereas other cells form similar response properties through more complex, non-linear processing.

8.2 The Vocal Repertoire of Bats

Before turning to the neural mechanisms by which acoustic signals are processed, the variety of communication calls bats use are considered first, as illustrated by the vocal repertoire of Brazilian free-tailed bats, *Tadarida brazilensis*. These bats are common in the Southwestern United States where they live in caves with populations that often number in the millions. Here males use vocal signals to establish dominance hierarchies, maintain territories, garner females into harems, and defend their harems against intruding males. Females use vocal signals for recognition of and bonding with their pups, among other behaviors (Schwartz et al. 2007; Bohn et al. 2008).

To give a flavor of the variety of calls emitted by these animals, a sample of calls is shown in Figure 8.1, together with a notation about the behaviors the bats displayed during the emission of each call type. Each call is composed of one or more repetitions of a syllable or note. Each syllable is composed of multiple harmonics with spectral components that change in amplitude, and often in frequency, throughout its duration. The syllables range not only in duration (from 2

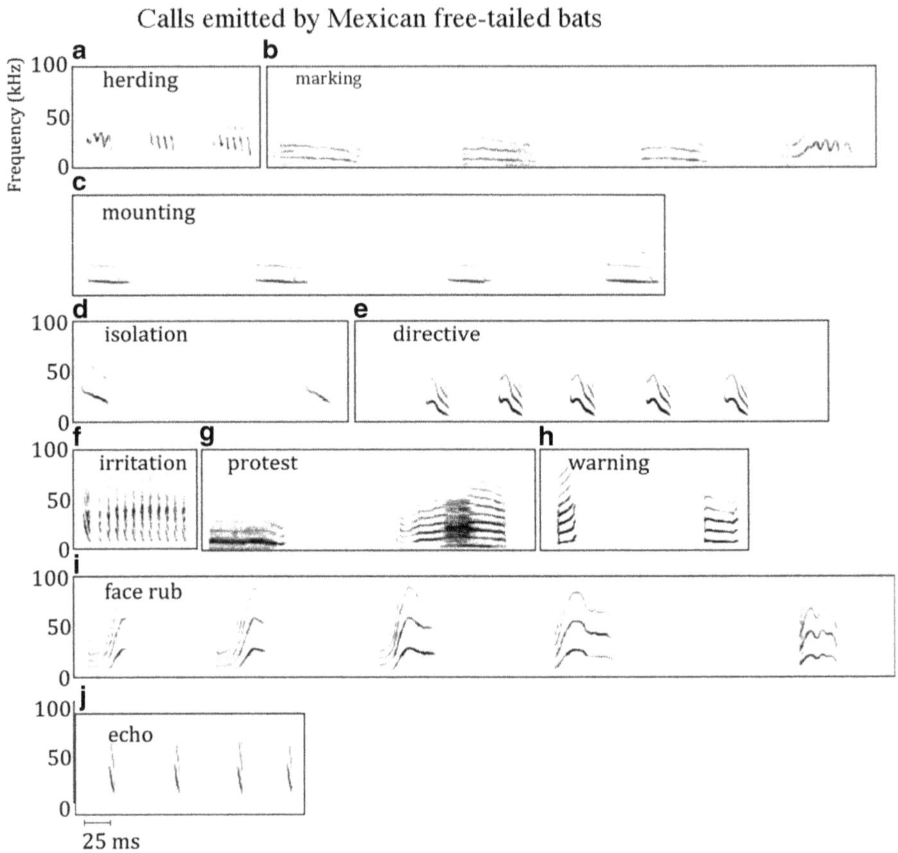

Fig. 8.1 Spectrograms that show the various communication calls emitted by Mexican free-tailed bats. (**a**) *Herding Calls* were emitted while males forcefully pushed one or more females with muzzle or wing into his territory. (**b**) *Marking Calls* were emitted by dominant males while rubbing their faces and gular glands on the surfaces of their territories. (**c**) *Mounting Calls* were emitted by males to convey dominance, when males would mount females and forcefully push their muzzles repeatedly between their shoulders. (**d**) *Isolation Calls* were emitted by pups immediately after birth and throughout development; pups called when they were isolated or hungry. (**e**) *Directive Calls* were emitted by females while giving birth and throughout pup development when females approached pups or in response to their pups' isolation calls. (**f**) *Irritation Calls* were emitted when bats were jostled by other bats. (**g**) *Protest Calls* were emitted in response to aggressive behaviors by other bats. (**h**) *Warning Calls* were emitted prior to aggressive encounters. (**i**) *Face-Rubbing Calls* were used for social bonding; they were emitted in roost sites while approaching another bat and rubbing their muzzles across the body of the bat. (**j**) Echolocation calls were emitted for orientation while the bats were flying (Reproduced from Bohn et al. 2008 with permission; ©Acoustical Society of America)

to 3 ms to over 100 ms) but also in their spectral structures. For example, some syllables are simply brief downward sweeping frequency modulations (FMs) (e.g., irritation call in Figure 8.1f), whereas others are more complex with both upward and downward FMs (e.g., directive, Figure 8.1e, and herding calls, Figure 8.1a), and yet others have only harmonic stacks of constant frequencies (marking, Figure 8.1b, and mounting calls, Figure 8.1c). The temporal sequence in which the syllables are emitted is also an important feature that varies with behavioral context (Bohn et al. 2008). The syllables produced in several different calls associated with completely different contexts are indistinguishable except for differences in temporal intervals or the repetition rate at which the syllables are emitted.

The Brazilian free-tailed bats not only emit the simpler types of calls illustrated in Figure 8.1, they also sing elaborate "songs." During the breeding season, free-tailed bats, as well as many other animals, emit simple repetitions of one or a few syllables that are generally referred to as mating or advertisement calls. In a few exceptional cases, such as songbirds (Catchpole and Slater 1995), whales (Payne and McVay 1971), and some bat species (Behr and von Helversen 2004; Bohn et al. 2009), these advertisement signals can be more complex vocalizations termed songs. The major difference between mating "calls" and "songs" is that songs are longer and more complex and contain multiple types of elements (e.g., syllables or notes) that are combined in a stereotypical manner (Catchpole and Slater 1995; Marler 2004). Therefore, songs have an added dimension of complexity in the form of syntax, the patterns by which elements are ordered and combined. Indeed, in most songs, element ordering is not random but instead is highly structured, with individual, regional, and/or species-specific patterns (Balaban 1988).

A remarkable feature of Brazilian free-tailed bats is that the ways in which phrases are combined to form songs follow broad syntactical rules, yet males dynamically vary phrase order from one rendition to the next. During the breeding season dominant males sing their courtship songs (Figure 8.2A) when females approach their territories (Schwartz et al. 2007; Bohn et al. 2009). Their courtship songs are composed of three types of phrases. A phrase is composed of one or more syllable types that form a distinct and reproducible unit, and the phrases are combined to form songs. The three phrases are chirps, trills, and buzzes (Figure 8.2A). Chirps are phrases composed of two types of syllables: "A" and "B" syllables. The A syllables are short (~5 ms) downward sweeping FMs. The B syllables are longer (~15 ms) and more complex than A syllables. The B syllables often begin with an upward FM followed by a longer downward FM, and some signals end with a second upward FM. Thus, their spectral contours often have multiple inflection points. Several A syllables always precede each B syllable, and the sequence of several A syllables followed by a B syllable is then repeated to form the chirp phrase.

The second type of phrase is the trill. Trills are composed of short (3–4 ms), downward FM syllables that are sometimes connected, resulting in sinusoidal patterns (Figure 8.2Ac). Trill syllables, whether discrete or connected, are produced as a distinct phrase or burst with durations of approximately 25 ms. Sequential trill phrases are often emitted in songs, but they are highly distinctive since each phrase is separated from the next by a silent interval that is much greater than the interval between syllables within each trill phrase.

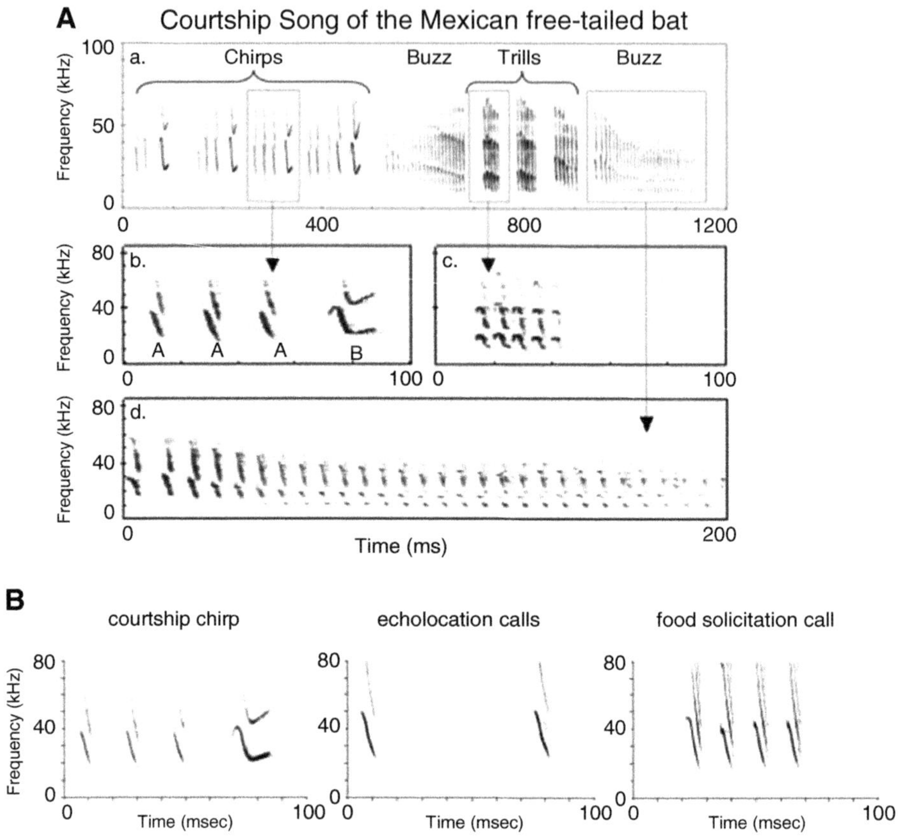

Fig. 8.2 Communication and echolocation calls emitted by Brazilian free-tailed bats. (**A**) The courtship song of a Brazilian free-tailed bat: (**a**) one complete song showing the three types of phrases: a chirp, buzzes, and trills; (**b**) Expanded section of a chirp phrase showing the A and B syllables; (**c**) Expanded section of a trill; (**d**) Expanded section of a buzz (Adapted from Bohn et al. (2009)). (**B**) The FM sweeps emitted in courtship chirps, echolocation calls, and food solicitation calls are strikingly similar

The third phrase in this song is the buzz (Figure 8.2Ad). Buzzes are also composed of short (3-ms) downward FM syllables that are always separated by a few milliseconds. Although the acoustical structures of trill and buzz syllables are similar, the phrases are distinguished by the number of syllables they contain: Trills have only 3–4 four syllables, whereas buzzes have on average 35 syllables. They are also distinguished by the spectral structure of the syllables. The initial FM syllables in each buzz have relatively high beginning and end frequencies and are followed by 5–10 syllables with progressively lower beginning and end frequencies (Figure 8.2Ad).

A significant feature of some of the FMs in the communication calls is that they are strikingly similar to the FMs these bats emit in their echolocation calls (Figure 8.2B) (Schwartz et al. 2007; Bohn et al. 2009). The similarity is in both the spectral and temporal structure of the FMs emitted in the two different contextual situations. An example is the similar structures of the FMs of the A syllables of the courtship songs, the FM sweeps of the food solicitation call, and the echolocation FMs emitted during certain phases of the echolocation cycle (Figure 8.2B). In all cases, the FMs are only about 1–3 ms in duration with fundamental frequencies that sweep downward from about 30–15 kHz. It follows then that the auditory system must process the two types of signals in the same way. The similarity in the structural features also underscores the hypothesis presented earlier, that the processing of communication signals is one of the primary tasks of the bats' auditory system, and adaptations required for echolocation were subsequently added or co-opted from the basic processing of the elements in communication calls. In the sections below, how the auditory system in bats processes and represents the various communication calls and songs they emit are discussed first and then attention is turned to the processing of brief FM sweeps.

8.3 Responses in the Colliculus Are Selective

When a series of echolocation and conspecific communication calls is presented to a group of isofrequency IC neurons (i.e., neurons tuned to the same frequency), most neurons respond to only a subset of the calls and not to others, and thus most neurons express response selectivity (Andoni et al. 2007; Andoni and Pollak 2011). Selectivity of this sort is seen in the IC of all mammals that have been studied (Holmstrom et al. 2007; Portfors et al. 2009). Selectivity is illustrated in Figure 8.3, which shows a suite of ten species-specific communication and echolocation calls from a Brazilian free-tailed bat and the responses that were evoked from four IC cells that were all tuned to about the same frequency. Each call had a different and unique spectrotemporal structure, was broadband with multiple harmonics, and was presented at an intensity that was at least 20 dB above the neuron's threshold at the frequency to which the neuron was most sensitive (its best frequency, BF). Thus, each call had suprathreshold energy that encroached upon each neuron's excitatory tuning curve. The differential responses to each of the calls showed that IC cells are not only selective but also that their selectivities are diverse in that the particular subset of calls that evoke discharges varied from neuron to neuron, even though the neurons were all tuned to the same frequency.

The selectivity for calls is shaped in the IC by the interaction of the excitatory and the inhibitory innervation that plays upon each IC cell (Klug et al. 2002;

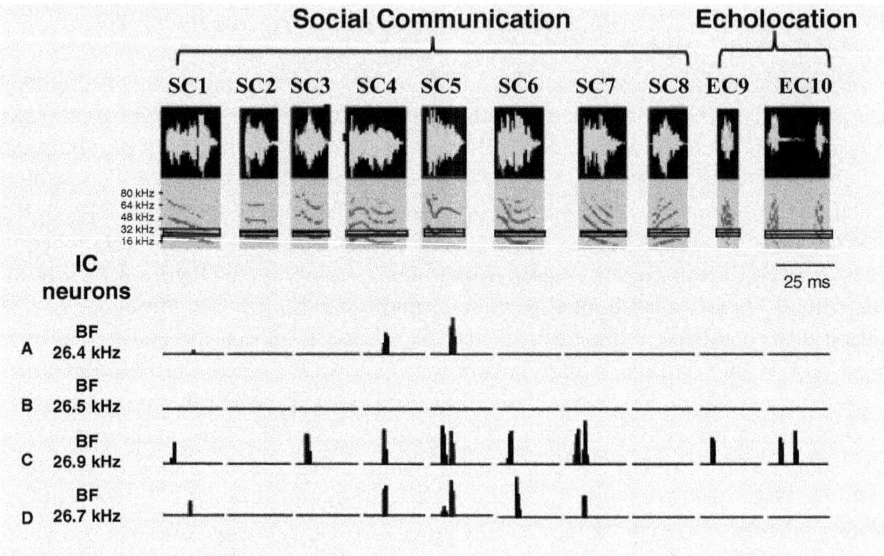

Fig. 8.3 Responses of four IC neurons to ten species-specific calls. Eight of the calls (SC1–SC8) are social communication calls and two others (EC9–EC10) are echolocation calls. The four IC cells are isofrequency and all tuned to about 26 kHz. The IC cells were selective in that each fired to only a subset of the ten calls although each of the calls had suprathreshold energy that swept through each neuron's excitatory tuning curve. The selectivity was also heterogeneous in that each cell fired to a particular subset of calls that was different from the subset to which the other cells fired. One cell failed to fire to any of the calls (Adapted from Klug et al. 2002)

Xie et al. 2005). Although each IC neuron receives both excitatory and inhibitory innervation, the excitatory inputs are either non-selective or only slightly selective. Indeed, inhibition is the major sculptor in the IC, where inhibition shapes a selective response property out of a far less selective excitatory input.

The profound impact of inhibition on the selective responses evoked by communication calls is illustrated in Figure 8.4, which shows the responses of nine IC neurons to two different calls, social communication call 4 (SC4) and social communication call 6 (SC6). The responses evoked by the two calls were recorded before and while inhibition was blocked by the iontophoretic application of bicuculline, a drug that selectively blocks $GABA_A$ receptors, and/or strychnine, which selectively blocks glycine receptors. Each of the nine neurons had a different BF and the cells are arranged from low to high, which corresponds to the tonotopic organization of the IC. Note that the calls had similar spectrotemporal features but evoked different responses among the population. Before inhibition was blocked, the nine neurons expressed different selectivities, since only three of the nine neurons responded to call SC4 and four different neurons responded to call SC6.

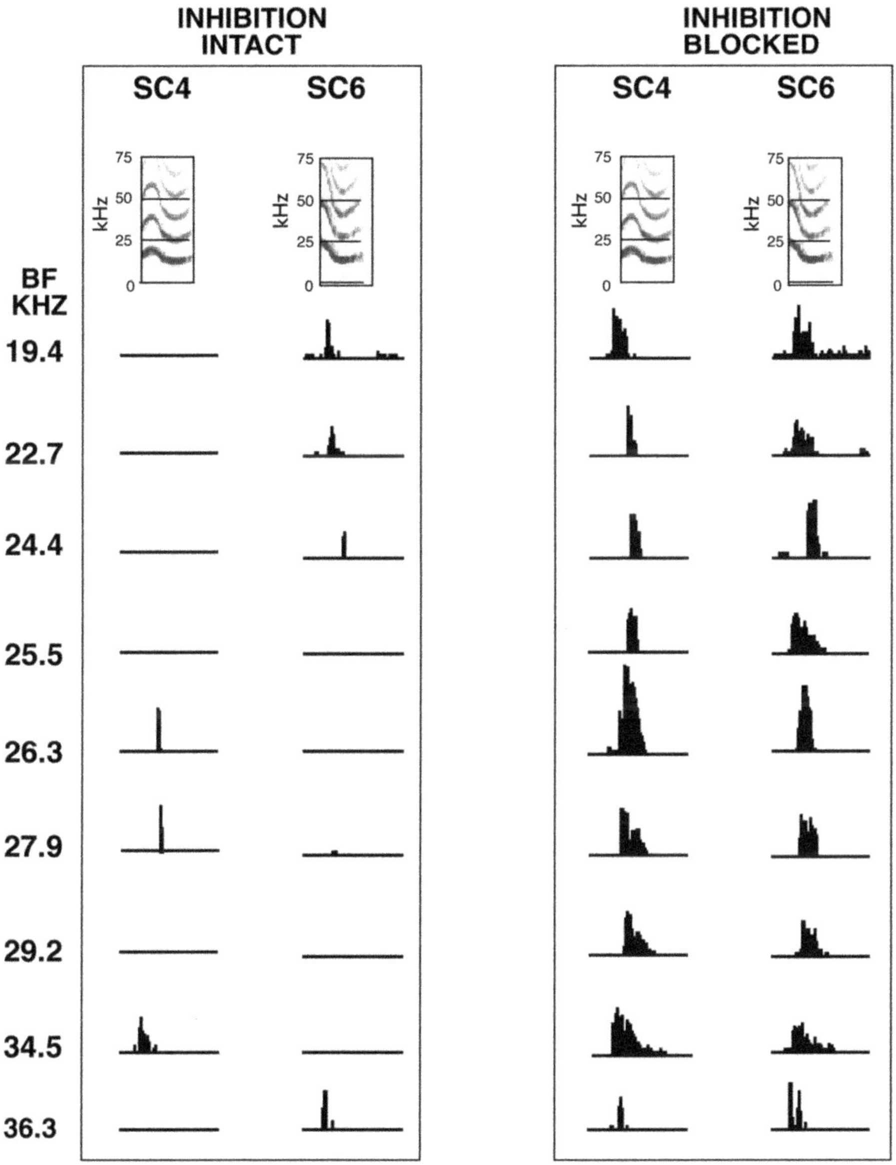

Fig. 8.4 Responses of nine IC neurons to two conspecific communication calls, SC4 and SC6, before and while inhibition was blocked. The BFs of the neurons are arranged from low to high, which corresponds to the tonotopic organization of the IC. Note that the calls had similar spectrotemporal features but evoked different responses among the population. Before inhibition was blocked, the selectivities of the two neurons were different, the three neurons that responded to call SC2 did not respond to SC4, and the four neurons that responded to SC6 did not respond to SC4. Blocking inhibition eliminated selectivities and all neurons responded to both signals (Adapted from Klug et al. 2002)

Blocking inhibition virtually eliminated selectivity and allowed all nine neurons to respond to both calls.

8.4 Spectrotemporal Receptive Fields Reveal the Importance of Sideband Inhibition

But what are the features of inhibition that act to shape response selectivity? The most important feature is the temporal and spectral structure of each neuron's sideband inhibition. Sideband, or surround inhibition as it is sometimes called, is composed of the frequencies that flank the excitatory frequency region of a neuron's tuning curve and evoke inhibition. Consistent with the results in Figure 8.4, other studies have also shown that when sideband inhibition is eliminated by the iontophoretic application of bicuculline and/or strychnine, IC neurons responded to many more calls than they did before inhibition was blocked or even responded to all of the calls presented (Andoni et al. 2007; Holmstrom et al. 2007; Mayko et al. 2012). Specifically, it must be the timing and magnitude of inhibition relative to excitation that underlies selectivity, but exactly how those features are expressed in each IC cell and how they differ among IC cells to create the diverse selectivities among isofrequency cells could not be determined from blocking inhibition alone.

To obtain a more detailed picture of both the excitatory and inhibitory fields in IC cells, a large number of complex signals called "moving ripples" were presented. Moving ripples are complex signals that contain a broad range of both spectral and temporal modulations that have been used by numerous investigators to generate spectrotemporal receptive fields (STRFs) (Kowalski et al. 1996). These signals were used to generate STRFs by a process analogous to spike-triggered averaging of the signals that preceded each spike (Andoni et al. 2007). The STRF derived from one IC cell is shown in Figure 8.5. The idea is that each ripple stimulus is a signal with a broad spectrum but unique spectrotemporal structure. When the different rippled stimuli are presented, frequencies in each ripple stimulus that are always present prior to a discharge are summed and thereby form the red region in the STRF. Frequencies that are rarely or never present prior to a discharge form the blue regions in the STRF. The frequencies represented in the red colors are presumed to be excitatory, whereas the frequencies in blue colors are presumed to be inhibitory. Moreover, whenever the neuron fires, some frequencies will be present in a random fashion, thereby generating the green background color in the STRF in Figure 8.5. Given these assumptions, the STRF provides a picture of relative magnitudes and temporal relationships of the excitation and inhibition that plays upon the cell.

If the STRF provides an accurate representation of the spectrotemporal arrangement of the excitatory and inhibitory innervation, then convolving the STRF with a suite of communication calls should yield predicted responses for that cell, and the predicted responses should be in close agreement with the responses that are actually evoked by each call. Convolution is a calculation in which two matrices are slid

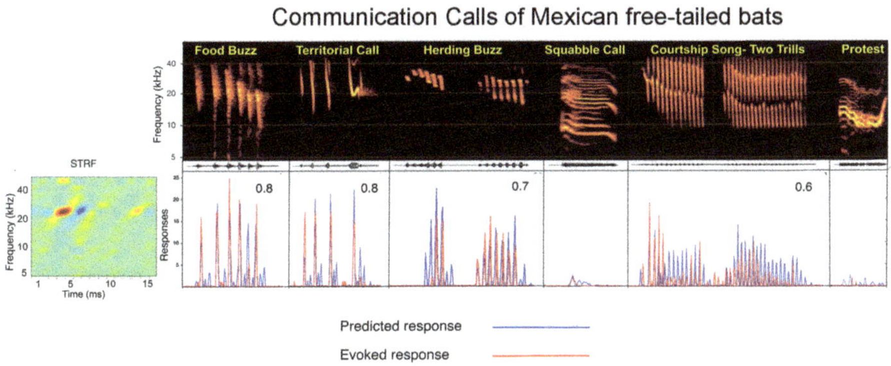

Fig. 8.5 The spectrotemporal receptive field (STRF) provides accurate predictions of responses to species-specific calls. The STRF derived from one cell in the inferior colliculus is shown on the far left. Spectrograms of each species-specific vocalization are shown in the top, with the evoked responses (*red*) and the responses predicted from the STRF (*blue*) displayed below each call. Convolving the STRF with the spectrogram of each call generated predicted responses. The correlations between the predicted and actual responses are shown in *top right* of each panel. Convolutions predicted the call selectivity of the neuron because they predicted high response magnitudes for those calls that evoked strong responses, but they also predicted very low response magnitudes for the calls that evoked little or virtually no responses (Adapted from Andoni et al. 2007)

past each other: one constructed from the cell's STRF and the other from the spectrogram of each call. At each point in time, the values of the two matrices are multiplied. If, at a point in time, a portion of the excitatory portion of the STRF overlaps with a part of the call spectrogram, a positive number is obtained. The positive number results from the multiplication of the energy in the call frequency at that point in time and the value of the corresponding excitation in the STRF. The value resulting from the multiplication indicates how strongly the neuron should have responded at that point in time. However, if the response of the STRF at that time point is inhibitory, the resultant value is negative and is recorded as zero, since there can be no negative number of spikes. Thus, the convolution takes into account not only whether the sound contains frequencies that enter the cell's excitatory or inhibitory response regions but also the magnitudes of the excitation and inhibition at times when the spectral components of the sound and the excitatory and inhibitory regions of the STRF overlap. Assuming the neuron's responses to complex calls are determined largely by the linear sum of responses evoked by the component frequencies in the call, the convolution of the STRF with each call provides a prediction of whether the neuron should respond to the call and, if so, how the neuron should respond in terms of relative response magnitude, latency, and temporal discharge pattern, i.e., it should predict the neuron's discharge profile.

The STRF is, in essence, a linear filter that represents the optimal signal to which the neuron is tuned (Klein et al. 2000). Thus, the prediction is that the strongest responses should be evoked by stimuli that are most similar to the filter, the spectrotemporal features of the neuron's STRF, and the more that the spectrotemporal features of the signal differ from the STRF, the weaker the predicted response. The responses predicted by the convolutions can then be compared or correlated with the discharge profile that is actually evoked by the same call, thereby providing a numerical value of correspondence.

In about 25 % of the IC cells, the responses evoked by the calls were accurately predicted by the convolutions (Andoni et al. 2007). An example is shown in Figure 8.5. The convolutions not only accurately predicted the calls to which the neurons responded, they also predicted the temporal discharge pattern evoked by each call. Equally important, they also predicted the calls to which the neurons did not respond. In short, the STRF in these cells captured the essential features of the cell and provided a picture of the relative magnitude and timing of excitation and inhibition, which in turn predicted how the cell would respond to any of the communication calls or to any other stimulus.

8.4.1 Spectrotemporal Receptive Fields Explain FM Directional and Velocity Selectivities

A basic response feature tested by FM sweeps is whether the neuron responds equally well to both upward and downward frequency sweeps or whether it responds only to, or most strongly to, one FM direction and thus expresses a directional preference or selectivity. Directional selectivity for FM sweeps is strongly influenced by inhibition, since blocking inhibition greatly reduces directional preferences in IC neurons (Casseday et al. 1997; Fuzessery et al. 2011). However, it is not inhibition per se that shapes directional selectivity. The important features are the relative timing of the excitation and inhibition. The timing is reflected in the tilting of the inhibitory fields along the spectrotemporal axis of the STRF (Andoni et al. 2007). Tilted inhibitory fields enhance directional preferences, or even create them, because signals sweeping in the non-preferred direction simultaneously evoke both excitation and inhibition, thereby suppressing responses to that FM direction (Figure 8.6, top panel). In contrast, signals sweeping in the preferred direction activate excitation and inhibition at different times, excitation first followed by inhibition, thereby allowing the cell to respond to the preferred direction (Figure 8.6, lower panel). This interpretation is supported by results obtained when inhibition was blocked by the iontophoretic application of bicuculline and/or strychnine (Andoni et al. 2007). Blocking inhibition not only reduced or even eliminated the inhibitory fields in their STRFs, but it also reduced FM directional selectivities in the IC.

The degree of tilt in the receptive field shapes both the neuron's directional selectivity and the FM velocity that evokes the strongest response (Andoni et al. 2007; Andoni and Pollak 2011). The response strength is determined by the correspondence between the tilt in the excitatory field and the rate of frequency sweep or FM velocity. Thus neurons with strong tilts are most sensitive to high FM velocities, whereas neurons with lesser tilts are most sensitive to lower FM velocities. Based

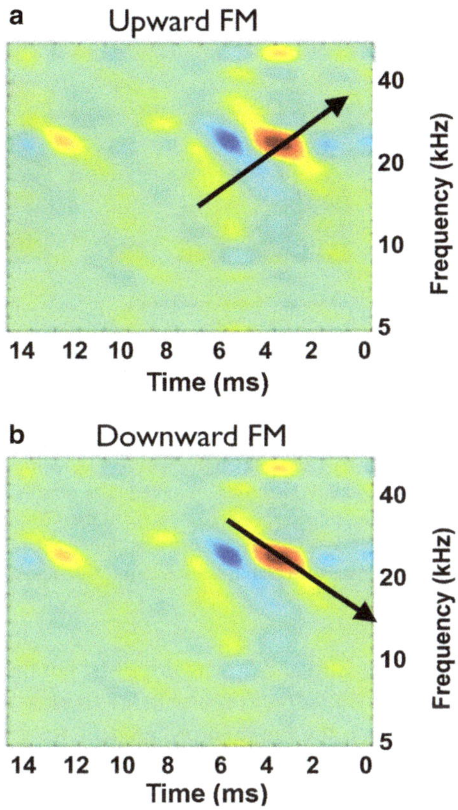

Fig. 8.6 Tilted spectrotemporal receptive fields impart directional selectivity for FM sweeps. Cell with a tilted (*inseparable*) receptive field. *Arrows* indicate how an *upward* (**a**) and *downward* (**b**) FM sweep would traverse the STRF at one point in time. The key feature is that at some point in time the downward FM will only sweep through the excitatory portion of the STRF without encroaching upon the inhibitory portion and thereby excite and drive the cell. In contrast, the upward FM will never encroach only upon the excitatory part of the STRF but rather will sweep through both its excitatory and inhibitory portions, which will suppress excitation thereby preventing the neuron from firing. This is the same STRF shown in Figure 8.5 but flipped in time (STRF is adapted from Andoni et al. 2007)

on these features, it was estimated that most IC neurons had best velocities between 5 and 100 octaves/s, with a mean of ~60 octaves/s (Figure 8.7c).

Of particular importance is the close agreement between the FM features in their conspecific communication sounds and the tuning for those FM features among the IC population (Andoni and Pollak 2011; Pollak et al. 2011a, b). As can be seen in the spectrograms of the various calls in Figures 8.1, 8.2, and 8.3, all echolocation and most communication signals emitted by Mexican free-tailed bats contain FMs. At least a portion of the FMs in almost all calls sweeps downward at velocities ranging from 0 to 250 octaves/s. Consistent with these signal features, the IC of all bats

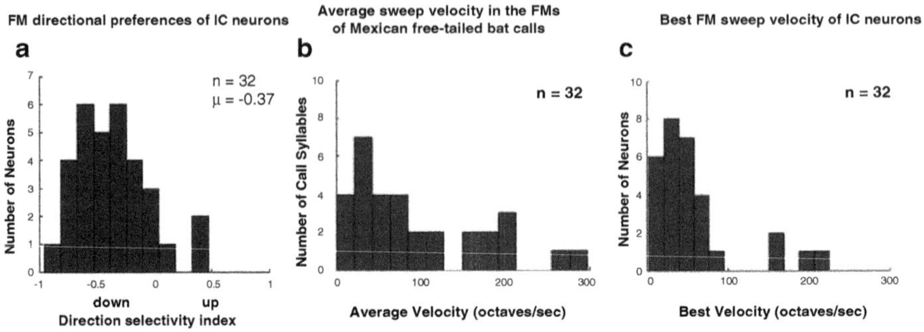

Fig. 8.7 Directional selectivities of IC cells, call velocities, and best velocities of IC cells. (**a**) Distribution of directional selectivity indices in IC cells. Downward preferring cells have negative indices, upward preferring cells have positive indices, and non-directional cells have an index of 0. A small number of cells preferred upward sweeps or were non-directional. The vast majority, however, preferred downward FMs. (**b**) Distribution of the FM velocities found in 21 calls. (**c**) Distribution of best velocities to which IC neurons are tuned. The distributions in B and C are well correlated (r = 0.7), showing the close correspondence between the FM velocities in their communication calls and the FM velocities to which IC neurons are tuned. The STRFs of all cells shown yielded good predictions for responses to communication calls (Adapted from Andoni et al. (2007))

has cells selective for both upward and downward FMs, but the majority of cells are selective for the downward direction (Figure 8.7a) (Razak and Fuzessery 2006; Andoni et al. 2007). Moreover, the range of preferences for sweep velocities corresponds closely to the sweep velocities in the signals these animals emit (Figure 8.7b, c) (Andoni et al. 2007). Thus the structure of their excitatory and inhibitory fields biases many IC neurons for downward direction selectivity and shapes their responsiveness to the FM velocities and other features present in their vocalizations.

8.4.2 Predictive Spectrotemporal Receptive Fields Found in Minority of IC Neurons

The STRFs of cells in which the convolutions accurately predicted responses and explained response selectivities present a comprehensive view of the quantitative features of excitation and inhibition in both frequency and time. The cells that yielded predictive STRFs must have linearly added the response of inhibitory and excitatory frequencies evoked by the rippled stimuli. Since STRFs reflect the average signal generated by such linear additions, the average representation of the excitatory and inhibitory fields generated by ripple stimuli was appropriate for predicting responses to other complex stimuli, such as the communication calls.

The neuronal population in the IC, however, is heterogeneous (Li et al. 2010; Pollak 2012) and most IC cells did not behave in the relatively simple way that the IC cells described above did. Specifically, predictive STRFs were only found in only 25 % of IC cells; the STRFs in most cells (~75 %) provided poor predictions or were non-interpretable (Andoni et al. 2007). Those cells apparently had either static or dynamic non-linear response properties that were stronger than the linear response properties extracted by the STRFs generated by ripples. Stated differently, there was no linear relationship between the magnitudes of the excitation and inhibition in time and frequency that would apply to every complex signal. Therefore, the STRFs computed for those cells could not predict the response to a new complex signal such as the conspecific calls. The reason for the lack of predictability is that the nonlinear interactions of excitation and inhibition would be different than the average STRF derived from the linearly summed ripples.

8.4.3 Most Neurons Had More Than One Spectrotemporal Filter

Neurons in which the STRF (generated by spike-triggered averaging) yielded poor predictions for calls had multiple spectrotemporal features of the stimulus that defined the neuron's overall receptive field (Andoni and Pollak 2011). In these neurons, the nonlinear combination of multiple spectrotemporal features, or filters, predicted the neuron's spiking responses. To evaluate how multiple filters influenced the responses to communication calls, a computation was used that was a spike-triggered covariance procedure somewhat similar to principal component analysis (Rust et al. 2005). This method yielded two or three relevant filters in most of the IC cells, where the first spectrotemporal filter captured the most information of the stimulus-response relationship of each neuron.

In a study by Andoni and Pollak (2011), the set of relevant spectrotemporal filters was not computed from rippled stimuli but rather was computed from the responses evoked by a large number of conspecific communication calls. Natural calls were used because previous studies showed that STRFs derived from natural stimuli in both the IC of songbirds (Woolley et al. 2005; Gill et al. 2006) and in the cortex of ferrets (David et al. 2009) are significantly different than the STRFs derived with synthetic stimuli, such as ripples or noise. Most importantly, the receptive fields derived with natural stimuli provided far better predictions of responses to natural calls than did the receptive fields derived with synthetic stimuli (Woolley et al. 2005).

Predicted responses for both electronically generated FMs and conspecific calls were then calculated using either the first most informative spectrotemporal filter alone or the two most informative spectrotemporal filters (Figure 8.8). The most significant finding was that the predicted responses were poor when only the first filter was used but improved significantly when two filters were used (Andoni and Pollak 2011). The correlation coefficient between the predicted and the evoked responses for calls had a mean of 0.46 with only one filter but increased to a mean of 0.61 when two filters were used. This showed that these neurons did indeed have two or more spectrotemporal filters that determined the responses to calls.

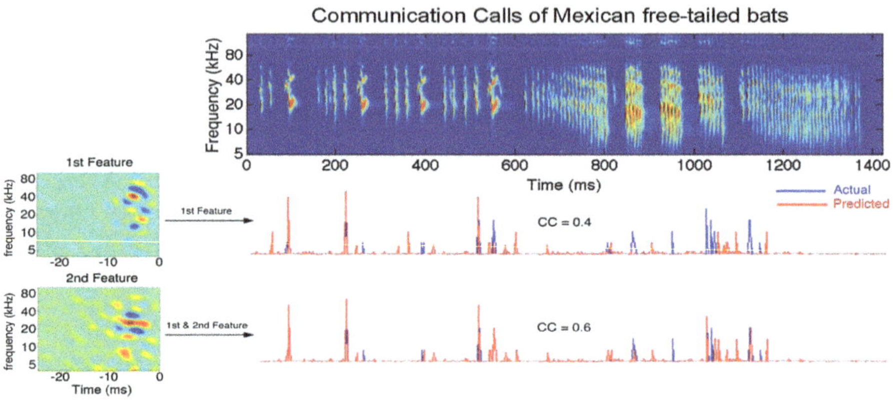

Fig. 8.8 Convolution with only first filter or feature and with both first and second features. Responses predicted from spectrotemporal features improve when multiple stimulus features are considered. The two most informative features of an IC neuron are shown. Convolving the calls shown with only the first feature yielded poor predictions, with an average correlation coefficient between the predicted responses and those evoked by the calls of only 0.4. When both the first and second features were used to calculate the predicted responses, the correlation coefficient increased to 0.6. This shows that this IC neuron is tuned for multiple spectrotemporal features of natural calls (Adpated from Andoni and Pollak 2011)

The relevance of the two filters was further supported by the near-perfect agreement between the responses evoked by electronically generated FMs and the responses predicted with the non-linear combination of the two most relevant filters.

8.5 The Importance of Frequency Modulations for Call Selectivity

As was shown above, an important acoustic feature in the calls of bats is the structure of the FM components (Andoni et al. 2007; Andoni and Pollak 2011). FMs are not only prominent in both the echolocation and social communication calls emitted by bats (Bohn et al. 2008, 2009), FMs are also important components of communication signals in most animals, including humans (Doupe and Kuhl 1999). The structure of the FM component is important because IC cells are tuned for FM features and the velocity of the sweep and its direction, whether it sweeps upward or downward (Andoni et al. 2007; Fuzessery et al. 2011; Gittelman et al. 2012). Indeed, the specific tuning of IC cells for FM features is one of the principal features that determines the response selectivity for calls in the IC of bats (Andoni and Pollak 2011).

8.6 Directional Preferences for FMs Measured with In-Vivo Whole Cell Recordings

The previous sections emphasized the roles of inhibition in shaping directional preferences for FM sweeps. However, inhibition cannot be measured directly with extracellular recordings because extracellular electrodes only record spikes, which are evoked by excitation. With extracellular recordings, inhibition has to be inferred from the suppressive effects of some stimulus manipulation on the excitation evoked by another signal. Furthermore, with extracellular recordings there is an uncertainty about whether the observed spike suppression was due to inhibition at the IC or whether suppression was inherited from the inhibition that occurred in a lower nucleus that projects to the IC. To obtain a more direct and detailed view of sound-evoked inhibition, and to evaluate how the temporal features of inhibition interact with excitation to shape responses to complex signals, intracellular recordings were obtained with patch electrodes from the IC in awake bats in response to FM sweeps.

With intracellular recordings using patch electrodes, as with extracellular recordings, the discharges evoked in most IC cells exhibited a preference for downward sweeping FMs (Gittelman et al. 2009; Gittelman and Pollak 2011). With patch recordings, however, both the inputs to the cells (expressed in the amplitudes of post-synaptic potentials, PSPs) and their outputs, their discharges, are obtained. The selectivity differences of the inputs can be quantified by computing a PSP directional index (PSP amplitude evoked by the downward FM minus PSP amplitude evoked by the upward FM divided by the sum of the two amplitudes). Similarly, the selectivity differences of the outputs (discharges) are quantified by computing a discharge directional index (DSI) based on spike counts rather than PSP amplitudes. Thus the directional preferences of the inputs can be quantitatively compared to the directional preferences of the outputs.

In most IC cells, the differences in the discharge vigor evoked by the two FMs are substantially greater than the differences in the magnitudes of the excitatory postsynaptic potentials (EPSPs) evoked by the same signals (Gittelman et al. 2009). The discharge output of cell 2 in Figure 8.9, for example, was selective for the preferred (downward) FM; it fired to every presentation of the preferred FM but only fired 50 % of the time to the null (upward) FM and had a spike directional index of 0.33. In marked contrast, the EPSP amplitudes evoked by the two signals were very similar in amplitude. The PSP directional index was only 0.06. The disparity in the higher spike selectivity compared to the low EPSP selectivity is due to the nonlinear influence of spike threshold, where the larger EPSP evoked by the preferred FM in this cell was above threshold and evoked a discharge on every presentation, whereas the EPSP evoked by the null FM was only slightly smaller and just at threshold level. This cell illustrates the general finding that the inputs (PSPs) were less selective than the outputs (spikes). On average, the spike DSI was more than twice as large as the PSP-DSI among the IC population (Gittelman et al. 2009).

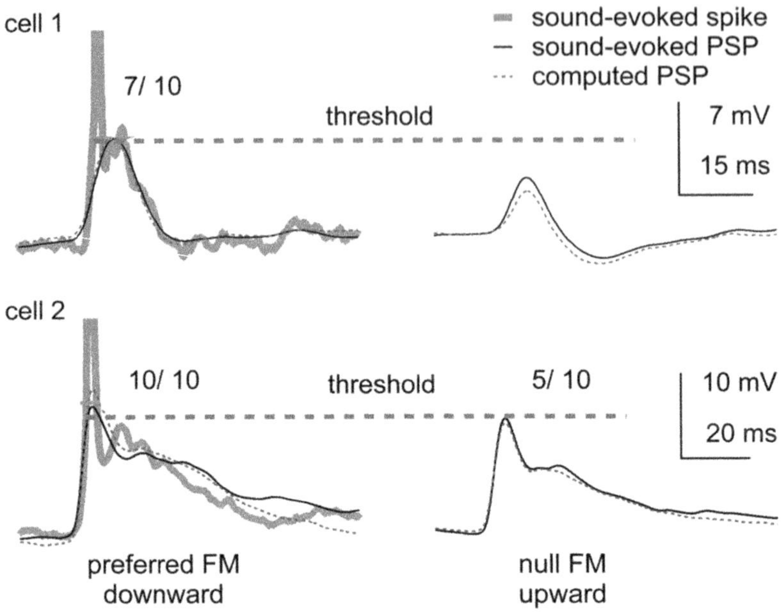

Fig. 8.9 Two directionally selective cells. Black traces are the measured postsynaptic potentials (PSPs) (mean of ten trials; spikes were removed by filtering), dashed traces are PSPs computed from derived conductance waveforms, and gray traces illustrate spiking with a single sweep response. Dashed line is spike threshold (Adapted from Gittelman and Pollak 2011)

8.7 The Role of Spike Timing for Creating Directional Selectivity

While the intracellular recordings with patch electrodes, like the recordings with extracellular electrodes, showed that most IC cells express directional preferences for FM sweeps, the comparison of EPSPs and spikes did not show by itself how the interactions of excitation and inhibition shaped the directional preferences of the cells. Previously, it was shown that for cells whose responses to calls could be predicted by their STRFs, the directional preferences were formed by the relative timing of the excitatory compared to the inhibitory inputs evoked by an FM sweep. The wide acceptance of this explanation is based on two principal observations. The first is that neurons selective for downward (or upward) FM sweeps have inhibitory fields that are lower (or higher) in frequency than the frequencies that activate their excitatory fields. These features were confirmed by the excitatory and inhibitory response fields in the linear STRFs, as illustrated by the cell in Figure 8.5, and these features were shown in a large number of previous studies in a variety of mammals (Razak and Fuzessery 2006; Kuo and Wu 2012). The second observation is that blocking inhibition reduces or eliminates directional preferences (Koch and Grothe 1998; Razak and Fuzessery 2009).

The timing hypothesis, which follows from the results of those experiments, posits that downward FM signals first sweep through the excitatory field, thereby evoking an initial excitation, and slightly later in time, the signal sweeps through the inhibitory field (Covey and Casseday 1999; Zhang et al. 2003). With upward sweeping FMs, on the other hand, inhibition is activated first and the initial inhibition quenches the subsequent excitation. This is exactly the result obtained from the STRF of the IC neuron shown in Figures 8.5 and 8.6. The same arguments apply for upward preferring cells, but the frequencies of the excitatory and inhibitory fields are reversed.

But there is an additional implicit assumption in this explanation. Specifically, the explanation assumes that the inputs behave in a linear manner, where the excitatory and inhibitory inputs are evoked in synchrony with the spectrotemporal features of the signals. Thus, the same excitation and inhibition are evoked by downward and upward sweeping FMs, but the timing of excitation and inhibition is reversed because the temporal features of the signals are reversed. Moreover, it assumes that the cells are sensitive to even small differences in the arrival times of excitation and inhibition. These assumptions were used to explain the directional selectivity based on the STRF shown in Figure 8.5 and for all the other IC neurons that had linear STRFs. The strengths and relative timings of excitation and inhibition should simply be reversed as the FM direction is changed from upward to downward. In short, there is strong evidence from a variety of different studies in a variety of mammals to support the hypothesis that spectral arrangement of the excitatory and inhibitory fields generates FM directionality and that hypothesis also explains why blocking inhibition eliminates directionality.

8.7.1 FM Directional Selectivity Formed by Timing Disparities of Excitation and Inhibition Does Not Apply to All IC Cells

The IC is heterogeneous, as was pointed out previously, and a particular response property is formed in different ways among its neuronal population (Li et al. 2010; Pollak et al. 2011a, b). With regard to the formation of FM directional preferences, sensitivity for small differences in the timing of excitation and inhibition requires that such cells have features suitable for fast temporal processing, such as low input resistances and fast time constants. Such features would generate brief sound-evoked responses, where even small changes in the arrival of excitation and inhibition would change the amplitude of the sound-evoked response and hence the evoked firing rate. In cells with high input resistances and long time constants, in contrast, small changes in timing would be ineffective. The sound-evoked membrane potentials in these cells would change far more slowly with the arrival of each input, and these cells would be suited to integrate inputs over longer time periods. A recent study of IC cells in bats showed that about half of the cells in the IC do indeed have low input resistances that range from 40 to 100 megohms and fast time constants (Xie et al. 2008). Presumably these are the cells whose FM preferences are formed by the relative timing of excitation and inhibition. The other side of the finding is that about half of the IC population has high input resistances and long time constants, features that are inappropriate for sensitivity to small changes in the timing of excitation and inhibition. It may well be that the non-linear cells with multiple filters had high input resistances and slow time constants, although there is no direct proof of this correspondence.

8.7.2 *The Timing of Excitation and Inhibition Explored with Whole Cell Recordings*

To evaluate the role of the timing of excitation and inhibition in IC cells with high input resistances, the excitatory and inhibitory conductances that generated the responses evoked by an upward and by a downward FM sweep were computed in a subset of IC neurons (Gittelman et al. 2009; Gittelman and Pollak 2011). Conductances reflect the number of ligand-gated receptors that are opened during excitation and inhibition. Thus, the excitatory conductance indicates the number of excitatory receptors opened in the IC cell by the release of glutamate from the presynaptic axons, whereas the inhibitory conductance reflects the number of inhibitory receptors opened in the IC cell by the release of GABA and/or glycine. Additionally, the latency and shape of the excitatory conductance waveform indicates the latency of excitatory innervation, the length of time the receptors are open, and the strength or magnitude of the excitation. The same applies to the inhibitory conductance waveform. The response of the membrane potential, the postsynaptic potential (PSP), is generated by the flow of currents through the conductive receptors. Those experiments showed several important features of the conductances evoked by the preferred and null FMs, as well as several other features of the EPSPs that would occur when the timing of inhibition is advanced or delayed (Gittelman and Pollak 2011).

The first important feature is that the excitatory conductances evoked by both the preferred and null FMs by themselves would have evoked a suprathreshold response. This finding is consistent with the general finding from extracellular studies that blocking inhibition, which simply eliminated inhibition and left only excitation, reduces or eliminates the directional preferences in almost all IC cells, allowing the cells to fire to both FMs. The second finding is that the amplitudes of the excitatory and inhibitory conductances evoked by the preferred FM and null FMs are almost always different. In other words, even though the spectral composition of the preferred and null FMs are identical but reversed in time, each signal does not evoke the same-but-time-reversed excitatory and inhibitory conductance waveforms. Rather, the excitatory and inhibitory conductance waveforms evoked by the preferred FM differ in either waveform shape or amplitude, or both shape and amplitude, from the conductances evoked by the null FM. The third finding is that there was no consistent relationship between the timing of the excitatory and inhibitory conductances evoked by the preferred FM compared to the null FM.

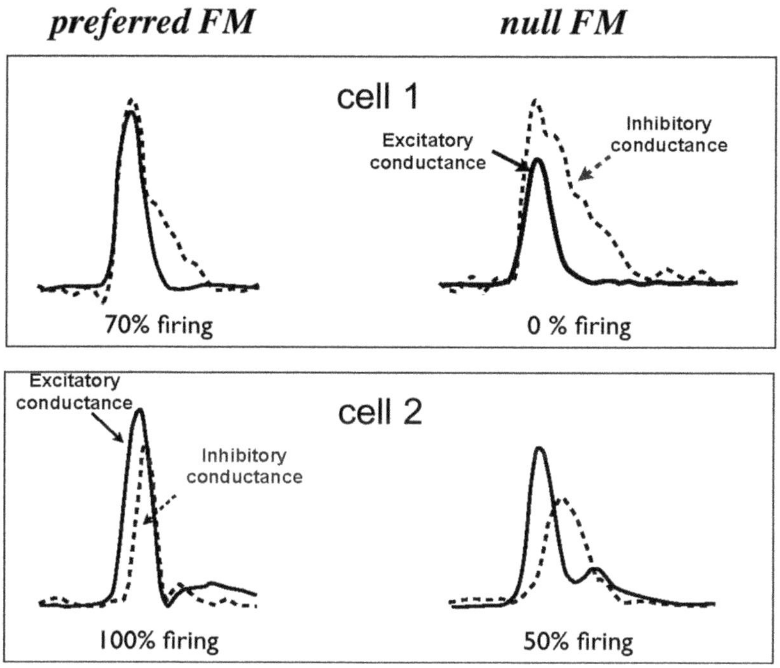

Fig. 8.10 Timing of excitation and inhibition provides no information about directional preferences. The calculated excitatory (*solid lines*) and inhibitory (*dashed lines*) conductances for a downward (*preferred*) and upward (*null*) FM sweep in two IC neurons. These are the same cells whose responses to the FMs are shown in Figure 8.9. In cell 1 (*top panel*), the timing of the excitatory and inhibitory conductances was virtually simultaneous for both the preferred and null FMs. In cell 2 (*bottom panel*), the excitatory conductance led the inhibitory conductance for both the preferred and the null FMs. However, the lead time of excitation was even greater for the null than the preferred. In both cells, the differences in the responses to the two FMs are due largely to the differences in the shapes and amplitudes of the excitatory compared to the inhibitory conductances rather than to their relative timing (Adapted from Gittelman and Pollak 2011)

The two cells in Figures 8.9 and 8.10 illustrate two of the three features. Although not shown in Figure 8.9, the EPSPs of the preferred and null FMs, computed only from the excitatory conductances shown in Figure 8.10, were above threshold for both cells. In addition, the waveforms of the excitatory and inhibitory conductances of the preferred and null FMs in Figure 8.10 differed in shape and in peak amplitude (the exception is the peak amplitudes of the inhibitory conductances for cell 1, which were about the same). Finally, excitation and inhibition in cell 1 were virtually coincident for both the preferred and null FMs. In cell 2, in contrast, excitation led inhibition in the response to the preferred FM, but excitation led by an even greater amount of time in the response to the null FM. For these cells at least, the notion that the preferred FM evokes an excitation that leads inhibition, whereas the null FM always evokes an inhibition that either leads or is coincident with the excitation, is not supported by the relative timing of the excitatory and inhibitory conductances. Since the relative timings of the excitatory and inhibitory conductances evoked by the preferred and null FMs differed from cell to cell, the relative timing of excitation and inhibition by itself provides little or no information about the preferences of these cells for the direction of an FM sweep. While the timing of inhibition and excitation contributed to the FM directional preference, in that inhibition and excitation had to overlap in time, the principal features that shaped directional selectivity in these cells were the magnitudes of inhibition compared to excitation coupled with the shapes of the conductance waveforms (Gittelman and Pollak 2011).

8.8 Combination Sensitivity

The results of the studies on FM directional selectivity illustrate that selective response properties, FM sweep directionality in this case, are formed in a variety of ways in the IC. Moreover, since FMs occur prominently in both echolocation and communication calls, the various ways in which FM directionality is formed applies to both echolocation and communication signals.

The idea of shared mechanisms for processing acoustic features of echolocation and communication calls is further illustrated by combination sensitive neurons in the auditory system of mustache bats (*Pteronotus parnellii*). Mustache bats emit biosonar signals composed of an initial long constant-frequency (CF) component and a terminal brief, downward sweeping FM component (Figure 8.11) (Pollak and Casseday 1986; Suga et al. 1987; Wenstrup and Grose 1995). The duration of the CF component can be up to 30 ms in duration, whereas the terminal FM is only 2–4 ms. Each call is emitted with a fundamental frequency and three harmonics, but the second harmonic always contains the most energy. Thus, the first CF harmonic, or fundamental, is emitted at about 30 kHz, and the terminal FM component sweeps downward by about 7 kHz, from 30 kHz to about 23 kHz. The CF of the second harmonic, the dominant harmonic, is emitted at about 60 kHz, and the second harmonic FM sweeps downward from 60 kHz to about 45 kHz. The CF of the third harmonic is emitted at about 90 kHz and the CF of the fourth at about 120 kHz, each with initial and terminal FM components as shown in Figure 8.11.

Suga was the first to discover combination sensitive neurons in the mustache bat's cortex (O'Neill and Suga 1979; Suga and O'Neill 1979). Combination sensitive neurons are distinguished by their far more vigorous responsiveness to two tones that have a specific-frequency relationship and temporal relationship, as compared to their responsiveness to a single tone at their BF. Hence the name "combination sensitive neurons."

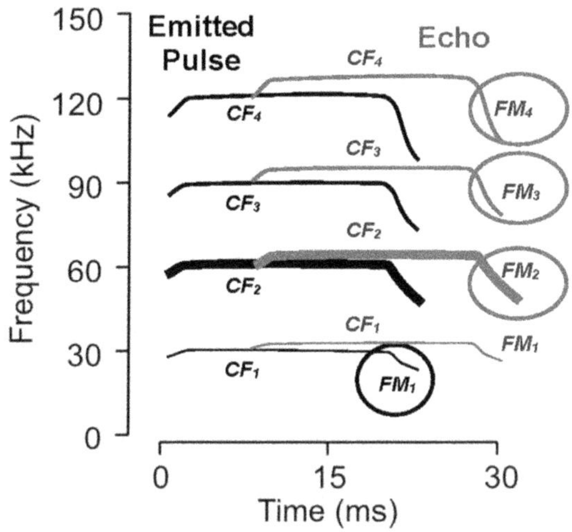

Fig. 8.11 Schematic sonogram of the sonar signals of mustache bats with the emitted pulse (*black lines*) and returning echo (*gray lines*). Line thickness indicates the relative strength of each signal harmonic. FM-FM neurons express response facilitation when presented with a frequency in the FM of the fundamental (FM1, *black circle*) followed in time by a frequency in the FM of either the second, third, or fourth harmonics (FM2, 3, 4, *gray ovals*) (Adapted from Portfors and Wenstrup 1999)

There are several types of combination sensitive neurons in the mustache bat's cortex (Suga et al. 1983). The most thoroughly studied are the so-called FM-FM neurons, which are topographically segregated from other types and are localized in the FM-FM region that is just adjacent to the primary auditory cortex. These neurons respond best to a combination of the first-harmonic, frequency-modulated (FM1) sweep in the emitted pulse and the frequency-modulated component of a higher harmonic (FM2, FM3, or FM4) in the returning echoes (Figure 8.12). In other words, FM-FM neurons are facilitated by the FM1 component in the simulated pulse and a higher harmonic FM component in the simulated echoes but only when there is a specific delay between the two signals.

One of the functional attributes assigned to FM-FM neurons is the coding of range or distance between the bat and its target (O'Neill and Suga 1979; Portfors and Wenstrup 1999). Target range is conveyed by the time interval between the FM of the emitted pulse and the FM of the returning echo (Simmons et al. 1974). The rationale follows from the finding that each FM-FM neuron has a best delay, the delay that produces the largest facilitation. The best delays of FM-FM neurons vary from about 1–20 ms, which correspond to the timing differences between the pulses and echoes that the bats receive during echolocation. There is then a striking concordance between the highly specified spectral and temporal requirements of the signals that drive these neurons optimally, and the spectral and temporal features of the biosonar signals that mustache bats emit and receive. The combinatorial proper-

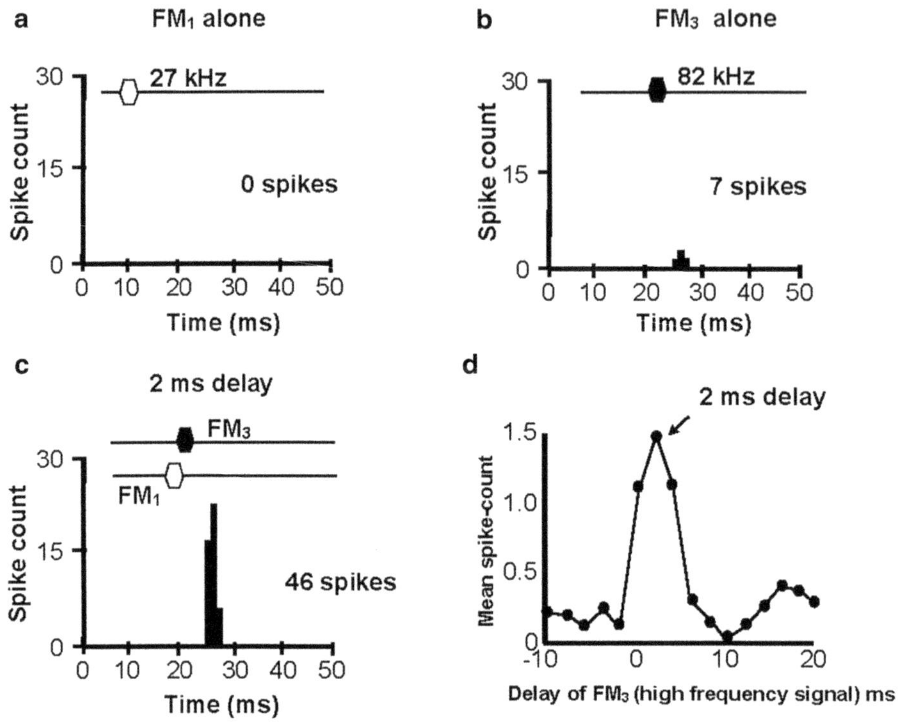

Fig. 8.12 Responses of a facilitated FM-FM neuron to single tone bursts and to the combination of a low- and high-frequency tone. (**a**) The neuron did not respond to a 27.0 kHz (FM1) tone burst presented alone. (**b**) The neuron displayed a weak response to an 82.7 kHz (FM3) tone (time of stimulus onset was 20 ms, and the response latency was 4 ms). (**c**) A facilitated response was elicited by delaying the onset of the high-frequency sound from the onset of the low-frequency sound by 2 ms. Response latency of the facilitated response was 7 ms from the onset of the low-frequency signal (5 ms from the onset of the high-frequency signal). (**d**) Delay tuning curve obtained by presenting two tone bursts at best facilitating frequencies (82.7 and 27.0 kHz), then varying the relative timing between the two tones. Zero milliseconds represents simultaneous presentation of the two signals. The best delay for this neuron was 2.0 ms (Adapted from Portfors and Wenstrup 1999)

ties of these neurons are tailored to the mustache bat's biosonar signals, and thus their particular features are unique to this animal.

This concordance led to the implicit hypothesis that the mustache bat's cortex is super-specialized for processing biosonar signals. It, therefore, came as a surprise when, in later studies, neurons in the FM–FM area were found to respond vigorously to a variety of signals used by mustache bats for social communication (Kanwal et al. 1994). Indeed, the neurons were selective in that they responded best to calls having particular spectral and temporal features that were similar to the spectral and temporal features of the FM-FM neurons, and they responded poorly or not at all to other signals. Thus, FM-FM neurons, as well as the other combination-sensitive types, express combinatorial properties that impart selectivity for both biosonar signals and for communication signals.

8.8.1 Combination Sensitive Neurons Are Created in the IC

Although combination-sensitive neurons were first discovered in the cortex and thus were thought to be an emergent or new response property resulting from cortical processing, subsequent studies by Wenstrup and his colleagues showed that combinatorial neurons are initially created in the mustache bat's IC (Mittmann and Wenstrup 1995; Portfors and Wenstrup 1999). Indeed, the combination-sensitive neurons in the mustache bat's IC have properties strikingly similar to those that Suga and his colleagues found in the cortex. In an elegant series of studies, Wenstrup and his colleagues worked out the mechanisms that generate combinatorial facilitation (Sanchez et al. 2007, 2008; Wenstrup et al. 2012). Remarkably, the mechanism is due entirely to inhibition rather than a summation of excitation evoked by the two frequencies. The role of inhibition in creating combinatorial facilitation once again underscores the importance of inhibition for generating selective response properties in the IC.

The evidence for the role of inhibition is that when the appropriate stimuli are presented and generate facilitation, blocking all excitatory receptors (both AMPA- and NMDA- type receptors) eliminates the response to a best frequency tone but has no effect at all on facilitation (Sanchez et al. 2008; Wenstrup et al. 2012). Indeed, even blocking GABAergic inhibition has no influence on facilitation. However, facilitation is eliminated completely when glycinergic inhibition is blocked by the iontophoretic application of strychnine. This led to the hypothesis that an initial low-frequency tone activates a glycinergic inhibitory input that produces an inhibition. The hypothesis proposed by Wenstrup et al. (2012) is shown in Figure 8.13. The key feature is that the inhibition evoked by the first low-frequency tone is followed by a post-inhibitory rebound that is excitatory but sub-threshold. The following higher frequency tone at the cell's BF also generates a glycinergic inhibition but one that is brief and has a fixed latency. Thus the BF also generates a short inhibition with a sub-threshold post-inhibitory rebound. The idea is that when the BF tone is delayed by an appropriate amount, the post-inhibitory rebound evoked by the first, low-frequency tone and the rebound evoked by the following BF tone coincide and summate to evoke a supra-threshold post-inhibitory rebound, which generates discharges and is the facilitated response. The facilitation is due entirely to glycinergic inhibition or to the rebounds from the summation of the two inhibitions. Thus, blocking AMPA or NMDA receptors has no influence on the facilitation, whereas blocking glycinergic inhibition completely eliminates it.

8.8.2 Combination Sensitivity also Imparts Selectivity for Communication Calls in the IC

One of the features that originally led to the idea that combination-sensitive neurons were specialized for echolocation was that the frequencies that evoked the facilitated responses correspond to the frequencies in the first and higher

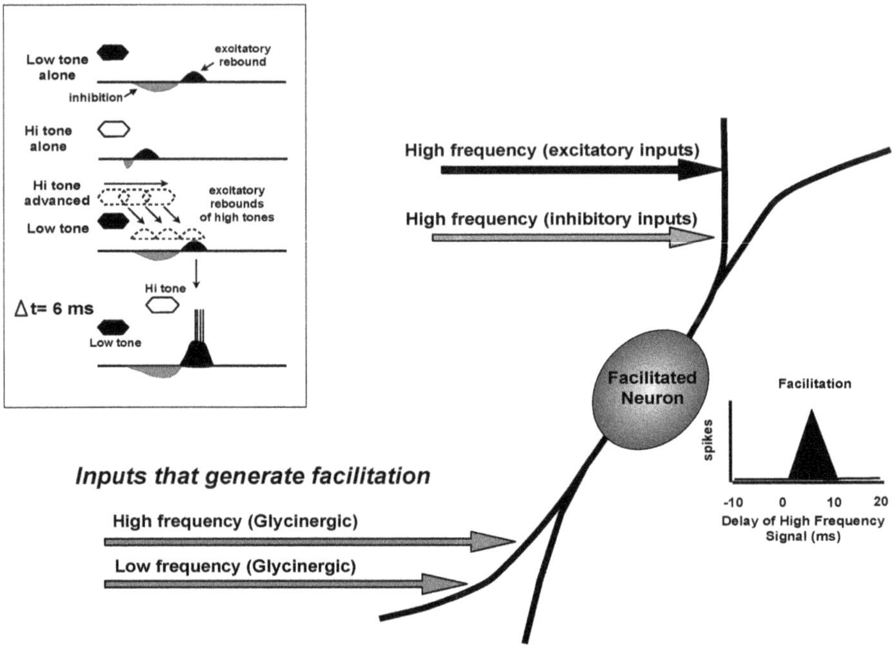

Fig. 8.13 Schematic diagram of mechanisms and circuitry underlying combination-sensitive facilitation in an IC neuron. Inset shows mechanism of post-inhibitory rebound. The high-frequency tone alone evokes a short duration inhibition of fixed latency followed by the post-inhibitory rebound. The low-frequency tone evokes a more prolonged inhibition with a longer latency followed by a rebound. When the high-frequency tone (representing the FM of the echo) follows the low-frequency tone (representing the FM of the emitted pulse) at an appropriate delay, the inhibitory rebounds evoked by both tones coincide in time and summate to generate the facilitation. The neuron receives a variety of high-frequency excitatory and inhibitory inputs tuned to its best frequency (*upper right*) that do not interact with the glycinergic inputs related to facilitation (*lower left*) (Adapted from Wenstrup et al. 2012)

harmonics of the mustache bat's echolocation calls. However, subsequent studies by Wenstrup, Portfors, and their colleagues (Portfors and Wenstrup 2002; Portfors 2004; Holmstrom et al. 2007) showed that many of the combination-sensitive neurons in the IC were tuned to non-echolocation frequencies, frequencies that were not contained in either the emitted calls or the echoes. Moreover, the best delays of many of the "non-echolocation" neurons were at or around 0 ms, intervals so short that they would never occur between the emitted pulse and echo.

These mismatches between the combinatorial properties of some IC neurons and the spectrotemporal features of their echolocation calls indicate that the non-echolocation neurons, as well as the other combinatorial neurons whose features correspond to the echolocation calls, may be important for encoding communication signals. Thus, Portfors (2004) suggested that since bats evolved from a non-echolocating ancestor and those ancestors most likely used acoustic signals for communication, it seems reasonable to suppose that similar neural mechanisms underlie the processing of echolocation and communication calls in the auditory midbrain. To be more specific, it seems reasonable to suppose that combinatorial neurons could also generate response selectivity for communication calls.

In subsequent studies, Portfors and her colleagues showed that combination sensitivity imparts response selectivity for communication calls in the IC, allowing the cells to respond only to some calls but not to others (Portfors 2004). Indeed, the role of combination sensitivity for the processing of communication calls in the IC cells is similar to the role combination sensitivity plays in creating response selectivity for communication calls in the cortex.

8.8.3 Combination Sensitivity also Occurs in the Auditory Systems of Other Animals

Combination-sensitive neurons have been intensively studied in mustache bats, but whether they also occur in other bats, or the degree to which they occur in other bats, is unclear, mainly because it has not been studied in other species. It is significant, however, that combinatorial neurons also have been found in the IC of mice (Portfors and Felix 2005) and songbirds (Schneider and Woolley 2011), suggesting that such neurons are likely to occur in the IC of other bats as well. It should be noted, however, that there are fewer combination-sensitive neurons in the IC of mice and birds than there are in the IC of mustache bats. Additionally, and importantly, in both mice and birds the combination-sensitive neurons have been shown to be important for creating response selectivity for communication calls, as was shown previously in mustache bats.

8.9 Summary and Concluding Thoughts

The results of the studies reviewed here illustrate at least three general features of processing in the IC. The first is the dominant role that inhibition plays in shaping the responses of IC neurons. The roles of inhibition are illustrated by the marked change in the response selectivity for communication calls when inhibition is blocked and by the prominent roles of sideband inhibition for shaping FM directionality. The second feature is the heterogeneity of mechanisms that shape the response properties. There is not a single mechanism that the IC employs to form a given response property, but rather there are multiple ways in which the same response property is formed among the IC population. The third feature is the close correspondence between neural tuning and acoustic properties of conspecific communication signals. In bats at least, this correspondence suggests that IC neurons are specifically encoding features of these signals through the neural computations that generate FM selectivity. Moreover, it is clear that the various selectivities expressed by IC neurons for communication calls are a consequence of the multiple ways in which their selective responses for features of acoustic signals, such as the direction of FM sweeps, are created.

These findings provide further support for the idea presented in the beginning of this chapter: The auditory systems of bats are not distinguished by novel mechanisms but rather that common mechanisms and features are far more pronounced in their auditory systems than in other mammals. This is also well illustrated by several higher order features that were first reported in the IC of bats, FM directional selectivity (Suga 1965), combination sensitivity (Mittmann and Wenstrup 1995), and duration tuning (Casseday et al. 1994), features that were subsequently seen in the auditory systems of other mammals (Brand et al. 2000; Portfors and Felix 2005). While all of these features are seen in the auditory systems of other mammals, they occur in different proportions than occur in bats. There is, for example, a pronounced over-representation of directional selectivity for downward FMs in the auditory systems of bats (Andoni and Pollak 2011), whereas in rats, upward and downward FMs are about equally prevalent (Zhang et al. 2003; Kuo and Wu 2012). Similarly, combination sensitivity is seen in about half of the auditory neurons in the mustache bat's IC (Wenstrup et al. 2012), whereas combination sensitivity occurs in only about 15 % of neurons in the mouse IC (Portfors and Felix 2005).

In short, the adaptations in the brain stem auditory nuclei are primarily, although not exclusively, a matter of quantity, where a species expresses certain features that are shared by other species but to a greater degree or in a more pronounced form rather than expressing wholesale qualitative changes in the mode of processing. The principal mechanisms for processing acoustic information are conserved among mammals, and thus the processing of sound in the brain stem auditory nuclei in a bat that is listening passively, as far as I can tell, is accomplished in the same way as it is in other mammals.

If the brain stem auditory systems of bats are so similar to other mammals, why don't all mammals echolocate? Perhaps the answer is that they possess the capability for echolocation, as do humans, as exemplified by Daniel Kish and his students, but they have never learned to express that ability. The argument is that the processing of sound in an animal that is passively listening must be different from one engaged in the active process of echolocation. Indeed, there is evidence that shows that sound-evoked responses change markedly when a bat is passively listening to a sound compared to when it hears the same sound during echolocation (Schuller 1979; Rubsamen and Betz 1986).

Exactly why and how active echolocation changes the responses evoked by sound is unclear. It is significant, in this regard, that there are massive descending projections from the auditory cortex that not only innervate the IC but also extend to almost all lower nuclei (Weedman and Ryugo 1996; Winer et al. 1998, 2002), and descending projections can have a profound influence on acoustically evoked responses (Suga et al. 2002; Xiao and Suga 2002). In addition, there is strong innervation by serotonergic as well as other neuromodulatory systems along the entire auditory system (Kossl and Vater 1989; Motts and Schofield 2009; Hurley and Sullivan 2012). Those systems are almost surely activated during vocalizations and during the active listening associated with vocalizations, as occurs during echolocation.

These features do not explain echolocation. But whatever role these systems play that enable echolocation, they may well function in an analogous way in most mammals. If so, the operation of these systems, together with the basic mechanisms of acoustic processing in the ascending auditory system, would provide the framework for echolocation, not only in bats and cetaceans, but in humans as well.

Acknowledgments I would like to express my sincerest appreciation to the National Institutes of Health for their generous support of my research over the past 40 years. I also thank Carl Resler for his expertise with computers and his invaluable technical support throughout my career. Finally, I thank my graduate students, postdoctoral fellows, and all of my colleagues who have been so supportive and have taught me so much. Supported by NIH Grant DC 0078.

References

Andoni, S., & Pollak, G. D. (2011). Selectivity for spectral motion as a neural computation for encoding natural communication signals in bat inferior colliculus. *Journal of Neuroscience,* 31(46), 16529–16540.

Andoni, S., Li, N., & Pollak, G. D. (2007). Spectrotemporal receptive fields in the inferior colliculus revealing selectivity for spectral motion in conspecific vocalizations. *Journal of Neuroscience,* 27(18), 4882–4893.

Balaban, E. (1988). Bird song syntax: Learned intraspecific variation is meaningful. *Proceedings of the National Academy of Sciences of the USA,* 85(10), 3657–3660.

Behr, O., & von Helversen, O. (2004). Bat serenades—complex courtship songs of the sac-winged bat (*Saccopteryx bilineata*). *Behavioral Ecology and Sociobiology*, 56, 106–115.

Bohn, K. M., Schmidt-French, B., Ma, S. T., & Pollak, G. D. (2008). Syllable acoustics, temporal patterns, and call composition vary with behavioral context in Mexican free-tailed bats. *Journal of the Acoustical Society of America*, 124(3), 1838–1848.

Bohn, K. M., Schmidt-French, B., Schwartz, C., Smotherman, M., & Pollak, G. D. (2009). Versatility and stereotypy of free-tailed bat songs. *PLoS ONE*, *4*(8), e6746.

Boughman, J. W. (1998). Vocal learning by greater spear-nosed bats. *Proceedings of the Royal Society of London B: Biological Sciences*, 265(1392), 227–233.

Brand, A., Urban, R., & Grothe, B. (2000). Duration tuning in the mouse auditory midbrain. *Journal of Neurophysiology*, 84(4), 1790–1799.

Casseday, J. H., Ehrlich, D., & Covey, E. (1994). Neural tuning for sound duration: Role of inhibitory mechanisms in the inferior colliculus. *Science*, 264, 847–850.

Casseday, J. H., Covey, E., & Grothe, B. (1997). Neural selectivity and tuning for sinusoidal frequency modulations in the inferior colliculus of the big brown bat, *Eptesicus fuscus*. *Journal of Neurophysiology*, 77(3), 1595–1605.

Casseday, J. H., Fremouw, T., & Covey, E. (2002). The inferior colliculus: A hub for the central auditory system. In D. Oertel, A.N. Popper, & R. R. Fay (Eds.), *Integrative functions in the mammalian auditory pathway* (pp. 238–318). New York: Springer.

Catchpole, C. K., & Slater, P. J. B. (1995). *Bird song: Biological themes and variations*. Cambridge: Cambridge University Press.

Covey, E., & Casseday, J. H. (1999). Timing in the auditory system of the bat. *Annual Review of Physiology*, 61, 457–476.

David, S.V., Mesgarani, N., Fritz, J.B., & Shamma, S.A. (2009) Rapid synaptic depression explains nonlinear modulation of spectro-temporal tuning in primary auditory cortex by natural stimuli. *Journal of Neuroscience*, 29, 3374–3386.

Doupe, A. J., & Kuhl, P. K. (1999). Birdsong and human speech: Common themes and mechanisms. *Annual Review of Neuroscience*, 22, 567–631.

Fuzessery, Z. M., Razak, K. A., & Williams, A. J. (2011). Multiple mechanisms shape selectivity for FM sweep rate and direction in the pallid bat inferior colliculus and auditory cortex. *Journal of Comparative Physiology A: Neuroethology, Sensory, Neural, and Behavioral Physiology*, 197, 615–623.

Gill, P., Zhang, J., Woolley, S. M., Fremouw, T., & Theunissen, F. E. (2006). Sound representation methods for spectro-temporal receptive field estimation. *Journal of Computational Neuroscience*, 21, 5–20.

Gittelman, J. X., & Pollak, G. D. (2011). It's about time: How input timing is used and not used to create emergent properties in the auditory system. *Journal of Neuroscience*, 31, 2576–2583.

Gittelman, J. X., Li, N., & Pollak, G. D. (2009). Mechanisms underlying directional selectivity for frequency modulated sweeps in the inferior colliculus revealed by in vivo whole-cell recordings. *Journal of Neuroscience*, 29, 13030–13041.

Gittelman, J. X., Wang, L., Colburn, H. S., & Pollak, G. D. (2012). Inhibition shapes response selectivity in the inferior colliculus by gain modulation. *Frontiers of Neural Circuits*, *6*, 67.

Griffin, D. R. (1944). Echolocation by blind men and bats. *Science*, 100, 589–590.

Griffin, D. R. (1958). *Listening in the dark*. Ithaca, NY: Cornell University Press.

Holmstrom, L., Roberts, P. D., & Portfors, C. V. (2007). Responses to social vocalizations in the inferior colliculus of the mustached bat are influenced by secondary tuning curves. *Journal of Neurophysiology*, 98, 3461–3472.

Hurley, L. M., & Sullivan, M. R. (2012). From behavioral context to receptors: Serotonergic modulatory pathways in the IC. *Frontiers of Neural Circuits*, doi: 10.3389/fncir.2012.00058

Kanwal, J. S. (1999). Processing species-specific calls: Combination-sensitive neurons in an echolocating bat. In M. D. Hauser & M. Konishi (Eds.), *The design of animal communication* (pp. 133–156). Cambridge, MA: The MIT Press.

Kanwal, J. S., Matsumura, S., Ohlemiller, K., & Suga, N. (1994). Analysis of acoustic elements and syntax in communication sounds emitted by mustached bats. *Journal of the Acoustical Society of America*, 96, 1229–1254.

Klein, D. J., Depireux, D. A., Simon, J. Z., & Shamma, S. A. (2000). Robust spectrotemporal reverse correlation for the auditory system: Optimizing stimulus design. *Journal of Computational Neuroscience*, 9, 85–111.

Klug, A., Bauer, E. E., Hanson, J. T., Hurley, L., Meitzen, J., & Pollak, G. D. (2002). Response selectivity for species-specific calls in the inferior colliculus of Mexican free-tailed bats is generated by inhibition. *Journal of Neurophysiology*, 88, 1941–1954.

Knornschild, M., Behr, O., & von Helversen, O. (2006). Babbling behavior in the sac-winged bat, *Saccopteryx bilineata*. *Naturwissenschaften*, 93, 451–454.

Koch, U., & Grothe, B. (1998) GABAergic and glycinergic inhibition sharpens tuning for frequency modulations in the inferior colliculus of the big brown bat. *Journal of Neurophysiology*, 80, 71–82.

Kossl, M., & Vater, M. (1989). Noradrenaline enhances temporal auditory contrast and neuronal timing precision in the cochlear nucleus of the mustached bat. *Journal of Neuroscience*, 9, 4169–4178.

Kowalski, N., Depireux, D. A., & Shamma, S. A. (1996). Analysis of dynamic spectra in ferret primary auditory cortex. II. Prediction of unit responses to arbitrary dynamic spectra. *Journal of Neurophysiology*, 76, 3524–3534.

Kuo, R. I., & Wu, G. K. (2012). Generation of direction selectivity in the auditory system. *Neuron*, 73, 1016–1027.

Li, N., Gittelman, J. X., & Pollak, G. D. (2010). Intracellular recordings reveal novel features of neurons that code interaural intensity disparities in the inferior colliculus. *Journal of Neuroscience*, 30, 14573–14584.

Marler, P. (2004). Science and birdsong: The good old days. In P. Marler & H. Slabbekoorn (Eds.), *Nature's music: The science of birdsong*.Vol. 1. Amsterdam: Academic Press.

Mayko, Z. M., Roberts, P. D., & Portfors, V. (2012). Inhibition shapes selectivity for vocalizations in the inferior colliculus of awake mice. *Frontiers of Neural Circuits*, doi: 10.3389/fncir.2012.00073

Mittmann, D. H., & Wenstrup, J. J. (1995). Combination-sensitive neurons in the inferior colliculus. *Hearing Research*, 90, 185–191.

Motts, S. D., & Schofield, B. R. (2009). Sources of cholinergic input to the inferior colliculus. *Neuroscience*, 160, 103–114.

Neuweiler, G. (1990). Auditory adaptations for prey capture in echolocating bats. *Physiological Reviews*, 70, 615–641.
O'Neill, W. E., & Suga, N. (1979). Target range-sensitive neurons in the auditory cortex of the mustache bat. *Science*, 203, 69–73.
Payne, R. S., & McVay, S. (1971). Songs of humpback whales. *Science*, 173, 585–597.
Pollak, G. D. (2012). Circuits for processing dynamic interaural intensity disparities in the inferior colliculus. *Hearing Research*, 288, 47–57.
Pollak, G. D. & Casseday, J. H. (1986) *The neural basis of echolocation in bats*. New York: Springer-Verlag.
Pollak, G. D., Winer, J. A., & O'Neill, W. E. (1995). Perspectives on the functional organization of the mammalian auditory system: Why bats are good models. In A. N. Popper & R. R. Fay (Eds.), *Hearing by bats* (pp. 481–498). New York: Springer.
Pollak, G. D., Gittelman, J. X., Li, N., & Xie, R. (2011a). Inhibitory projections from the ventral nucleus of the lateral lemniscus and superior paraolivary nucleus create directional selectivity of frequency modulations in the inferior colliculus: A comparison of bats with other mammals. *Hearing Research*, 273, 134–144.
Pollak, G. D., Xie, R., Gittelman, J. X., Andoni, S., & Li, N. (2011b). The dominance of inhibition in the inferior colliculus. *Hearing Research*, 274, 27–39.
Portfors, C. V. (2004). Combination sensitivity and processing of communication calls in the inferior colliculus of the moustached bat *Pteronotus parnellii. Anais da Academia Brasoleira de Ciencias, 76*, 253–257.
Portfors, C. V., & Wenstrup, J. J. (1999). Delay-tuned neurons in the inferior colliculus of the mustached bat: Implications for analyses of target distance. *Journal of Neurophysiology*, 82, 1326–1338.
Portfors, C. V., & Wenstrup, J. J. (2002). Excitatory and facilitatory frequency response areas in the inferior colliculus of the mustached bat. *Hearing Research*, 168, 131–138.
Portfors, C. V., & Felix, II, R. A. (2005). Spectral integration in the inferior colliculus of the CBA/CaJ mouse. *Neuroscience*, 136, 1159–1170.
Portfors, C. V., Roberts, P. D., & Jonson, K. (2009). Over-representation of species-specific vocalizations in the awake mouse inferior colliculus. *Neuroscience*, 162, 486–500.
Razak, K. A., & Fuzessery, Z. M. (2006). Neural mechanisms underlying selectivity for the rate and direction of frequency-modulated sweeps in the auditory cortex of the pallid bat. *Journal of Neurophysiology,* 96, 1303–1319.
Razak, K. A., & Fuzessery, Z. M. (2009). GABA shapes selectivity for the rate and direction of frequency modulated sweeps in the auditory cortex. *Journal of Neurophysiology*, 102, 1366–1378.
Rubsamen, R., & Betz, M. (1986). Control of echolocation pulses by neurons of the nucleus ambiguus in the rufous horseshoe bat, *Rhinolophus rouxi*. I. Single unit recordings in the ventral motor nucleus of the laryngeal nerves in spontaneously vocalizing bats. *Journal of Comparative Physiology A: Neuroethology, Sensory, Neural, and Behavioral Physiology*, 159, 675–687.
Rust, N. C., Schwartz, O., Movshon, J. A., & Simoncelli, E. P. (2005). Spatiotemporal elements of macaque v1 receptive fields. *Neuron*, 46, 945–956.
Sanchez, J. T., Gans, D., & Wenstrup, J. J. (2007). Contribution of NMDA and AMPA receptors to temporal patterning of auditory responses in the inferior colliculus. *Journal of Neuroscience*, 27, 1954–1963.
Sanchez, J. T., Gans, D., & Wenstrup, J. J. (2008). Glycinergic "inhibition" mediates selective excitatory responses to combinations of sounds. *Journal of Neuroscience*, 28, 80–90.
Schneider, D. M., & Woolley, S. M. (2011). Extra-classical tuning predicts stimulus-dependent receptive fields in auditory neurons. *Journal of Neuroscience*, 31, 11867–11878.
Schuller, G. (1979). Coding of small sinusoidal frequency and amplitude modulations in the inferior colliculus of 'CF-FM' bat, *Rhinolophus ferrumequinum. Experimental Brain Research*, 34, 117–132.
Schwartz, C., Tressler, J., Keller, H., Vanzant, M., Ezell, S., & Smotherman, M. (2007). The tiny difference between foraging and communication buzzes uttered by the Mexican free-tailed bat,

Tadarida brasiliensis. Journal of Comparative Physiology A: Neuroethology, Sensory, Neural, and Behavioral Physiology, *193*, 853–863.

Simmons, J. A., Lavender, W. A., Lavender, B. A., Doroshow, C. A., Kiefer, S. W., Livingston, R., et al. (1974). Target structure and echo spectral discrimination by echolocating bats. *Science*, 186, 1130–1132.

Simmons, J. A., Howell, D. J., & Suga, N. (1975). Information content of bat sonar echoes. *American Scientist*, 63, 204–215.

Suga, N. (1965) Analysis of frequency-modulated sounds by auditory neurons of echolocating bats. *Journal of Physiology*, 179, 26–53.

Suga, N., & O'Neill, W.E. (1979) Neural axes representing target range in the auditory cortex of the mustached bat. *Science*, 206, 351–353.

Suga, N., O'Neill, W. E., Kujirai, K., & Manabe, T. (1983). Specificity of combination-sensitive neurons for processing of complex biosonar signals in auditory cortex of the mustached bat. *Journal of Neurophysiology*, 49, 1573–1626.

Suga, N., Niwa, H., Taniguchi, I., & Margoliash, D. (1987). The personalized auditory cortex of the mustached bat: Adaptation for echolocation. *Journal of Neurophysiology*, 58, 643–654.

Suga, N., Xiao, Z., Ma, X., & Ji, W. (2002). Plasticity and corticofugal modulation for hearing in adult animals. *Neuron*, 36, 9–18.

Thaler, L., Arnott, S.R., & Goodale, M.A. (2011) Neural correlates of natural human echolocation in early and late blind echolocation experts. *PLoS ONE*, 6 (5) e20162.

Weedman, D. L., & Ryugo, D. K. (1996). Projections from auditory cortex to the cochlear nucleus in rats: Synapses on granule cell dendrites. *Journal of Comparative Neurology*, 371, 311–324.

Wenstrup, J. J., & Grose, C. D. (1995). Inputs to combination-sensitive neurons in the medial geniculate body of the mustached bat: the missing fundamental. *Journal of Neuroscience*, 15, 4693–4711.

Wenstrup, J. J., Nataraj, K., & Sanchez, J. T. (2012). Mechanisms of spectral and temporal integration in the mustached bat inferior colliculus. *Frontiers of Neural Circuits*, doi:10.3389/fncir.2012.00075

Winer, J. A., Larue, D. T., & Pollak, G. D. (1995). GABA and glycine in the central auditory system of the mustache bat: Structural substrates for inhibitory neuronal organization. *Journal of Comparative Neurology*, 355, 317–353.

Winer, J. A., Larue, D. T., Diehl, J. J., & Hefti, B. J. (1998). Auditory cortical projections to the cat inferior colliculus. *Journal of Comparative Neurology*, 400, 147–174.

Winer, J. A., Chernock, M. L., Larue, D. T., & Cheung, S. W. (2002). Descending projections to the inferior colliculus from the posterior thalamus and the auditory cortex in rat, cat, and monkey. *Hearing Research*, 168, 181–195.

Woolley, S. M., Fremouw, T. E., Hsu, A., & Theunissen, F. E. (2005). Tuning for spectro-temporal modulations as a mechanism for auditory discrimination of natural sounds. *Nature Neuroscience*, 8, 1371–1379.

Xiao, Z., & Suga, N. (2002). Modulation of cochlear hair cells by the auditory cortex in the mustached bat. *Nature Neuroscience*, 5, 57–63.

Xie, R., Meitzen, J., & Pollak, G. D. (2005). Differing roles of inhibition in hierarchical processing of species-specific calls in auditory brainstem nuclei. *Journal of Neurophysiology*, 94, 4019–4037.

Xie, R., Gittelman, J. X., Li, N., & Pollak, G. D. (2008). Whole cell recordings of intrinsic properties and sound-evoked responses from the inferior colliculus. *Neuroscience*, 154, 245–256.

Zhang, L. I., Tan, A. Y., Schreiner, C. E., & Merzenich, M. M. (2003). Topography and synaptic shaping of direction selectivity in primary auditory cortex. *Nature*, 424, 201–205.

Chapter 9
Behavioral and Physiological Bases for Doppler Shift Compensation by Echolocating Bats

Shizuko Hiryu, Emanuel C. Mora, and Hiroshi Riquimaroux

Keywords Acoustic glints • CF-FM pulse • CF_2 • DSCF area • Echo intensity compensation • Foveal neurons • Reference frequency • Resting frequency (F_{rest})

9.1 Introduction

Unlike the large majority of echolocating bats studied to date, some bats (Rhinolophidae, Hipposideridae, and a few species of Mormoopidae) are high duty cycle (HDC) echolocators. They produce long calls dominated by a single frequency (constant frequency, CF) and separated by brief periods of silence. They separate pulse and echo in frequency while virtually all other echolocators, including most bats, separate pulse and echo in time (low duty cycle, LDC) (see review in Fenton et al. 2012). The species of bats using this approach to echolocation were initially referred to as CF-FM bats, reflecting the general structure of the call. To date only one HDC species, the East Asian tailless leaf-nosed bat (*Coelops frithii*, Hipposideridae) has been shown to use a low duty cycle echolocation strategy even when approaching fluttering targets (Ho et al. 2013).

HDC echolocation depends largely on Doppler shift compensation (DSC). HDC bats compensate for flight-induced Doppler shifts in echoes by adjusting the CFs in their outgoing calls, thereby stabilizing the CFs in returning echoes. By DSC, the echo CF can be maintained within the range of the "acoustic fovea," which allows fine-frequency analysis in the auditory system of HDC bats.

S. Hiryu (✉) • H. Riquimaroux
Faculty of Life and Medical Sciences, Doshisha University,
Kyotanabe 610-0321, Japan
e-mail: shiryu@mail.doshisha.ac.jp; hrikimar@mail.doshisha.ac.jp

E.C. Mora
Research Group in Bioacoustics and Neuroethology, Havana University,
Street 25 No. 455 between J & I, Vedado, Havana, CP 10400, Cuba
e-mail: emanuel_mora@yahoo.com

M.B. Fenton et al. (eds.), *Bat Bioacoustics*, Springer Handbook of Auditory Research 54, DOI 10.1007/978-1-4939-3527-7_9

9.2 General Principles of Doppler Shift Compensation

9.2.1 Doppler Effect

Figure 9.1a illustrates a sound source producing a constant frequency f_s (Hz) while approaching a stationary observer at a constant moving velocity v_s (m/s). The emitted sound waves propagate at a rate of c (m) per second (sound velocity c) and are accompanied by the movement of v_s (m) of the sound source itself in the same direction per second; thus, the waves emitted from the sound source per second, f_s, are distributed along a distance of $c - v_s$ (m). Therefore, the wavelength, λ_o, at the observer is given by

$$\lambda_o = \frac{c - v_s}{f_s} \tag{9.1}$$

The observed frequency at the observer f_o is

$$f_o = \frac{c}{\lambda_o} = f_s \frac{c}{c - v_s} > f_s \tag{9.2}$$

As a result, the observer will receive the sound waves at a frequency higher than the original f_s emitted by the sound source. In contrast, when the sound source moves away from the observer, the frequency at observer f_o is lower than f_s

$$f_o = f_s \frac{c}{c + v_s} < f_s \tag{9.3}$$

In a case of an observer with constant moving velocity, v_o, approaching a stationary sound source emitting a f_s (wavelength λ_s) (Figure 9.1b), the observed frequency by the observer f_o is

$$f_o = \frac{c + v_o}{\lambda_s} = f_s \frac{c + v_o}{c} > f_s \tag{9.4}$$

Figure 9.1c depicts a situation where a flying bat is emitting a sound with f_s and receives echoes from a large stationary object located in front of the bat. In this case, Eq. (9.2) and (9.4) should be combined. The frequency of echoes, f_e, received at the bat is given by

$$f_e = f_s \frac{c + v_b}{c - v_b} \tag{9.5}$$

where v_b is the moving velocity of the bat. When the object is also moving, v_b can be replaced by the relative moving velocity between the bat and the object. Because

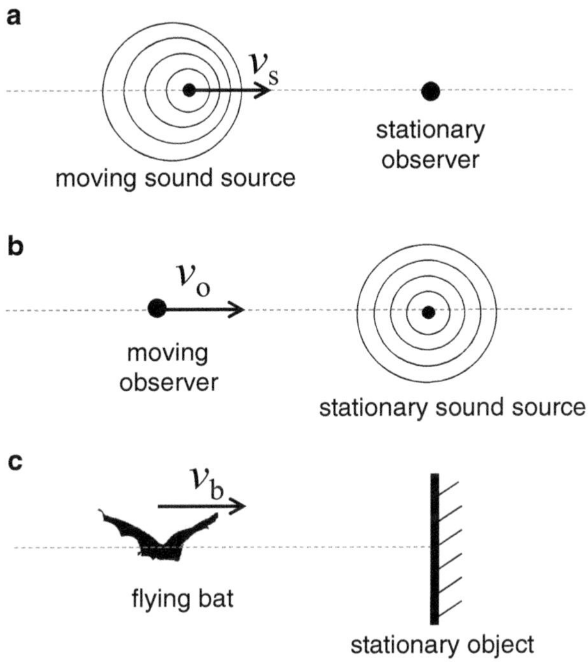

Fig. 9.1 Conceptual diagrams of the Doppler effect. (**a**) A moving sound source approaches a stationary observer. (**b**) A moving observer approaches a stationary sound source. (**c**) A flying bat approaches a stationary object

v_b can be considered to be small compared to the sound velocity c, the Doppler shift Δf can be estimated by

$$\Delta f = f_e - f_s = f_s \frac{2v_b}{c - v_b} \simeq f_s \frac{2v_b}{c} \tag{9.6}$$

Equation (9.6) indicates that the relative moving velocity can be obtained from the observed Doppler shift, which can be applied to various measurement techniques of target velocity in the field of engineering.

9.2.2 *Ecology of Doppler Shift Compensation*

9.2.2.1 High Duty Cycle Echolocation in Bats

The approximately 165 species of HDC bats use calls with a dominant, long CF component that begins and/or ends with a brief frequency-modulated (FM) component, referred to as the CF-FM pulse. The greater horseshoe bat, *Rhinolophus ferrumequinum* (10–50 ms), and Parnell's mustached bat, *Pteronotus parnellii* (7–30 ms), use rather long CF component pulses, whereas hipposiderid bats emit shorter pulses (5–10 ms). HDC bats generally produce calls whose signal durations are ≥25 % of the time between the onset of successive calls (Fenton et al. 2012).

Therefore, HDC echolocation results in overlap of the returning echo with the emitted pulse. To avoid masking of echoes by outgoing long-duration calls, HDC bats use DSC to separate the dominant CF components of the call and the echo in frequency. In contrast, LDC bats produce short-duration calls with long intervals between calls, allowing the bats to avoid forward masking by separating pulse and echo in time.

The CF-FM pulses emitted by HDC bats typically consist of harmonics in which the second harmonic has the highest energy because the fundamental component is attenuated by vocal tract filtering (Hartley and Suthers 1988, 1990). The frequency of the second CF component (CF_2) of calls produced when at rest (resting frequency, F_{rest}, e.g., when the bat is roosting and not compensating for Doppler-shifted echoes) differs among subspecies and among individuals (e.g., 81–85 kHz for *R. ferrumequinum*; 59–64 kHz for *P. parnellii*). In addition to physical constitution, sex, age, geography, and morphometrics, the F_{rest} of an adult HDC bat also shows a slight but continual individual drift over several months or seasons or throughout its lifetime (Jones and Ransome 1993; Hiryu et al. 2006).

Among mormoopid bats, only *Pteronotus parnellii*, the Paraguayan mustached bat (*P. paraguayensis*), and the Mesoamerican mustached bat (*P. mesoamericanus*) use HDC echolocation, although Wagner's moustached bat *P. personatus* (but not other mormoopids) performs DSC (Smotherman and Guillén-Servent 2008). The lesser bulldog bat (*Noctilio albiventris*) and the greater bulldog bat (*N. leporinus*) are LDC echolocators that sometimes use short, narrowband (quasi-CF) pulses, and the latter is considered to partially exhibit DSC (Wenstrup and Suthers 1984). For details, see Section 9.5 of this chapter.

9.2.2.2 Discovery of Doppler Shift Compensation

In 1968, Schnitzler reported that *Rhinolophus ferrumequinum* lowered the CF_2 of the emitted pulse (pulse CF_2) when flying from one place to another in a flight chamber. The bats maintained the CF_2 of returning echoes (echo CF_2) around the F_{rest}. The Doppler shifts induced by the bat's flight speed were compensated by lowering the pulse frequency. Schnitzler (1968) called this Doppler shift compensation (Figure 9.2). DSC was also confirmed in flying *Pteronotus parnellii* (then called *Chilonycteris rubiginosa*) (Schnitzler 1970). Schnitzler (1973) later demonstrated that *R. ferrumequinum* flying in a wind tunnel compensated for the Doppler shifts based on the ground speed, not the airspeed. Bats flying in a He-O_2 gas mixture exhibited DSC based on the change in sound speed manipulated by the mixture rate of the gas (Schnitzler 1973). These findings demonstrated that the bats use feedback control for DSC involving the echo frequency, triggering a change in the pulse CF_2 so that echo CF_2 is maintained at a constant value.

In flight experiments, the pulse and echo frequencies determined from remote recordings by a stationary microphone were corrected to eliminate flight-induced Doppler shifts. This required measurement of the bat's flight speed with appropriate accuracy. Later experiments used a pendulum on which a stationary bat was

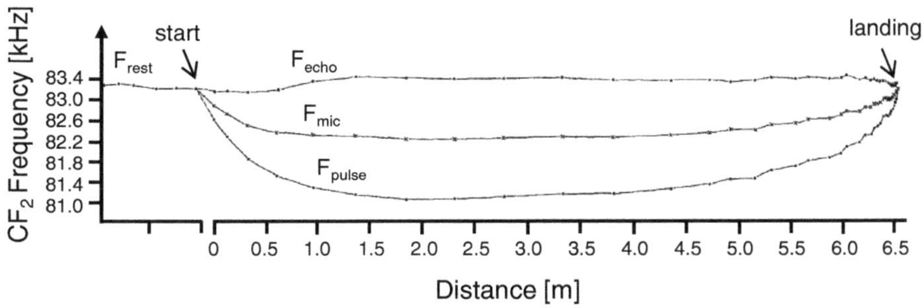

Fig. 9.2 Doppler shift compensation of a bat during flight in a flight chamber. Before the flight, the CF_2 of the emitted pulse is maintained at a constant (F_{rest}). During flight, the bat lowers the CF_2 of the emitted pulse (F_{pulse}) so that the of the returning echo (F_{echo}) remains constant at about F_{rest}. F_{mic} is the CF_2 of the emitted pulse detected by the microphone. F_{pulse} and F_{echo} are determined based on the values of F_{mic} and the flight speed of the bat measured by a photoelectric detector (Adapted from Schnitzler 1968)

mounted and swung toward a large target. Henson et al. (1980) first demonstrated that *P. parnellii* held on a pendulum lowered its pulse CF_2 to keep the echo CF_2 within a narrow frequency band near the frequency with the lowest threshold in a cochlear microphonic (CM) audiogram.

To evaluate the detailed responses of bats over a wide range of positive and negative Doppler shifts, the emitted pulses were electronically shifted in frequency so that artificial echoes could be played back to a stationary bat in real time (Schuller et al. 1974; Simmons 1974). Playback experiments offer substantial advantages for the quantitative analysis of DSC because arbitrary target motion can generate Doppler shifts. Some relevant findings are described in detail in Section 9.4.

9.2.2.3 Discovery of the Auditory Fovea

Schnitzler (1968) had pointed out that bats maintained the echo CF_2 at a frequency approximately 150 Hz higher than the F_{rest}. Schuller et al. (1974) referred to the echo CF_2 maintained by DSC as the reference frequency. The compensated frequency difference between the F_{rest} and the reference frequency varies among bat species and among studies, but it is usually 150–200 Hz above the F_{rest}. The compensation offset is considered to be the DSC threshold, the point at which bats begin to exhibit DSC when the change in the echo CF_2 exceeds the DSC threshold (Schuller et al. 1974; Smotherman and Metzner 2003b).

The most remarkable physiological features of HDC bats are specializations of the auditory system for fine-frequency analysis of the CF component dominating their echolocation sounds, particularly the CF_2 used in DSC. As an analogy to the fovea in the visual system (maximal visual sharpness due to a high concentration of cones in the retina), Schuller and Pollak (1979) called this specialization in HDC bats the "auditory fovea" (or acoustic fovea). This specialization begins in the frequency place map of the basilar membrane (BM) in the cochlea where there are widely expanded representations around the echo CF_2 (for a review see Kössl and Vater 1995), and an over-representation of sharply tuned neurons around the echo CF_2 occurs in all stations along the auditory pathway from the peripheral to the central auditory system. Thus, the majority of neurons in the Doppler-shifted CF processing area (DSCF area) are tuned to frequencies between 61 and 63 kHz, corresponding to the frequency range of the echo CF_2 modified by DSC in *P. parnellii* (for reviews see Suga 1984, 1990).

Schnitzler et al. (1976) demonstrated that the threshold minimum of the summated neural potentials (the N1-on response audiogram) in *R. ferrumequinum* was tuned to the individual reference frequency and that the individual reference frequency was 30–500 Hz higher than the F_{rest}. In addition, the sharp minima of CM audiograms in rufous horseshoe bat *Rhinolophus rouxi* and *P. parnellii* are 200 Hz above the F_{rest} (Henson et al. 1980). The specialized frequency in the auditory receiver of HDC bats corresponds to the reference frequency at which the echo CF_2 is maintained by DSC. Anatomical and neurophysiological specializations found in HDC bats are described in detail in Section 9.3.

9.2.2.4 Impact of Doppler Shift Compensation on High Duty Cycle Echolocation

HDC echolocating bats are thought to hunt in clutter where their ability to detect fluttering targets permits them to detect and track a flying insect. A long CF component transmits fluttering information of insect prey because both periodic fluctuations of amplitude and frequency (acoustical glints) are encoded in the echo CF component (Figure 9.3). From an acoustical perspective, these acoustic glints are easily detected from background echoes in the cluttered environment. HDC bats use this acoustical information about fluttering insects while foraging. Bats that use HDC echolocation have better flutter detection ability than LDC bats (Lazure and Fenton 2011).

What is the impact of DSC on echolocation by the HDC bats? The auditory receiver of HDC bats is highly sensitive to frequencies around the biologically most important frequency range, the reference frequency, where the echo CF_2 is maintained by DSC (see Section 9.3). In contrast, in an echolocation pulse the CF_2 is lower than the frequency of the sharp threshold minima in the audiograms. Thus the auditory receiver of HDC bats is sensitive to the compensated echo CF_2 but relatively insensitive to the pulse CF_2, suggesting that DSC reduces masking of weak echoes by intense emitted pulses. Furthermore, neurons highly tuned to the best frequency (BF) in the auditory fovea facilitate encoding information about fluttering insect prey. HDC bats increase the duration of a CF component or repetition rate when exposed to a fluttering target, further increasing the duty cycle so that they can repeatedly obtain the fluttering information within single echoes or over several successive echoes. DSC is a unique and important behavioral and physiological adaptation that supports flutter detection as a foraging strategy in

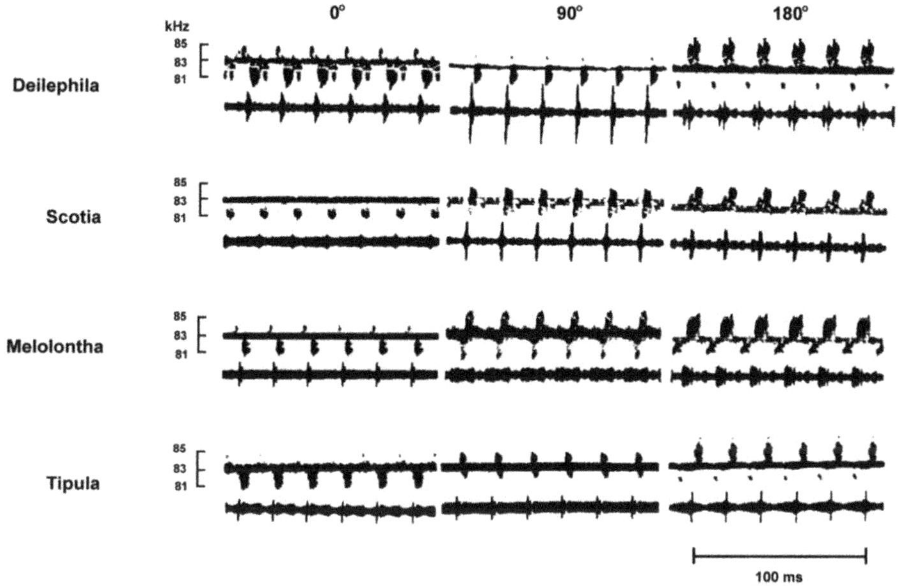

Fig. 9.3 Spectrograms (*above*) and oscillograms (*below*) of acoustic frequency/amplitude glints generated by four different fluttering insects ensonified with an 83 kHz constant-frequency tone, which represents the main component of the echolocation call of *Rhinolophus ferrumequinum*. Each insect is facing in three different directions: 0°, 90°, and 180°. Notice that all insects are fluttering at 50 Hz, but glint structure and echo structure between the glints are species specific and orientation specific. Deilephila: sphingid moth, *Deilephila elpenor*, Lepidoptera; Scotia: noctuid moth, *Scotia exclamations*, Lepidoptera; Melolontha: scarabid beetle, *Melolontha melolontha*, Coleoptera; Tipula: crane fly, *Tipula oleracea*, Tipulidae, Diptera (Adapted from von der Emde and Schnitzler 1990)

HDC bats. Details of the neurophysiological aspects related to the processing of flutter information appear in the next Sec. 9.3.

9.3 Adaptations for Doppler Shift Compensation in the Auditory Receiver

The ears and the auditory pathway of HDC bats are an integral part of the echolocation system for bats that hunt flying insects in highly cluttered spaces. To process flutter information, the auditory receiver of HDC bats is strikingly specialized in the CF range that dominates the echolocation calls. The first evidence of such specializations was observed in the audiograms of HDC bats (Grinnell 1967).

Behavioral audiograms or neuronal audiograms (measured from single neurons of the auditory nerve or the auditory brain stem) are arguably the most reliable measurements of the threshold of hearing. In these audiograms, sharply tuned

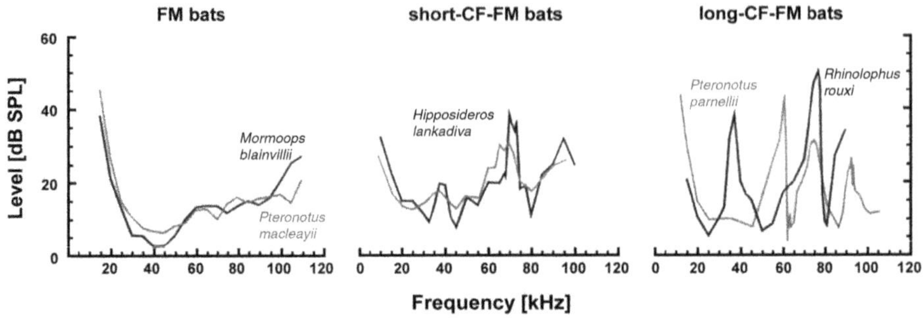

Fig. 9.4 Mechanical audiogram of the cochlea in five bat species measured with Distortion Products Otoacoustic Emissions (DPOAEs). The DPOAE threshold curves represent the level of the f1 tone necessary to elicit a 2f1–f2 distortion product of −10 dB SPL; the level of f2 was 10 dB below that of f1 (for details see Kössl 1994; Kössl et al. 1999). The FM bat *M. blainvillii* and the short CF-FM bat *P. macleayii* do not employ DSC. *H. lankadiva* is a short CF-FM bat with incomplete DSC; *P. parnellii* and *R. rouxi* are CF-FM bats with DSC. Note the narrow threshold minimum and the distinct maximum a few kHz below in the threshold curves of the CF bats. Maxima and minima in the CF_2 ranges of the DPOAE audiogram in the hipposiderid are less pronounced than in the rhinolophids and the moustached bat. The audiograms in *M. blainvillii* and *P. macleayii* are relatively smooth (Adapted from Foeller and Kössl 2000)

threshold minima and contrasting response maxima in the CF_2 region were found for *P. parnellii* (Kössl and Vater 1996), for rhinolophids (Neuweiler et al. 1971; Long and Schnitzler 1975), and for hipposiderids (Neuweiler et al. 1984) (Figure 9.4). Maxima and minima in the CF_2 ranges of neuronal audiograms in the hipposiderids, Schneider's round-nosed bat (*Hipposideros speoris*) and bicoloured roundleaf bat (*H. bicolor*), however, are less pronounced than in rhinolophids and *P. parnellii* (Schuller 1980; Rübsamen et al. 1988). The sharp tuning of behavioral and neuronal audiograms in HDC bats is already apparent at the level of the CM and the N1 audiograms, and must be based on properties of the cochlea.

9.3.1 Auditory Fovea in the Cochlea of High Duty Cycle Echolocating Bats

Although the bat cochlea follows the common mammalian "bauplan" in structure and function, laryngeal echolocating bats have cochleae that are unusually large relative to body weight, in keeping with the importance of hearing for Chiroptera (Davies et al. 2013). The cochleae of HDC bats are larger relative to skull size than those of LDC bats (Habersetzer and Storch 1992). Among *Pteronotus* (Mormoopidae), the LDC echolocators have smaller cochleae than the HDC species. Among the HDC CF-FM bats (rhinolophids, hipposiderids, and *P. parnellii*) the smallest cochleae are found in hipposiderids, which use the shortest CF components in their echolocation (Fenton et al. 2011).

Among HDC bats, the morphology of the basilar membrane (BM) shows two common features: (1) abrupt changes or discontinuities in thickness and width that might play a role in enhancing tuning in a narrow-frequency band and (2) expanded areas with very little change in morphology and probably a very slight stiffness gradient, thus leading to expanded frequency mapping (Kössl and Vater 1995). These two special features of the BM both filter and largely over-represent the biologically important frequency range for flutter detection, the CF_2 component in the stabilized echo, the reference frequency.

Cochlear frequency maps in HDC bats clearly show that a narrow-frequency range around the reference frequency is expanded to about 30 % of the BM length (Kössl and Vater 1995). The area on the BM representing the reference frequency has the highest afferent innervation. The abrupt thickening of the BM could provide a reflection zone for incoming waves, allowing standing waves to be set up in the region between the BM discontinuity and the stapes, which would then implement a passive and highly tuned resonator. The resonator would ensure the high sensitivity and sharp tuning apical to the BM discontinuity and into the reference frequency region (reviewed in Kössl and Vater 1995; Neuweiler 2003).

In addition to low threshold and sharp tuning, an active cellular component also may account for spontaneous otoacoustic emissions in the region of the reference frequency. Still under investigation, the cellular force generator (electromotility) that amplifies the sound energy of the CF echo could be established by fast movements of the bodies or stereocilia of outer hair cells (OHCs) (review in Kössl and Vater 1995). The electromotility of OHCs has been found up to at least 79 kHz (Frank et al. 1999). Interestingly, a number of the observed macro and micromechanical properties of the cochlear fovea differ among species of HDC bats (Kössl and Vater 1995; Vater 1998).

9.3.2 *Auditory Fovea in the Higher Auditory Nuclei*

Foveal areas with overrepresented neurons with best frequencies near the reference frequency characterize the entire auditory system of HDC bats. These correspond with the cochlear frequency expansion in the cochlear nucleus of HDC bats, where about half of all recorded auditory neurons are tuned to frequencies around the species-specific CF_2 component. Sharp tuning of foveal neurons is evidenced in extremely narrow tuning curves with a Q_{10}dB value (best frequency divided by bandwidth of the tuning curve at 10 dB above minimal threshold) well above 20 and often as high as 400 (Covey and Casseday 1995). The frequency ranges expanded at the cochlea and cochlear nucleus are further expanded at the level of the superior olivary complex (SOC) and the lateral lemniscus (LL).

The inferior colliculus (IC) of horseshoe bats and *P. parnellii* shows the typical tonotopic organization but with a distorted general arrangement of the isofrequency layers due to the overrepresentation of the CF_2 range (Pollak and Park 1995). In the hipposiderid bat *H. speoris*, there is a less developed foveal area in the IC (Rübsamen et al. 1988; Fu et al. 2010).

The overrepresentation of a very narrow frequency band around CF_2 also characterizes the tonotopically organized primary auditory cortex (AC) (O'Neill 1995). In *P. parnellii*, one-third of the tonotopic region within the primary AC represents frequencies between 60 and 63 kHz. In *R. ferrumequinum*, there is also significant magnification of the CF_2 representation in the primary AC relative to the cochlear representation (Ostwald 1984). The foveal area in the primary AC is "personalized" in that the expanded frequencies vary among individual F_{rest} and reference frequency in *P. parnellii* (Suga et al. 1987). This is called the "Doppler-shifted CF" area.

9.3.3 The Processing of Flutter Information in the Auditory Pathway

Both in the IC and in the AC of *R. ferrumequinum*, neurons processing information about fluttering accurately encode natural species-specific glint patterns (Schuller 1984; Ostwald 1988), perhaps allowing HDC bats to precisely identify prey. Because natural echoes are complex, in the laboratory sinusoidally amplitude-modulated (SAM) and frequency-modulated (SFM) stimuli were used to simulate flutter information from flying insects. In studies with *P. parnellii* and with horseshoe bats, foveal neurons show response selectivity to specific parameters of the modulating waveform, such as carrier frequency, modulation rate, modulation depth, and intensity. Already in the peripheral auditory system of HDC bats, foveal cochlear neurons show clear phase-locked responses to frequency modulations as small as ±0.01 to ±0.02 % of the carrier frequency.

Modulation rate reflects the wingbeat frequencies of different insects. Unlike peripheral neurons, filter neurons in the central auditory pathway respond preferentially to a limited range of modulation frequencies. In AC neurons, synchronization occurred up to 100–150 Hz with the range of maximal activity between 40 and 70 Hz (Ostwald 1988). The activity of most filter neurons in the higher auditory centers covers the wingbeat frequencies of the insects that HDC bats perhaps preferred as prey. There are also high sensitivity and selectivity for specific ranges of amplitude modulations in the foveal areas of the central auditory nuclei. Neurons sensitive to small amplitude variations of 10–20 % are able to encode the fine structure of the echoes created by wingbeat patterns (Vater 1982; Reimer 1987).

Many SFM-sensitive foveal units exhibit the most vigorous response and sharpest locking at low intensities. They reduce or lose their modulation encoding capabilities for stimuli with sound pressure levels above 50–70 dB SPL. This may be an adaptation for detecting faint echoes (Figure 9.5) (Pollak and Schuller 1981). In the auditory cortex, the DSCF area of *P. parnellii* and the CF_2 area of *R. ferrumequinum* are populated with flutter processing neurons segregated by their best amplitudes. This may support insect discrimination tasks according to echo strength. The influence of sound pressure level on the processing of flutter information is of relevance also in light of echo intensity compensation. HDC bats maintain the intensity of the echoes returning from approaching targets at an optimal range (Kobler et al. 1985; Hiryu et al. 2008).

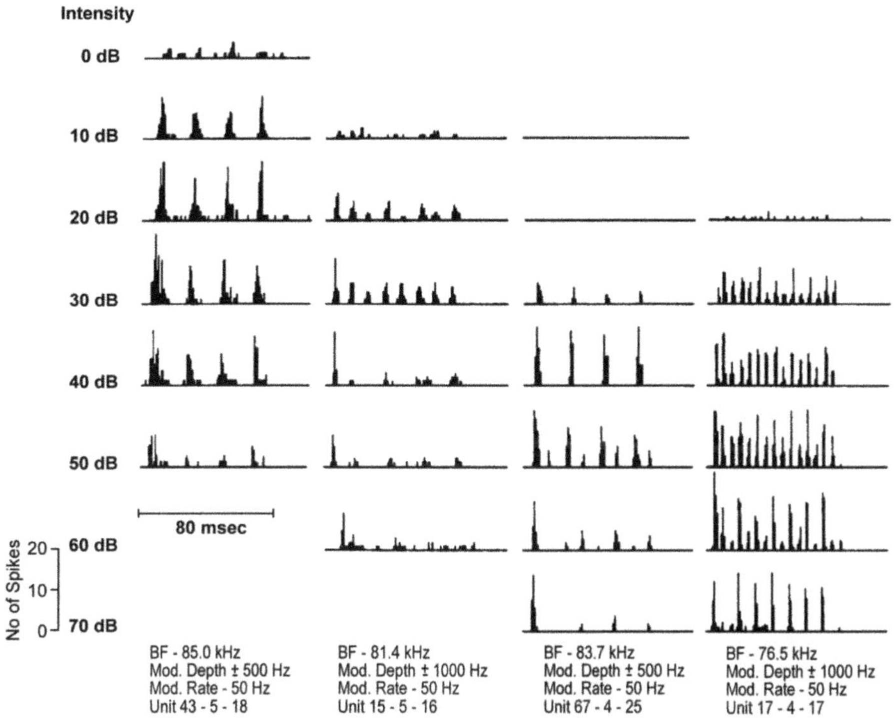

Fig. 9.5 Effect of stimulus intensity on the locked discharges to sinusoidal frequency-modulated signals in four neurons of the inferior colliculus of *Rhinolophus ferrumequinum*. The neuron on the right shows tightly locked firings at all intensities above threshold; the three other units each locked best to only a small range of low intensities. Stimulus frequency was set at the neuron's best frequency (BF) as indicated. All signals were 80 ms long (Adapted from Pollak and Schuller 1981)

Combination-sensitive neurons in nontonotopic areas show response selectivity to flutter information in the CF_2 range. CF_1/CF_2 sensitive neurons, for example, are sensitive to small periodic modulations in the CF_2 echo-frequency range if there is also stimulation in the CF_1 range (Suga et al. 1983).

Foveal neurons all along the auditory pathway show preferences for selective ranges of frequency and intensity as well as modulation depth and rate. These foveal neurons may play a significant role in the dynamic neural representation of target attributes due to changes in position, orientation, and speed of either the bat or its prey.

9.4 Ethology of Doppler Shift Compensation

DSC is the result of behaviors as well as specialized anatomical and neurophysiological functions. Bats using HDC echolocation use DSC primarily to detect fluttering target prey. In this section, additional significant features of DSC are discussed in the context of ethology.

9.4.1 Acoustical Measurements of Doppler Shift Compensation Behaviors

Schuller et al. (1974) pointed out that the observed maximum compensation for positive Doppler shift in playback echoes ranges from 4,400 to 6,000 Hz, corresponding to the Doppler shift induced by flight at approximately 9 m/s in *R. ferrumequinum* (Schnitzler 1973). Later, Schnitzler (1978) reported that *R. ferrumequinum* compensated for positive Doppler shift of up to 8,000 Hz. Playback experiments have revealed that bats do not respond to negative Doppler shifts (downward frequency shifts) in echoes that would occur when a simulated target moves away from the bat (Figure 9.6) (Gaioni et al. 1990). This significant difference in behavioral responses between positive and negative Doppler shifts suggests that DSC is more important when bats approach targets (prey) than when the distance between the bat and the target is increasing.

Metzner et al. (2002) used playbacks to demonstrate that *R. ferrumequinum* increased the pulse CF_2 response to negative Doppler shifts, although the magnitude of compensation was small compared to the response to positive Doppler shifts. Negative Doppler shifts may also occur when the flying bats slow down. Then the echo CF_2 will fall below the reference frequency, requiring the bats to increase the pulse CF_2.

In *R. ferrumequinum*, the returning echoes always overlap with outgoing pulses because of the long pulse duration (Tian and Schnitzler 1997). Pulse-echo overlap is a prerequisite for DSC (Schuller 1974, 1977). In contrast, hipposiderids (the

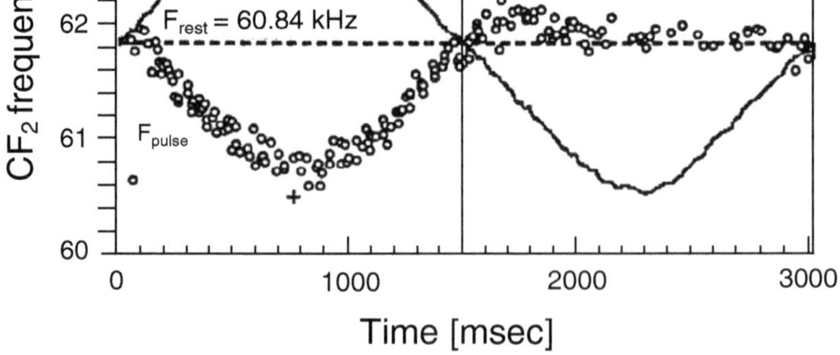

Fig. 9.6 Changes in the pulse CF_2 during pendulum experiments in *Pteronotus parnellii*. The solid line indicates estimated Doppler shift of the echo CF_2 if the bats do not exhibit DSC. The bats lower the pulse CF_2 on the forward swing of the pendulum but do not compensate echoes on the backward swing (Adapted from Gaioni et al. 1990)

Taiwanese leaf-nosed bat, *Hipposideros terasensis*, and trident leaf-nosed bat, *Asellia tridens*) do not exhibit this overlap because they compensate for flight-induced frequency shifts in echoes by emitting pulses of short duration (Gustafson and Schnitzler 1979; Hiryu et al. 2005), suggesting a fundamental difference between rhinolophids and hipposiderids.

Hipposiderids are considered to have lower DSC abilities than horseshoe bats, and *P. parnellii*, *Hipposideros speoris*, and *Hipposideros bicolor* showed incomplete DSC, decreasing the pulse CF_2 by only about half of the full Doppler shift (Habersetzer et al. 1984). However, some hipposiderid bats compensate for Doppler shifts in echoes during free flight (Gustafson and Schnitzler 1979; Hiryu et al. 2005). This suggests that certain experimental conditions, such as being retained on a swinging pendulum with an unnatural, greater rate of change in echo frequency, may result in lower DSC than under free-flight conditions (Schnitzler and Denzinger 2011). In fact, when the rate of change in the frequency of playback echoes is very fast, bats cannot keep pace with it and DSC becomes incomplete (Smotherman and Metzner 2003a). In sum, compensation performance varies among bat species and among studies, which may in part be due to unnatural conditions of some experimental procedures to induce Doppler shifts without flight.

9.4.2 Telemetry Recordings of Bats During Flight

For precise acoustical measurements of the calls of flying bats, microphones should be attached to the bats themselves. One option is a telemetry device consisting of a microphone, transmitter, and battery that is light enough to be mounted on a bat's head or body. Henson and his colleagues developed a telemetry device for *P. parnellii* (~11 g body mass) so that the echolocation signals emitted by flying bats were recorded without correcting for flight-induced Doppler shift. Their recordings demonstrated that flying bats lowered their pulse CF_2; the echo CF_2 was estimated to be maintained within 150 Hz around the reference frequency (Lancaster et al. 1992).

Riquimaroux and Watanabe (2000) developed an onboard telemetry microphone (Telemike), and confirmed DSC in free-flying *Hipposideros terasensis* (Hiryu et al. 2005). In another study, the CF_2 of returning echoes was observed directly and was compensated by DSC in flying Japanese horseshoe bats *R. ferrumequinum nippon* (Hiryu et al. 2008) (Figure 9.7). In that study, the echo CF_2 was maintained at the reference frequency, which is approximately 60 Hz higher than the F_{rest} of each individual (with the standard deviation of 80–90 Hz). This indicates that flying bats compensate for the echo CF_2 with an accuracy of regulation equivalent to bats at rest (Hiryu et al. 2008).

In addition to acoustical measurements, Henson and his colleagues used telemetry to record CM potentials from flying bats (Henson et al. 1982, 1987). Interestingly, telemetry-recorded CM responses of echoes are usually greater than pulse-evoked CM responses, although the pulses are considerably louder than the returning echoes (Henson 1967; Henson et al. 1982). In one study, when tethered

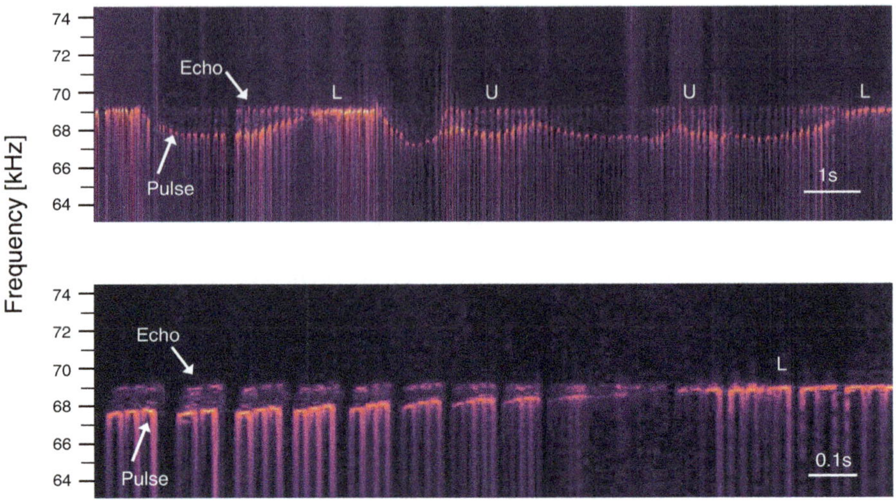

Fig. 9.7 Echolocation signals of *Rhinolophus ferrumequinum nippon* recorded by the telemetry microphone (Telemike) mounted on the back of the bat during free flight in an experimental flight chamber. Spectrograms show the sequence of pulse-echo pairs during U-turn (U) and landing (L) events (*top*) and the magnified view of recorded sounds before landing (*bottom*). The bat dynamically changes the pulse CF_2 while the echo CF_2 remains relatively stable (Adapted from Hiryu et al. 2008)

fluttering moths were presented to bats restrained on a swinging pendulum, the recorded sounds of echoes from insects did not show a prominent amplitude pattern, whereas the CM potentials were often prominent with acoustic glints caused by the fluttering moths (Henson et al. 1987). Such "amplified" echo-evoked CM potentials may indicate specialization of the auditory periphery in HDC bats, which likely plays an important role in detecting weak echoes from fluttering insects. An important challenge for future biosonar research is to combine telemetry recording of physiological data with acoustical measurements.

9.4.3 Flutter Detection by Doppler Shift Compensation

Some horseshoe bats hunt from perches in the wild and make short flights out to intercept prey. Before and after takeoff, the bats extend the duration of the CF component of their emitted pulse. The increase in pulse duration has also been observed in HDC bats at the beginning of the approach phase in the capture sequence for a fluttering moth in laboratory recordings (Mantani et al. 2012). HDC bats extend the pulse duration to increase the number of temporal repetitions of fluttering information.

HDC bats maintain the echo CF_2 at the reference frequency range within their own acoustic fovea. Hence, they perform DSC for echoes from their prey. However,

Trappe and Schnitzler (1982) reported that *R. ferrumequinum* performs DSC not on insect echoes but rather on echoes from stationary objects in the surroundings. Telemetry recordings have also provided direct evidence for this (Figure 9.8a) (Mantani et al. 2012). In this scenario, the echo from a moving target would be above or below the reference frequency. Such differences could be used by the bat to perceive the direction of the moth's flight, either toward or away from the bat.

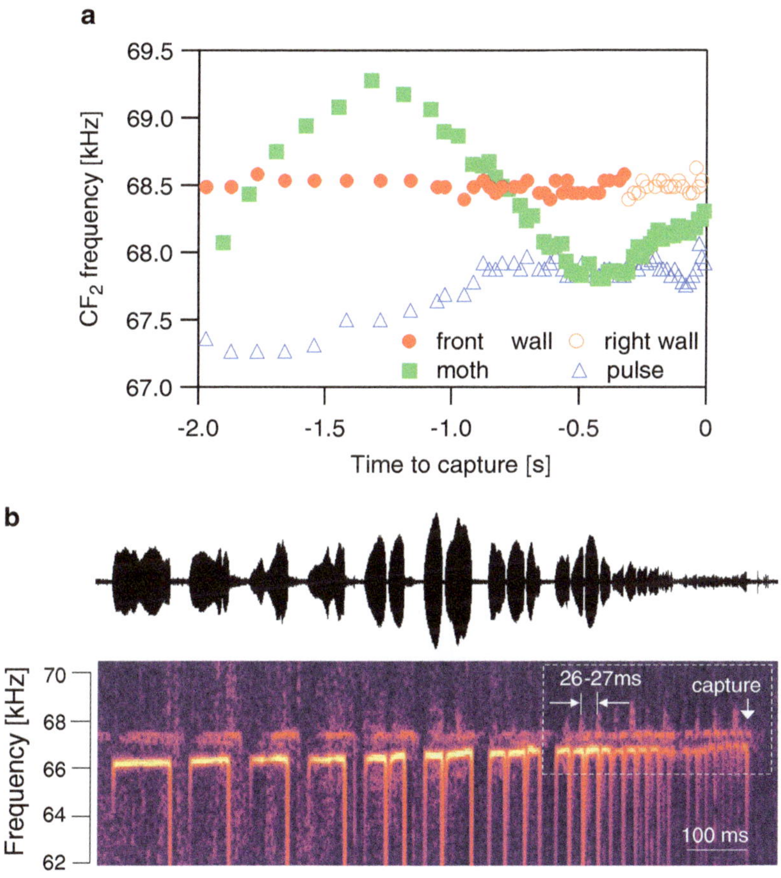

Fig. 9.8 Echolocation behavior of *Rhinolophus ferrumequinum nippon* in pursuit of fluttering prey. (**a**) Changes in the CF_2 of pulses and echoes as a function of time to capture during moth-capture flights in an experimental flight chamber. The bats compensate for echoes returned from the large static object in front of them (changing from the front wall to the right wall of the chamber during flight) but not for echoes from target moths, even though the bats were focused on capturing. (**b**) Sound sequence recorded by the Telemike while the bat was approaching a fluttering moth for capture; amplitude pattern (*top*) and spectrogram (*bottom*). Spectral glints caused by moth fluttering can be observed every 26–27 ms in the CF_2 component of echoes (*dashed box*) (Adapted from Mantani et al. 2012)

Doppler shifts in echoes from moving insects consist of flight-induced Doppler shift and acoustical glints caused by insect fluttering. In fact, telemetry recordings of flying bats capturing moths have revealed periodic spectral glints of 1–1.5 kHz that are synchronized with wing fluttering (Figure 9.8b) (Mantani et al. 2012). By exhibiting DSC on echoes from objects ahead of the bat's flight direction, the extent of Doppler shift in target insects is estimated as ±2–3 kHz from the reference frequency at a maximum. This indicates that fine-frequency analysis for fluttering information is necessary in the range of ±2–3 kHz from the reference frequency, which covers the acoustic fovea found in HDC bats (reviewed in Schnitzler and Denzinger 2011).

9.4.4 Effect of Echo Intensity on Doppler Shift Compensation

Schuller et al. (1974) reported that the ability to perform DSC is not affected by attenuation of playback echoes between 20 and 60 dB relative to emitted pulses. More recently, Smotherman and Metzner showed that the rapidity of DSC responses actually decreases with attenuation of playback echoes relative to that of emitted pulses (Metzner et al. 2002; Smotherman and Metzner 2003b).

Echolocating bats decrease the intensity of their emitted pulses as they approach a prey or an obstacle. This is considered to be echo intensity compensation, in which pulse intensity is adjusted with respect to the distance to a target, resulting in maintenance of echo intensity within the optimal sensitivity range (Kobler et al. 1985; Hiryu et al. 2007). Telemetry recordings of *R. ferrumequinum nippon* indicate that bats gradually decrease pulse amplitude as they approach a landing site so that observed echoes from the target are compensated for at a stable level (Hiryu et al. 2008). Thus, the bats compensate not only for increases in echo frequency but also for echo amplitude as the range to the target decreases.

The DSCF area of *P. parnellii* is tonotopic for the best frequency and amplitopic for the best amplitude in different axes. The delay tuning of FM-FM neurons in *P. parnellii* is affected by echo amplitude, suggesting that echo intensity compensation also helps to stabilize range estimations (Edamatsu and Suga 1993). HDC bats adjust their call frequency and amplitude together to maintain both within an optimal sensitivity range, which can help them to sustain consistent, fine analyses of returning echoes.

9.4.5 Jamming Avoidance Behavior of High Duty Cycle Echolocating Bats

The F_{rest} of HDC bats differs slightly among individuals. However, if the F_{rest} (or more precisely, the reference frequency) overlaps or comes into very close range with the calls of conspecifics, how would a HDC bat avoid or manage acoustic interference (i.e., a jamming avoidance response, JAR)?

By using telemetry, Furusawa et al. (2012) demonstrated that *R. ferrumequinum nippon* flying in pairs or flying alone made DSCs of identical accuracy. Interestingly, although the reference frequencies of individuals in that study were not significantly different, the bats did not shift their frequencies away from each other. Instead, most bat pairs actually shifted their frequencies slightly toward each other, decreasing the difference between them, the opposite of what is done by electric fish (Watanabe and Takeda 1963). Such paradoxical frequency shift was also observed in *Noctilio albiventris* during discrimination experiments; bats shifted the CF of emitted pulses toward that of the artificial jamming CF sounds (Roverud and Grinnell 1985). In contrast, non-DSC LDC bats adaptively change the characteristics of emitted FM signals to minimize acoustical interference from conspecific sounds (Habersetzer 1981; Chiu et al. 2009).

P. parnellii can detect frequency differences as small as 50 Hz in an echo CF_2 due to the high sensitivity of their auditory system (Suga 1984; Riquimaroux et al. 1991). Therefore, the inherent inter-individual variation in reference frequency may be sufficient to allow HDC bats to discriminate between each other without shifting their reference frequencies while flying in groups. In hipposiderids there is no strong evidence of an active shift of the frequencies in echolocation calls to avoid jamming (Jones et al. 1993, 1994).

9.5 Evolution of Doppler Shift Compensation

Wing morphology, cochlear size, and a variety of other characters clearly demonstrate that bats from the early Eocene already featured powered flight and echolocation (Habersetzer et al. 1992). Because no "pre-bats" have been found to answer the question about the timing of the origin of flight and echolocation, several hypotheses have been proposed to explain the sequence in which these two main bat traits have evolved (Fenton 2010). Some hypotheses agree that the putative first echolocation call used by bats may have been a short, broadband multi-harmonic call emitted with long inter-call intervals and low duty cycles. CF echolocation and HDC, both of which depend largely on DSC, are considered to be derived behaviors that evolved more recently from LDC bats (Fenton et al. 1995; Maltby et al. 2009).

9.5.1 *Doppler Shift Compensation in the Bat Phylogenetic Tree*

Within Yinpterochiroptera, DSC appears to characterize the echolocation behavior of about 77 species of horseshoe bats (Rhinolophidae) and about 81 species of round-leaf bats (Hipposideridae) (Altringham 2011). DSC is not known from other families in the suborder (e.g., Craseonycteridae, Rhinopomatidae, Megadermatidae). Species in Craseonycteridae (e.g., Kitti's hog-nosed bat, *Craseonycteris thonglongyai;* Surlykke et al. 1993) and in Rhinopomatidae (e.g., lesser mouse-tailed bat, *Rhinopoma hardwickei*; Habersetzer 1981) emit relatively long CF or narrowband signals with non-overlapping multiple harmonics of which the second harmonic is the most powerful. In the evolution of DSC, *R. hardwickei* may represent an intermediate evolutionary step because it emits long CF calls of about 50 ms and high duty cycles up to about 40 % (Habersetzer 1981). In addition, *R. hardwickei* shows a prominent sensitivity peak in its audiogram in the frequency of the dominant second harmonic (Simmons et al. 1984). The ancestor of *Rhinopoma*, *Rhinolophus*, and *Hipposideros* was probably in the process of evolving an acoustic fovea as a prerequisite for DSC (Neuweiler 1990).

In the suborder Yangochiroptera, just three species of Mormoopidae are HDC echolocators and one other, *Pteronotus personatus*, uses DSC (Smotherman and Guillén-Servent 2008). Recent phylogenetic evidence indicates that *P. parnellii* stems from the most basal node in the *Pteronotus* lineage and that *P. personatus* stems after *P. parnellii* from the second most basal node (Van den Bussche and Weyandt 2003). DSC has been reported in two species of *Noctilio*, suggesting that DSC may have occurred in the common ancestor of Noctilionidae and Mormoopidae.

9.5.2 Doppler Shift Compensation: CF and HDC in Bat Echolocation

It seems safe to state that the CF components of bat calls are a requisite to operate DSC. Long (>20 ms) CF components and calls are distinctive of rhinolophids, and within the family Mormoopidae, *P. parnellii* is the only species to use a particularly long CF component (Figure 9.9). Hipposiderids, the other recognized "DSC bats," emit short CF-FM calls. Among LDC bats, the two species of *Noctilio* and *Pteronotus personatus* employ DSC; the three of them show a short CF component in their calls. Outside the four bat families known to have "DSC species," CF components have been recorded in bats from Rhinopomatidae (Habersetzer 1981), Molossidae (Mora et al. 2004), and Phyllostomidae (Mora and Brinklov, personal observations). None of these species are known to employ DSC.

The same duty cycle implies the same amount of available information as prey-generated amplitude and frequency glints. Therefore, HDC due to longer call durations may have assisted the development of a more precise DSC. This last assumption seems difficult to prove. The DSC of *P. parnellii* is indeed more precise than that of hipposiderids, but it performs as well as that of *Pteronotus personatus*, a congeneric LDC species (Smotherman and Guillén-Servent 2008). The precise DSC behavior of *P. personatus* shows that HDC doesn't seem to be a requirement for the evolutionary acquisition of DSC. So far, narrowband calls appear to be a fact of life for echolocating bats while DSC and HDC are not.

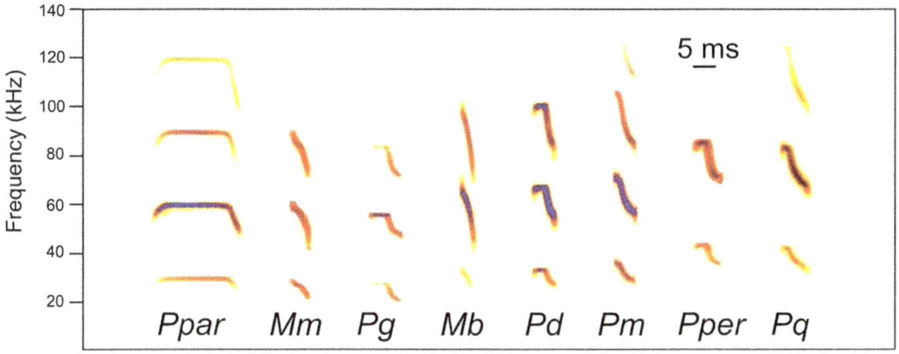

Fig. 9.9 Spectrograms of typical search echolocation calls of the eight bat species of the family Mormoopidae. Notice that despite the similarities in call design (signals with multiple harmonics without overlap in which most energy is concentrated in the 2nd harmonic), only one species (Pper) emits long CF calls at HDC and only two species (Ppar and Pper) perform DSC. Ppar, *Pteronotus parnellii*; Mm, *Mormoops megalophylla*; Pg, *Pteronotus gymnonotus*; Mb, *Mormoops blainvillei*; Pd, *Pteronotus davyi*; Pm, *Pteronotus macleayii*; Pper, *Pteronotus personatus*; Pq, *Pteronotus quadridens* (Adapted from Mora et al. 2013)

9.5.3 Ecological and Behavioral Factors in the Evolution of Doppler Shift Compensation

In both the Old and the New Worlds, several "DSC bat species" developed similar echolocation behaviors and auditory systems, which reveal similarities in early echolocation tasks. The hunting of flying insects in cluttered habitat was undoubtedly among the primeval tasks leading to DSC.

Because most airborne targets encountered by bats flying in the open are insects, there is no clear advantage for bats in the acquisition of a sophisticated echolocation based on DSC and an auditory fovea. On the other hand, DSC and flutter detection are of great value for hunting insects in cluttered environments. The ability to extract information from Doppler-shifted echoes of fluttering insects may have allowed pre-bats exploiting DSC to detect and approach prey in dense vegetation and thus forage in areas with little competition from other bats species without DSC (Lazure and Fenton 2011). There are several species without DSC (e.g., *Myotis nattereri*, *Murina* spp.) that effectively separate prey from background clutter, which is evidence that echolocation strategies based on FM calls can also support foraging in highly cluttered environments (Siemers and Schnitzler 2004; Lazure and Fenton 2011).

Outside of the forest understory, DSC and flutter detection also assist noctilionid bats in hunting over water. Both species of *Noctilio* produce pure CF signals interspersed with CF-FM signals. However, there are many other bat species known to capture prey from, or near, water surfaces; none of these perform DSC.

The specialized calling behavior and auditory receiver of "DSC bats" in both Yinpterochiroptera and Yangochiroptera are arguably the best examples of convergent evolution among echolocators. Because of large phylogenetic and geographic distances between Old World rhinolophids and New World mormoopids, the evolution of DSC clearly demonstrates that perceptual challenges imposed by the environment can override phylogenetic constraints.

9.6 Summary

DSC is achieved through behavioral and neurophysiological specializations in HDC bat species. These findings have advanced the understanding of biosonar systems considerably, and, therefore, DSC is among the most successful research topics in bat echolocation.

Man-made sonar systems are generally designed to transmit sonar sounds with fixed frequency and amplitude. Thus, target information, such as target velocity, is obtained by measuring deviations in frequency and amplitude of the echo. In contrast, HDC bats adjust the frequency of emitted sonar sounds to maintain the echo frequency within their auditory fovea. As a consequence, these bats can analyze the resulting echo within a narrow, sensitive range, allowing them to reduce computational effort by limiting the frequency and dynamic range being processed. To facilitate fine and stable analysis of fluctuating echoes, various compensation mechanisms may also underlie the fundamental processes of bat echolocation.

Doppler shift compensation may seem simple, but some of its behavioral and physiological features remain unexplained. The following are some open questions related to DSC:

1. In highly cluttered environments, it is difficult to detect the weak echo returning from small insect prey, even though DSC adjusts the carrier frequency of the echo to the foveal range of the auditory receiver. What are the acoustic characteristics of the compensated echoes to which the bats actually respond, and how do they change through DSC? Furthermore, as indicated by the early work by Henson, unrevealed specializations likely function in the auditory periphery to facilitate the extraction of information from target prey under cluttered conditions.
2. Thus far, the behavioral and physiological ontogeny of DSC has not been well studied. Furthermore, the evolution of DSC remains to be elucidated, as do the origin of flight and the origin of echolocation. Structured comparative studies of the mormoopids would help to reveal the evolutionary history of DSC. A more interesting question is why HDC bats diversified in the Old World and not in the New World. This also can give new leads to elucidate the evolution of DSC.
3. HDC bats can perform DSC under conspecific-jammed conditions. Further investigation is needed to understand how bats adapt their echolocation both behaviorally and physiologically to overcome unexpected jamming sounds while flying with conspecifics.

Again, we should consider the adaptive benefits of DSC, and the effects of DSC on echolocation (e.g., in finding prey) should also be examined experimentally. DSC, which is a unique strategy for echolocation in HDC bats, will provide new perspectives not only for animal neuroethology but also for various design concepts in the technology and engineering fields.

References

Altringham, J. D. (2011). *Bats: From evolution to conservation*. Oxford, UK: Oxford University Press.

Chiu, C., Xian, W., & Moss, C. F. (2009). Adaptive echolocation behavior in bats for the analysis of auditory scenes. *Journal of Experimental Biology*, 212, 1392–1404.

Covey, E., & Casseday, J. H. (1995). The lower brainstem auditory pathways. In A. N. Popper & R. R. Fay (Eds.), *Hearing by bats* (pp. 235–295). New York: Springer-Verlag.

Davies, K. T. J., Maryanto, I., & Rossiter, S. J. (2013). Evolutionary origins of ultrasonic hearing and laryngeal echolocation in bats inferred from morphological analyses of the inner ear. *Frontiers in Zoology*, 10, 1–15.

Edamatsu, H., & Suga, N. (1993). Differences in response properties of neurons between two delay-tuned areas in the auditory cortex of the mustached bat. *Journal of Neurophysiology*, 69, 1700–1712.

Fenton, M. B. (2010). Convergences in the diversification of bats. *Current Zoology*, 56, 454–468.

Fenton, M. B., Audet, D., Obrist, M. K., & Rydell, J. (1995). Signal strength, timing and self-deafening: The evolution of echolocation in bats. *Paleobiology*, 21, 229–242.

Fenton, M. B., Skowronski, M. D., McGuire, L. P., & Faure, P. A. (2011). Variation in the use of harmonics in the calls of laryngeally echolocating bats. *Acta Chiropterologica*, 13, 169–178.

Fenton, M. B., Faure, P. A., & Ratcliffe, J. M. (2012). Evolution of high duty cycle echolocation in bats. *Journal of Experimental Biology*, 215, 2935–2944.

Foeller, E., & Kössl, M. (2000). Mechanical adaptations for echolocation in the cochlea of the bat, *Hipposideros lankadiva. Journal of Comparative Physiology A: Neuroethology, Sensory, Neural, and Behavioral Physiology*, 186, 859–870.

Frank, G., Hemmert, W., & Gummer, A. W. (1999). Limiting dynamics of high-frequency electromechanical transduction of outer hair cells. *Proceedings of the National Academy of Sciences of the USA*, 96, 4420–4425.

Fu, Z. Y., Tang, J., Jen, P. H., & Chen, Q. C. (2010). The auditory response properties of singe-on and double-on responders in the inferior colliculus of the leaf-nosed bat, *Hipposideros armiger. Brain Research*, 1306, 39–52.

Furusawa, Y., Hiryu, S., Kobayashi, I. K., & Riquimaroux, H. (2012). Convergence of reference frequencies by multiple CF-FM bats (*Rhinolophus ferrumequinum nippon*) during paired flights evaluated with onboard microphones. *Journal of Comparative Physiology A: Neuroethology, Sensory, Neural, and Behavioral Physiology*, 198, 683–693.

Gaioni, S. J., Riquimaroux, H., & Suga, N. (1990). Biosonar behavior of mustached bats swung on a pendulum prior to cortical ablation. *Journal of Neurophysiology*, 64, 1801–1817.

Grinnell, A. D. (1967). Mechanisms of overcoming interference in echolocating animals. In R.-G. Bushel (Ed.), *Animal sonar systems* (pp. 451–481). Jouy-en-Josas, France: Laboratorie de Physiologie Acoustique.

Gustafson, Y., & Schnitzler, H.-U. (1979). Echolocation and obstacle avoidance in the hipposiderid bat *Asellia tridens. Journal of Comparative Physiology A: Neuroethology, Sensory, Neural, and Behavioral Physiology*, 131, 161–167.

Habersetzer, J. (1981). Adaptive echolocation in the bat *Rhinopoma hardwickei. Journal of Comparative Physiology A: Neuroethology, Sensory, Neural, and Behavioral Physiology*, 144, 559–566.

Habersetzer, J., & Storch, G. (1992). Cochlea size in extant Chiroptera and middle Eocene microchiropterans from Messel. *Naturwissenschaften*, 79, 462–466.

Habersetzer, J., Schuller, G., & Neuweiler, G. (1984). Foraging behavior and Doppler shift compensation in echolocating hipposiderid bats, *Hipposideros bicolor* and *Hipposideros speoris*. *Journal of Comparative Physiology A: Neuroethology, Sensory, Neural, and Behavioral Physiology*, 155, 559–567.

Habersetzer, J., Richter, G., & Storch, G. (1992). Bats: Already highly specialized insect predators. In S. Schaal & W. Ziegler (Eds.), *Massel: An insight into the history of life and of the earth* (pp. 179–191). Oxford, UK: Clarendon Press.

Hartley, D. J., & Suthers, R. A. (1988). The acoustics of the vocal tract in the horseshoe bat, *Rhinolophu hildebrandti*. *Journal of the Acoustical Society of America*, 84, 1201–1213.

Hartley, D. J., & Suthers, R. A. (1990). Sonar pulse radiation and filtering in the mustached bat, *Pteronotus parnellii rubiginosus*. *Journal of the Acoustical Society of America*, 87, 2756–2772.

Henson, O. W., Jr., Henson, M. M., Kobler, J. B., & Pollak, G. D. (1980). The constant frequency component of the biosonar signals of the bat, *Pteronotus parnellii parnellii*. In R.-G. Busnel & F. F. James (Eds.), *Animal sonar systems* (pp. 913–916). New York: Plenum Press.

Henson, O. W., Jr., Pollak, G. D., Kobler, J. B., Henson, M. M., & Goldman, L. J. (1982). Cochlear microphonic potentials elicited by biosonar signals in flying bats, *Pteronotus p. parnellii*. *Hearing Research*, 7, 127–147.

Henson, D. W., Jr., Bishop, A. L., Keating, A. W., Kobler, J. B., Henson, M. M., Wilson, B. S., & Hansen, R. (1987). Biosonar imaging of insects by *Pteronotus p. parnellii*, the mustached bat. *National Geographic Research*, 3, 82–101.

Henson, O. W. J. (1967). The perception and analysis of biosonar signals by bats. In R.-G. Busnel (Ed.), *Animal sonar systems* (pp. 949–1002). Jouy-en-Josas, France: Laboratorie de Physiologie Acoustique.

Hiryu, S., Hagino, T., Riquimaroux, H., & Watanabe, Y. (2007). Echo-intensity compensation in echolocating bats (*Pipistrellus abramus*) during flight measured by a telemetry microphone. *Journal of the Acoustical Society of America*, 121, 1749–1757.

Hiryu, S., Katsura, K., Lin, L. K., Riquimaroux, H., & Watanabe, Y. (2005). Doppler-shift compensation in the Taiwanese leaf-nosed bat (*Hipposideros terasensis*) recorded with a telemetry microphone system during flight. *Journal of the Acoustical Society of America*, 118, 3927–3933.

Hiryu, S., Katsura, K., Nagato, T., Yamazaki, H., Lin, L. K., Watanabe, Y., & Riquimaroux, H. (2006). Intra-individual variation in the vocalized frequency of the Taiwanese leaf-nosed bat, *Hipposideros terasensis*, influenced by conspecific colony members. *Journal of Comparative Physiology A: Neuroethology, Sensory, Neural, and Behavioral Physiology*, 192, 807–815.

Hiryu, S., Shiori, Y., Hosokawa, T., Riquimaroux, H., & Watanabe, Y. (2008). On-board telemetry of emitted sounds from free-flying bats: Compensation for velocity and distance stabilizes echo frequency and amplitude. *Journal of Comparative Physiology A: Neuroethology, Sensory, Neural, and Behavioral Physiology*, 194, 841–851.

Ho, Y.-Y., Fang, Y.-P., Chou, C.-H., Cheng, H.-C., & Chang, H.-W. (2013). High duty cycle to low duty cycle: Echolocation behaviour of the Hipposiderid bat *Coelops frithii*. *PloS ONE*, *8*, e62938.

Jones, G., & Ransome, R. D. (1993). Echolocation calls of bats are influenced by maternal effects and change over a lifetime. *Proceedings of the Royal Society of London B: Biological Sciences*, 252, 125–128.

Jones, G., Morton, M., Hughesand, P. M., & Buden, R. M. (1993). Echolocation, flight morphology and foraging strategies of some West African hipposiderid bats. *Journal of Zoology*, 230, 385–400.

Jones, G., Sripathi, K., Waters, D. A., & Marimuthu, G. (1994). Individual variation in the echolocation calls of three sympatric Indian hipposiderid bats, and an experimental attempt to jam bat echolocation. *Folia Zoologica*, 43, 347–361.

Kobler, J. B., Wilson, B. S., Henson, O. W., Jr., & Bishop, A. L. (1985). Echo intensity compensation by echolocating bats. *Hearing Research*, 20, 99–108.

Kössl, M. (1994). Otoacoustic emissions from the cochlea of the 'constant frequency' bats, *Pteronotus parnellii* and *Rhinolophus rouxi*. *Hearing Research*, 72, 59–72.
Kössl, M., & Vater, M. (1995). Cochlear structure and function in bats. In A. N. Popper & R. R. Fay (Eds.), *Hearing by bats* (pp. 191–234). New York: Springer-Verlag.
Kössl, M., & Vater, M. (1996). Further studies on the mechanics of the cochlear partition in the mustached bat. I. Ultrastructural observations on the tectorial membrane and its attachments. *Hearing Research*, 94, 63–77.
Kössl, M., Mayer, F., Frank, G., Faulstich, M., & Russell, I. J. (1999). Evolutionary adaptations of cochlear function in Jamaican mormoopid bats. *Journal of Comparative Physiology A: Neuroethology, Sensory, Neural, and Behavioral Physiology*, 185, 217–228.
Lancaster, W. C., Keating, A. W., & Henson, O. W., Jr. (1992). Ultrasonic vocalizations of flying bats monitored by radiotelemetry. *Journal of Experimental Biology*, 173, 43–58.
Lazure, L., & Fenton, M. B. (2011). High duty cycle echolocation and prey detection by bats. *Journal of Experimental Biology*, 214, 1131–1137.
Long, G. R., & Schnitzler, H.-U. (1975). Behavioural audiograms from the bat, *Rhinolophus ferrumequinum. Journal of Comparative Physiology, A: Neuroethology, Sensory, Neural, and Behavioral Physiology*, 100, 211–219.
Maltby, A., Jones, K. E., & Jones, G. (2009). Understanding the evolutionary origin and diversification of bat echolocation. In S. M. Brudzynski (Ed.), *Handbook of mammalian vocalization* (pp. 37–48). Oxford, UK: Academic Press.
Mantani, S., Hiryu, S., Fujioka, E., Matsuta, N., Riquimaroux, H., & Watanabe, Y. (2012). Echolocation behavior of the Japanese horseshoe bat in pursuit of fluttering prey. *Journal of Comparative Physiology A: Neuroethology, Sensory, Neural, and Behavioral Physiology*, 198, 741–751.
Metzner, W., Zhang, S., & Smotherman, M. (2002). Doppler-shift compensation behavior in horseshoe bats revisited: Auditory feedback controls both a decrease and an increase in call frequency. *Journal of Experimental Biology*, 205, 1607–1616.
Mora, E. C., Macias, S., Vater, M., Coro, F., & Kössl, M. (2004). Specializations for aerial hawking in the echolocation system of *Molossus molossus* (Molossidae, Chiroptera). *Journal of Comparative Physiology A: Neuroethology, Sensory, Neural, and Behavioral Physiology*, 190, 561–574.
Mora, E., Macías, S., Hechavarria, J., Vater, M., & Kössl, M. (2013). Evolution of the heteroharmonic strategy for target-range computation in the echolocation of Mormoopidae. *Frontiers in Physiology*, doi: 10.3389/fphys.2013.00141
Neuweiler, G. (1990). Auditory adaptations for prey capture in echolocating bats. *Physiological Reviews*, 70, 615–641.
Neuweiler, G. (2003). Evolutionary aspects of bat echolocation. *Journal of Comparative Physiology A: Neuroethology, Sensory, Neural, and Behavioral Physiology*, 189, 245–256.
Neuweiler, G., Schuller, G., & Schnitzler, H.-U. (1971). On- and off-responses in the inferior colliculus of the greater horseshoe bat to pure tones. *Zeitschrift für vergleichende Physiologie*, 74, 57–63.
Neuweiler, G., Singh, S., & Sripathi, K. (1984). Audiograms of a South Indian bat community. *Journal of Comparative Physiology A: Neuroethology, Sensory, Neural, and Behavioral Physiology*, 154, 133–142.
O'Neill, W. E. (1995). The bat auditory cortex. In A. N. Popper & R. R. Fay (Eds.), *Hearing by bats* (pp. 416–480). New York: Springer-Verlag.
Ostwald, J. (1984). Tonotopical organization and pure tone response characteristics of single units in the auditory cortex of the greater horseshoe bat. *Journal of Comparative Physiology A: Neuroethology, Sensory, Neural, and Behavioral Physiology*, 155, 821–834.
Ostwald, J. (1988). Encoding of natural insect echoes and sinusoidally modulated stimuli by neurons in the auditory cortex of the greater horseshoe bat, *Rhinolophus ferrumequinum*. In P. E. Nachtigall & P. W. B. Moore (Eds.), *Animal sonar processes and performance* (pp. 483–487). New York: Plenum Press.
Pollak, G. D., & Schuller, G. (1981). Tonotopic organization and encoding features of single units in inferior colliculus of horseshoe bats: Functional implications for prey identification. *Journal of Neurophysiology*, 45, 208–226.

Pollak, G. D., & Park, T. J. (1995). The inferior colliculus. In A. N. Popper & R. R. Fay (Eds.), *Hearing by bats* (pp. 296–367). New York: Springer-Verlag.
Reimer, K. (1987). Coding of sinusoidally modulated acoustic stimuli in the inferior colliculus of the rufous horseshoe bat, *Rhinolophus rouxi. Journal of Comparative Physiology A: Neuroethology, Sensory, Neural, and Behavioral Physiology*, 161, 305–313.
Riquimaroux, H., & Watanabe, Y. (2000). Characteristics of bat sonar sounds recorded by a telemetry system and a fixed ground microphone. *Seventh Western Pacific Regional Acoustics Conference (WESTPRACVII)*, 233–238.
Riquimaroux, H., Gaioni, S. J., & Suga, N. (1991). Cortical computational maps control auditory perception. *Science*, 251, 565–568.
Roverud, R. C., & Grinnell, A. D. (1985). Frequency tracking and Doppler shift compensation in response to an artificial CF/FM echolocation sound in the CF/FM bat, *Noctilio albiventris. Journal of Comparative Physiology A: Neuroethology, Sensory, Neural, and Behavioral Physiology*, 156, 471–475.
Rübsamen, R., Neuweiler, G., & Sripathi, K. (1988). Comparative collicular tonotopy in two bat species adapted to movement detection, *Hipposideros speoris* and *Megaderma lyra. Journal of Comparative Physiology A: Neuroethology, Sensory, Neural, and Behavioral Physiology*, 163, 271–285.
Schnitzler, H.-U. (1968). Die Ultraschallortungslaute der Hufeisen-Fledermäuse (Chiroptera-Rhinolophidae) in verschiedenen Orientierungssituationen [The ultrasonic sounds of horseshoe bats (Chiroptera-Rhinolophidae) in different orientation situations]. *Zeitschrift für Vergleichende Physiologie*, 57, 376–408.
Schnitzler, H.-U. (1970). Echoortung bei der Fledermaus Chlonycteris rubiginosa. *Zeitschrift für Vergleichende Physiologie*, 68, 25–38.
Schnitzler, H.-U. (1973). Control of Doppler shift compensation in the greater horsehoe bat *Rhinolophus ferrumequinum. Journal of Comparative Physiology A: Neuroethology, Sensory, Neural, and Behavioral Physiology*, 82, 79–92.
Schnitzler, H.-U. (1978). Die Detektion von Bewegungen durch Echoortung bei Fledermäusen. *Verhandlungen der deutschen Zoologischen Gesellschaft, 71*, 16–33.
Schnitzler, H.-U., & Denzinger, A. (2011). Auditory fovea and doppler shift compensation: adaptations for flutter detection in echolocating bats using CF-FM signals. *Journal of Comparative Physiology A: Neuroethology, Sensory, Neural, and Behavioral Physiology*, 197, 541–559.
Schnitzler, H.-U., Suga, N., & Simmons, J. A. (1976). Peripheral auditory tuning for fine frequency analysis by the CF-FM bat, *Rhinolophus ferrumequinum*. III Cochlear microphonic and N1 responses. *Journal of Comparative Physiology A: Neuroethology, Sensory, Neural, and Behavioral Physiology*, 82, 93–102.
Schuller, G. (1974). The role of overlap of echo with outgoing echolocation sound in the bat *Rhinolophus ferrumequinum. Naturwissenschaften*, 61, 171–172.
Schuller, G. (1977). Echo delay and overlap with emitted orientation sounds and Doppler-shift compensation in the bat, *Rhinolophus ferrumequinum. Journal of Comparative Physiology A: Neuroethology, Sensory, Neural, and Behavioral Physiology*, 114, 103–114.
Schuller, G. (1980). Hearing characteristics and Doppler shift compensation in South Indian CF-FM bat. *Journal of Comparative Physiology A: Neuroethology, Sensory, Neural, and Behavioral Physiology*, 139, 349–356.
Schuller, G. (1984). Natural ultrasonic echoes from wing beating insects are encoded by collicular neurons in the CF-FM bat, *Rhinolophus ferrumequinum. Journal of Comparative Physiology A: Neuroethology, Sensory, Neural, and Behavioral Physiology*, 155, 121–128.
Schuller, G., & Pollak, G. D. (1979). Disproportionate frequency representation in the inferior colliculus of Doppler-compensating greater horseshoe bats: Evidence for an acoustic fovea. *Journal of Comparative Physiology A: Neuroethology, Sensory, Neural, and Behavioral Physiology, 132*, 47–54.
Schuller, G., Beuter, K., & Schnitzler, H.-U. (1974). Response to frequency shifted artificial echoes in the bat *Rhinolophus ferrumequinum. Journal of Comparative Physiology A: Neuroethology, Sensory, Neural, and Behavioral Physiology*, 89, 275–286.
Siemers, B. M., & Schnitzler, H.-U. (2004). Echolocation signals reflect niche differentiation in five sympatric congeneric bat species. *Nature*, 429, 657–661.

Simmons, J. A. (1974). Response of the Doppler echolocation system in the bat, *Rhinolophus ferrumequinum*. *Journal of the Acoustical Society of America*, 56, 672–682.
Simmons, J. A., Kick, S. A., & Lawrence, B. D. (1984). Echolocation and hearing in the mouse-tailed bat, *Rhinopoma hardwickei*: Acoustic evolution of echolocation in bats. *Journal of Comparative Physiology A: Neuroethology, Sensory, Neural, and Behavioral Physiology*, 154, 347–356.
Smotherman, M., & Metzner, W. (2003a). Fine control of call frequency by horseshoe bats. *Journal of Comparative Physiology A: Neuroethology, Sensory, Neural, and Behavioral Physiology*, 189, 435–446.
Smotherman, M., & Metzner, W. (2003b). Effects of echo intensity on Doppler-shift compensation behavior in horseshoe bats. *Journal of Neurophysiology*, 89, 814–821.
Smotherman, M., & Guillén-Servent, A. (2008). Doppler-shift compensation behavior by Wagner's mustached bat, *Pteronotus personatus*. *Journal of the Acoustical Society of America*, 123, 4331–4339.
Suga, N. (1984). The extent to which biosonar information is represented in the bat auditory cortex. In G. M. Edelman, W. E. Gall & W. M. Cowan (Eds.), *Dynamic aspects of neocortical function* (pp. 315–373). New York: John Wiley & Sons.
Suga, N. (1990). Biosonar and neural computation in bats. *Scientific American*, 60–68.
Suga, N., Niwa, H., & Taniguchi, I. (1983). Representation of biosonar information in the auditory cortex of the mustached bat, with emphasis on representation of target velocity information. *Advances in Vertebrate Neuroethology*, 56, 829–867.
Suga, N., Niwa, H., Taniguchi, I., & Margoliash, D. (1987). The personalized auditory cortex of the mustached bat: Adaptation for echolocation. *Journal of Neurophysiology*, 58, 643–654.
Surlykke, A., Miller, L. A., Møhl, B., Andersen, B. B., Christensen-Dalsgaard, J., & Jørgensen, M. B. (1993). Echolocation in two very small bats from Thailand: *Craseonycteris thonglongyai* and *Myotis siligorensis*. *Behavioral Ecology and Sociobiology*, 33, 1–12.
Tian, B., & Schnitzler, H.-U. (1997). Echolocation signals of the greater horseshoe bat (*Rhinolophus ferrumequinum*) in transfer flight and during landing. *Journal of the Acoustical Society of America*, 101, 2347–2364.
Trappe, M., & Schnitzler, H.-U. (1982). Doppler-shift compensation in insect-catching horseshoe bats. *Naturwissenschaften*, 69, 193–194.
Van den Bussche, R. A., & Weyandt, S. E. (2003). Mitochondrial and nuclear DNA sequence data provide resolution to sister-group relationships within *Pteronotus* (Chiroptera: Mormoopidae). *Acta Chiropterologica*, 5, 1–13.
Vater, M. (1982). Single unit responses in cochlear nucleus of horseshoe bats to sinusoidal frequency and amplitude modulated signals. *Journal of Comparative Physiology A: Neuroethology, Sensory, Neural, and Behavioral Physiology*, 149, 369–388.
Vater, M. (1998). Adaptation of the auditory periphery of bats for echolocation. In T. H. Kunz & P. A. Racey (Eds.), *Bat biology and conservation* (pp. 231–245). Washington, DC: Smithonian Institution Press.
von der Emde, G., & Schnitzler, H.-U. (1990). Classification of insects by echolocating greater horseshoe bats. *Journal of Comparative Physiology A: Neuroethology, Sensory, Neural, and Behavioral Physiology*, 167, 423–430.
Watanabe, A., & Takeda, K. (1963). The change of discharge frequency by A. C. stimulus in a weakly electric fish. *Journal of Experimental Biology*, 40, 57–66.
Wenstrup, J. J., & Suthers, R. A. (1984). Echolocation of moving targets by the fish-catching bats, *Noctilio leporinus*. *Journal of Comparative Physiology A: Neuroethology, Sensory, Neural, and Behavioral Physiology*, 155, 75–89.

Chapter 10
Perceiving the World Through Echolocation and Vision

Annemarie Surlykke, James A. Simmons, and Cynthia F. Moss

Keywords Bats • Echolocation • Perception • Scene analysis • Sonar • Vision

10.1 Introduction

Spatial perception operates in diverse animal species, irrespective of habitat and dominant sensory modality, and supports the representation of objects in an animal's environment. Spatial perception guides behaviors such as (a) attending to object locations; (b) tracking object movements; (c) identifying objects by distinguishing features; (d) moving to intercept or avoid them; and (e) remembering both locations and identities while navigating through the environment. These tasks are carried out by echolocating bats, animals that use biological sonar to represent the environment. The natural behaviors of bats suggest that their spatial perception by echolocation is comparable in many ways to that of animals relying on vision.

The first stages of spatial perception involve the acquisition of stimuli by sensory organs. Animals then process and represent the information obtained by sensory signals and use it to guide actions. The representation derived from the stimuli is modulated by the animal's behavioral state, attention, and stored memories. Across different groups of animals, spatial perception usually is based on multisensory input, but the relative importance of the senses varies according to the species, its natural behaviors, and environmental conditions. For humans and other diurnal animals, vision is the primary sense for spatial perception. Other modalities dominate

A. Surlykke
Department of Biology, University of Southern Denmark, DK-5230 Odense, Denmark

J.A. Simmons
Department of Neuroscience, Brown University, Providence, RI 02912, USA
e-mail: james_simmons@brown.edu

C.F. Moss (✉)
Department of Psychological and Brain Sciences, The Johns Hopkins University, Baltimore, MD 21218-2686, USA
e-mail: cynthia.moss@jhu.edu

M.B. Fenton et al. (eds.), *Bat Bioacoustics*, Springer Handbook of Auditory Research 54, DOI 10.1007/978-1-4939-3527-7_10

for animals that are active under circumstances with little, no, or unpredictable light. The size of the animal's home range also influences which sense is primary. Animals operating over large areas use modalities well-suited for distal sensing. Vision and hearing are good examples. In contrast, animals operating in more restricted spaces, e.g., in burrows, depend more on proximal sensing, such as touch, smell, and taste. Some animals represent their environment through active sensing, generating signals that return information from the surroundings to the emitting individual. Electroreception by weakly electric fishes and echolocation by bats and toothed whales are examples of advanced active sensing systems that provide distal information about the environment in the absence of light.

This chapter highlights some principles and mechanisms of spatial perception by comparing bat echolocation and vision. Echolocation and vision are two different solutions to the problem of distal sensing. Nevertheless, perceptual tests reveal that the behavioral functions supported by echolocation and vision are comparable. Both visually guided animals and echolocating bats can detect individual objects, attend to them, follow their movements, identify them, and then intercept them while simultaneously detecting, localizing, and avoiding other objects in nearby space. Both visual and echolocating animals also can remember the objects' characteristics and locations for future reference. In measurements of performance, both systems support high-resolution spatial perception, even though the physical stimuli and early sensory processing are entirely distinct.

In this chapter, auditory scene analysis (Bregman 1990) by echolocation is explained with reference to the contemporary understanding of visual scene analysis. A comparison of scene segmentation and object recognition by echolocation and by vision reveals important constraints for spatial perception that would not be evident by examining just one system alone. Behaviorally equivalent solutions in vision and echolocation attest to the capacity of the two distal sensing systems to solve similar problems of scene analysis. Aerial-feeding birds (e.g., swifts) use vision to find flying insects by day. Bats that hunt flying insects by night are equally effective predators, often in the same locations, separated in time. This sensory comparison will highlight general rules that do not depend on modality but are dictated by the common function of spatial perception.

10.2 Essential Details About Echolocation Related to Spatial Perception

Echolocating bats are nocturnal and use a biological sonar, echolocation, to orient and hunt prey in little or no light. Bats emit trains of acoustic signals that result in corresponding trains of echoes from objects in the immediate environment (Griffin 1958; Neuweiler 1989; Thomas et al. 2004). There are over 1,200 species of echolocating bats, and their biosonar signals and behaviors reflect their diversity in habitat and lifestyle. While a few species of flying foxes (Pteropodidae) use tongue clicks for echolocation, this chapter focuses on bats using echolocation signals produced by the larynx. Bat sonar signals can be broadly grouped as frequency-modulated (FM) or constant-frequency (CF) signals or in various combinations of FM and CF. Virtually all species of bats make use of wideband sounds, either by covering a broadband with FM sounds or by emitting short (1–5 ms) nearly CF signals with several harmonic frequencies (Neuweiler 1989; Fenton et al. 2014). Echoes from broadband signals produce sharper sonar "pictures" of the surroundings than narrowband signals produce (Simmons and Stein 1980). The sharpness of the image governs the animal's accuracy in locating and identifying individual objects, as well as separating objects from the background. Accuracy improves as more frequencies are included in the sonar emission, and their spectral contributions are gathered into the echo image. FM sounds contain many frequencies spread out over the duration of each sonar sweep (typically from 0.5–1 ms to 10–15 ms). These short sounds have a low duty cycle (Fenton et al. 2014); they are "on" for less than 10 % of the time; there are much longer intervals of silence between the emitted sounds during which the bat is listening for echoes.

Echolocating bats that produce calls at low duty cycle separate pulse and echo in time. An alternative approach is to produce calls at high duty cycle and separate pulse and echo in frequency. Horseshoe bats in the families Rhinolophidae and Hipposideridae and the mustache bat *Pteronotus parnellii* (family Mormoopidae) emit long-duration calls separated by short periods of silence; they have a high duty cycle (Schnitzler and Kalko 2001; Fenton et al. 2012). Doppler shifts created by the bat's own flight velocity influence the frequency of the returning echo relative to the emitted call. Flying CF bats compensate for Doppler shifts introduced by their flight velocity by lowering the call frequency so that echoes are received at the frequency of the bat's highest sensitivity and selectivity, the reference frequency (Trappe and Schnitzler 1982). Listening in the reference frequency band, CF bats can distinguish between fluttering insects and the background based on acoustic glints (Doppler shifts and amplitude peaks) in echoes generated by wing movement (von der Emde and Schnitzler 1990). Moreover, the CF component of each sound is terminated by a short (1–3 ms) FM sweep, which bats use to estimate target distance (Simmons 1971, 1973). The low duty cycle of these FM components indicates that high duty cycle CF bats also sample the 3-D location of targets in a stroboscopic manner, similar to bat species that rely exclusively on FM signals for scene perception. Questions remain as to how target velocity information is integrated with the 3-D information extracted from FM echoes.

Typical FM bats are the many insectivorous species from the large, widely distributed families Vespertilionidae and Molossidae (Schnitzler and Kalko 2001; Fenton et al. 2014). They mostly hunt prey "on the wing" by detecting, tracking and intercepting flying insects using their sonar. Besides their well-documented aerial-feeding habits, some species of vespertilionids are more versatile, finding insects crawling on surfaces, such as leaves and branches, or the ground. The big brown bat, *Eptesicus fuscus*, is particularly flexible in its foraging behavior, and it is the bat example most often referred to in this chapter.

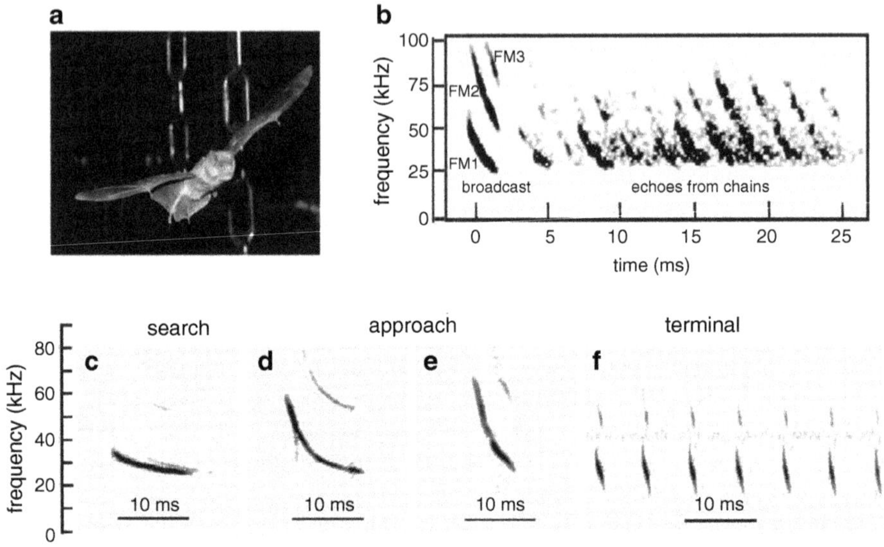

Fig. 10.1 (**a**) Big brown bat flying in a maze of chains. (**b**) Spectrogram of an FM broadcast and series of echoes returned by objects in the scene. The broadcast (*left*) is followed by a series of echoes that are reflected back to the bat's ears and used to guide flight. (**c**)–(**f**) Series of spectrograms illustrating adaptive changes that occur in the bat's broadcasts during an aerial interception of prey

Figure 10.1a shows an echolocating big brown bat flying through a complex environment, consisting of rows of vertically hanging plastic chains. Bats easily negotiate mazes of chains, even when the spacing of adjacent rows is only twice the bat's wingspan (Petrites et al. 2009; Barchi et al. 2013). During flight, the bat emits FM echolocation sounds and receives a cascade of echoes from a large sample of the chains (Figure 10.1b) because each broadcast signal is very intense and each chain is a strong reflector of sound. The room and its contents comprise a sonar scene, and the echoes from each emission comprise the auditory stimuli for perceiving the scene. To negotiate the passageway through the chains, the bat must avoid the chains just in front of it and determine a path empty of obstacles amid reverberating echoes from the chains located off to the sides (Petrites et al. 2009). The characteristics of the echo cascades provide the bat with information necessary to navigate and maneuver in complicated conditions. It would seem an insurmountable challenge for the bat to segregate acoustic objects in an almost continuous cascade of echoes, but the navigation performance of bats compares well to that of visually guided animals (e.g., birds), indicating that there may be considerable overlap in the degree to which visual and biosonar spatial perception yield similar representations of complex scenes.

Both the broadcast sounds and their echoes are processed by the bat's auditory system to produce a spatial representation of the bat's surroundings (Moss et al. 2011). The big brown bat's FM broadcasts cover frequencies from over ~100 kHz down to 25 kHz in two or three harmonic sweeps (e.g., FM1, FM2, FM3 in Figure 10.1b). The

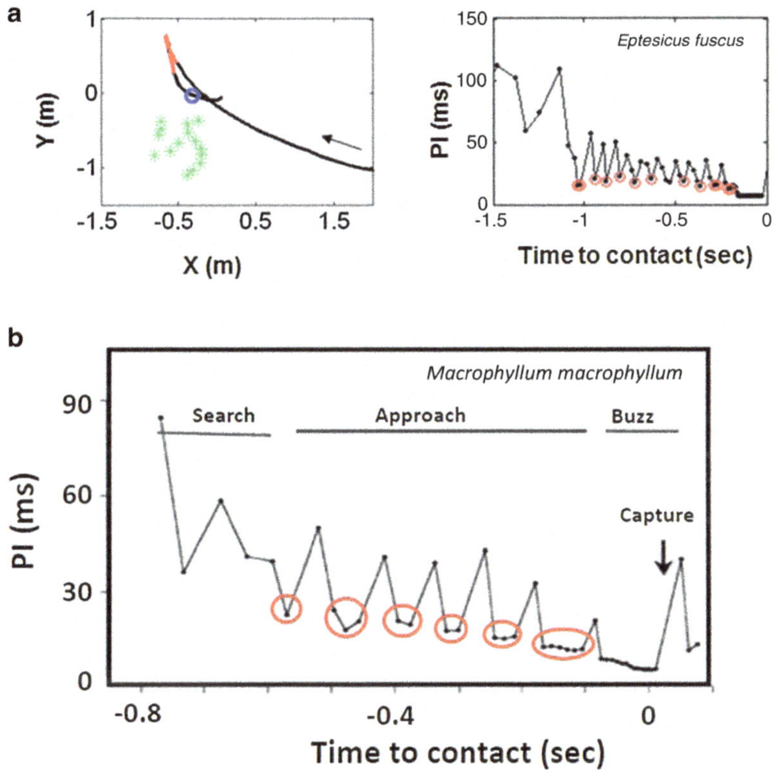

Fig. 10.2 Sonar calls are grouped when bats forage (**a**) in the lab (*Eptesicus fuscus*) and (**b**) in the field (*Macrophyllum macrophyllum*). Both examples illustrate the production of sound groups with relatively stable intervals, as the bats approach prey. (**a**) The left panel shows an overhead view of the flight path of the bat when it takes a tethered insect (*blue circle*) hanging ca. 40 cm from a plant shown in green. *Arrows* indicate the direction of the flight path. The *right panel* shows the pulse interval (PI) as a function of the time to contact with sonar sound groups circled in *red*. (**b**) PI as a function of time in a natural pursuit sequence. Sonar sound groups circled in *red*. (Adapted from data in Moss et al. 2006; Weinbeer and Kalko 2007)

bat adapts the features of the sounds: the duration, the frequency span, the sweep-rate, and the harmonic content across the phases of an insect pursuit (Griffin 1958) (Figure 10.1c–f). During insect pursuit, the bat's pulse interval (PI) between the onset of one call and the next decreases from over 200 ms in the search phase to about 6 ms in the terminal buzz phase. Simultaneously, there is a parallel decrease in pulse duration and change in time-frequency structure of calls. However, during active pursuit, when the bat is tracking a selected target, the PI does not change continuously with decreasing target distance. Instead, the bat emits sound groups (sonar "strobe groups") with short, often relatively stable intervals, interrupted by longer breaks. As the bat approaches the target, the intervals between sound groups decrease (Figure 10.2). The sound groups are produced in situations requiring precise positioning of objects, like prey pursuit, or when the bat is changing the direction of its flight path and is in close proximity to obstacles. The reduced signal intervals in the sound groups may sharpen the bat's representation of objects and support flight navigation in complex spaces (Moss et al. 2006; Kothari et al. 2014; Sändig et al. 2014).

In the search phase, big brown bats, like many other aerial-feeding insectivorous bats, extend the length of their FM sounds at the tail end to include an almost CF component (Figure 10.1c, d), often called "quasi-CF," QCF (Schnitzler and Kalko 2001). The QCF is most pronounced when bats fly in the open, away from background reflections. The extended duration of the QCF tail end of the sound is well suited for detecting a target. Energy in the QCF call is concentrated in a narrow spectral band, enlarging the sonar-operating range by increasing the signal to noise ratio and, in turn, facilitating detection of echo returns. Closer to a target in more or less cluttered environments, big brown bats emit a broadband signal, from over 100 kHz to about 25 kHz and change the duration of the individual sounds according to the distance to objects in the scene. Other species of bats in several families emit pulses in which the FM component is preceded, rather than followed, by short (1–5 ms) QCF (e.g., the greater bulldog bat, *Noctilio leporinus*) (Surlykke and Kalko 2008), components that appear to serve the same purpose as the QCF tail used by *E. fuscus* to facilitate target detection by concentrating energy in a narrow frequency band (Fenton et al. 2014). Because CF echoes are not well suited to carry information about object direction and distance (they only contain a narrow frequency band), bats that use CF signals always accompany them with FM signals or shift to emitting FM signals in contexts requiring spatial localization of targets and obstacles.

The bat's broadcasts attenuate with distance due to spatial spreading of the propagating wavefront. Furthermore, sounds at higher frequencies are attenuated more than lower frequencies because of atmospheric absorption (Lawrence and Simmons 1982b). When each sound impinges on ("ensonifies") an object, it interacts with the object's geometric structure and is scattered in different directions. The portion returning to the bat constitutes the echo that contributes to the animal's perception of its surroundings. Echo strength depends on the size of the object (its acoustic "cross-section") and distance (Wahlberg and Surlykke 2014), which together limit the operating range of bat echolocation. The bat's detection distance of insects is probably only 5–10 m, while large, extended background surfaces, such as screens of vegetation or the ground, may be detected at 20–40 m (Holderied and von Helversen 2003; Surlykke and Kalko 2008).

The big brown bat's echolocation sounds are directional, with most energy broadcast in the forward direction (Figure 10.3). In natural open habitats, insectivorous bats typically emit calls with on-axis intensities of 120–135 dB SPL at 0.1 m (Holderied and von Helversen 2003; Surlykke and Kalko 2008). In cluttered space in the lab, the beam is relatively broad with a half-amplitude angle at 35 kHz of around 60–70° (Ghose and Moss 2003; Jakobsen et al. 2013). The higher frequencies are beamed progressively more directionally (Figure 10.3c), reducing the bandwidth of echoes with off-axis angle of the object. Consequently, echoes are progressively low-pass filtered the further away or the more off-axis the echo objects are, which is different from the additional and complicated effects on the echo spectrum from the target's features. The echo spectrum has to be apportioned into its spatial, low-pass

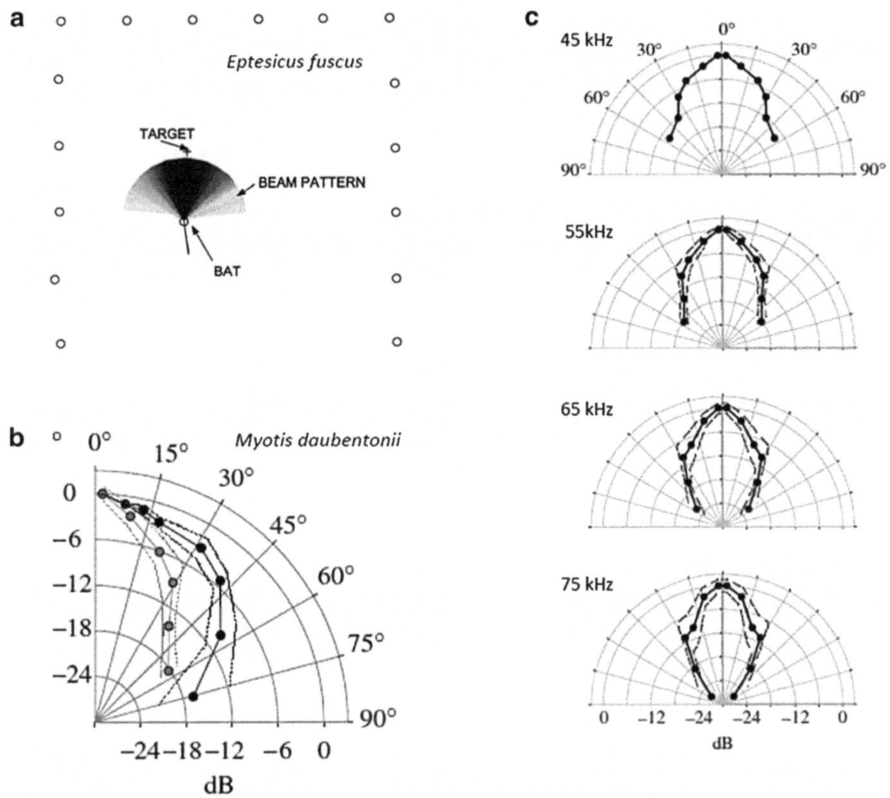

Fig. 10.3 Directionality of sonar calls depends on situation and frequency. (**a**) Reconstruction of the beam of *E. fuscus* hunting in the lab. The call is recorded by 16 microphones (*circles*) placed on three sides of the flight room. The thin line terminating at the bat's position (*arrow*) is the bat's trajectory up to that frame. The beam pattern is normalized such that the peak intensity is colored *black* with lighter shades of gray denoting progressively lower intensities. (**b**) The average directionality of the beam of *Myotis daubentonii* in the lab (*black*) and the field (*grey*) at 55 kHz. The beam is much wider in the confined space of the lab. *M. daubentonii* emits a broadband FM signal with energy from 40 to >90 kHz. (**c**) The directionality of *M. daubentonii* in the field, illustrating how beam width decreases with increasing frequency

components and its complex, target-related components due to the sound's interaction with the object's geometry (e.g., an insect's discrete reflecting points, such as the head, wings, or abdomen, or the many surfaces of complicated objects such as chains), which can provide the bat with information about the object's shape.

The sonar beam directionality has been determined in the lab for *E. fuscus,* but results from another vespertilionid FM bat, *Myotis daubentonii*, suggest that directionality is greatly increased in the field compared to the lab (Figure 10.3b) (Surlykke et al. 2009b). In the lab, the broad beam of vespertilionids ensonifies much of the scene around the target in the space in front of the bat. This scene contains multiple objects and returns multiple echoes (Figure 10.1b). However, in spite of the broad acoustic "field of ensonification," the bat aligns its head and external ears, along with the central axis of its sonar beam, precisely in the direction of the selected target when preparing to intercept prey or inspecting nearby objects for flight guidance (Ghose and Moss 2006; Surlykke et al. 2009a). By analogy, in a visually guided animal, gaze directs the fovea of each retina onto the target of interest to form a high-resolution visual image. In both the visual and biosonar systems, these tracking responses provide an empirical indicator of the specific objects in the scene to which the subject is focusing attention at each moment in time.

In vision, objective measures of a subject's attention include the direction of gaze, fixation of selected objects on the fovea, and accommodation of the lens to focus at a distance. In echolocation, the aim of the bat's head indicates its acoustic gaze, while time–frequency structure and emitted intensity of sonar calls reveal the bat's attention along the range axis (Surlykke et al. 2009a). Just before prey capture, aerial-hunting bats increase pulse repetition rate and progressively shorten individual calls, producing a feeding buzz (Figure 10.1f). The same patterns of sonar signal change are seen in trawling bats hunting prey on the surface of water (Surlykke et al. 2009b; Brinkløv et al. 2010). In contrast, gleaning bats take prey from surfaces of vegetation and often do not produce the terminal "buzz" (Ratcliffe et al. 2013). These bats emit their biosonar signals at a steady, higher rate when foraging than when flying in the open.

The bat's adaptive changes in sonar signal design as it detects, approaches, and intercepts food support task-specific perceptual requirements (Moss et al. 2011), providing researchers with a window into general solutions to the challenges of auditory scene analysis (Bregman 1990; Lewicki et al. 2014). Shorter calls with broader bandwidth produced by bats closer to targets increase sonar resolution by conveying more accurate time information and reducing call/echo and echo/clutter overlap. Importantly, the increased call rate provides more frequent images for faster updates of moving elements in the scene. Similarly, visual animals make shifts in gaze (smooth pursuit) to track moving objects to prevent losing sight of them.

10.3 Spatial Perception in Vision and Echolocation

In vision, stimuli consist of spatial patterns of light delivered to the 2-D receptor arrays of the retinas in the left and right eyes through refraction by the cornea and lens. These spatial patterns of light lead to corresponding spatial patterns of receptor excitation, which change over time through movement of objects in the scene, along with movements of the animal's eyes, head, and body. The initial process of object representation in vision occurs through topographic activation of the retina from stimuli located in different regions of space. Horizontal and vertical object position is directly registered by activation patterns on the retina, with binocular disparity coding for object distance (depth). Additional information about target distance is available in monocular cues, such as relative size and motion parallax. What is the equivalent process for space representation in echolocation?

In echolocation, the receptor surface cannot directly code for object direction and distance. Instead, the cochlea of each inner ear separates the frequencies of sounds and registers their amplitudes and times of occurrence (Kössl and Vater 1995). Because the nature of this representation differs from that of vision in that no explicitly spatial information is registered along the receptor array, central neural mechanisms must be used to compute the locations of objects reflecting the echoes. The stimuli for sonar localization consist of pairs of sounds: signal broadcasts and individual echoes reflected by different objects in the surrounding scene. Each broadcast sonar signal or echo contains a range of frequencies that stimulate the 1-D receptor arrays in the left and right inner ears. The spatial pattern of excitation in each inner ear receptor array reflects the momentary spectrum of the sound, which includes the temporal sequence of signal frequencies, as well as amplitudes at those frequencies. However, these features only convey information about the spatial scene itself after computations in the central auditory system.

In contrast to the topographic representation of object location in vision, time is the primary stimulus dimension used for spatial localization by sonar. Time carries information about the distance to each object (target range) and is represented by the delay of echoes from the object reflecting the sounds (Simmons 1973). Echo time delay is not encoded by the spatial distribution of excitation on the receptor arrays of the inner ears. All of the frequencies in the broadcast impinge on the inner ear's frequency-selective receptors at the moment of transmission. The resulting excitation generates neural responses that ascend the auditory pathways. Responses evoked by the broadcast start the neural equivalent of a clock, actually a series of clocks for the individual frequencies contained in the broadcast. These clocks give a reference time for interpreting the echoes that return shortly thereafter. As reflected versions of the broadcast itself, echoes necessarily contain a subset of the frequencies in the broadcast: acoustic interference from overlapping echoes, excess attenuation of high frequencies, and properties of the target and the external ear all add to filtering of the echo. Each echo's frequencies impinge on the same frequency-selective auditory receptors that have just received the broadcast evoking a similar pattern of responses to the broadcast but occurring later in time. The bat's inner ear thus registers two separate volleys of neural responses from sonar emissions and returning echoes, separated in time by the delay of each echo return. Target distance is estimated by pooling the neural delay estimates across responses to different sound frequencies.

When a single object, such as an insect flying in open space, returns a single echo for each sonar signal, a bat can compute distance to a target from the echo delay. The bat can compute azimuth from binaural differences in the timing, intensity, and spectrum of returning echoes (Batra et al. 1997; Holderied and von Helversen 2006; Sümer et al. 2009). The complex shape of the bat's external ears modifies the echo spectrum, which is used to determine elevation (Wotton and Simmons 2000; Aytekin et al. 2004). Presumably, the bat's sonar receiver combines information about target distance and direction to build a unified representation of the object's relative spatial location, which changes in the bat's dynamic as it flies.

The bat emits a series of sounds to follow changes in the echo scene over time. In natural situations, the bat tracks a target's horizontal and vertical position by aiming its head (Ghose and Moss 2006) to maintain a broad spectrum of the broadcast sounds impinging on the target. The bat tracks the target's range as it approaches by shortening call durations to prevent call-echo overlap (Surlykke et al. 2009a). Echoes arriving from an off-axis target differ binaurally in time, amplitude, and spectrum. The bat must determine whether to track left or right to keep its sonar beam centered on the target. A target straight ahead means minimal differences in the amplitude and timing of echoes arriving at the two ears. Small ear movements can enhance interaural cues and aid in the accuracy of target localization.

In more complicated surroundings, such as those encountered by a bat chasing an insect flying near vegetation, the scene consists not only of the insect but also of other sources of echoes (clutter) distributed around the bat at different distances. While tracking the target, the bat must also acquire some information about the background to prevent collisions with obstacles. In experiments involving the presence of several objects in its path, the bat scans the scene, locking its sonar beam axis sequentially on selected targets (Surlykke et al. 2009a; Falk et al. 2011). The bat also takes into account the relative distance of objects by keeping its broadcast durations shorter than the delay of the echoes from the closest object (Aytekin et al. 2011).

In simple cases, the objects are far enough apart in range that each echo arrives alone. Then, the target can be kept separate from the surrounding clutter by keeping track of its distance and direction. In more complex scenes, bats try to avoid call overlap with echoes from the target of interest, while accepting call-echo overlap from nearby objects in the scene (Surlykke et al. 2009a). Even then, when echoes from a close target and clutter merge at the bat's inner ears, it appears as if the sonar receiver can still segregate the target from clutter (Simmons et al. 1989; Simmons 2014). Sampling echo information over time and frequency surely contributes to the bat's ability to separate objects in a cluttered sonar scene, but for many species of bats, figure-ground segregation by echolocation remains poorly understood.

10.4 Inferences from Behavioral Data on 3-D Object Position in Vision and Echolocation

The primary visual input is organized as a 2-D image of horizontal and vertical excitation on the retina, whereas the primary input for echolocation is organized as a 1-D "image" consisting of excitation distributed along the frequency axis of each cochlea. Sound is processed with reference to the primary dimension, which is the time elapsed between broadcast and echo (Simmons 2012). For the bat, the relative distances of different objects are directly encoded in the time sequence of the returning echoes. However, to determine the absolute distance to each object, the bat must compute the time delay between the outgoing vocalization and returning echo (Simmons 1971). Echo delay-tuned neurons in the bat's midbrain

(Valentine and Moss 1997; Portfors and Wenstrup 1999), thalamus (Olsen and Suga 1991), and cortex (O'Neill and Suga 1979; Dear et al. 1993) are likely to provide at least part of the neural basis for target range estimation in bats.

Both vision and echolocation lack one or more spatial dimensions of scenes in the initial representation of their respective stimuli. In both systems, the missing coordinates in the physical 3-D spatial scene are determined through computations that take place at higher neural processing stations that receive direct and indirect inputs from the receptor organs. Spatial information processing involves comparisons of bilateral inputs from both eyes or from both ears. Although spatial perception by echolocation is based on fundamentally different physical stimuli and early neural processing, higher order representations of scenes in the two systems may generally be more comparable in the sense that one or more dimensions appear from neural computations instead of simply registering the excitation pattern distributed across the receptor array. In effect, the perceptual scene, whether sensed through echolocation or through vision, must be a functional representation of the spatial organization in the actual physical scene, which constrains goal-directed and navigation behaviors.

While maximum detection range by echolocation is orders of magnitude shorter than by vision, the accuracy of the two systems is not so different. Range discrimination in bats has been studied in psychophysical experiments, which give the bat a choice between a nearer or farther target (a shorter or a longer echo delay). At close range, at an absolute distance of 0.5–0.6 m, the big brown bat can discriminate the difference in distance between two objects with an accuracy on the order of 1 cm, corresponding to an echo delay difference of less than 60 μs (Simmons 1971; Simmons and Grinnell 1988; Surlykke 1992). These experiments are important for understanding spatial perception in biosonar because they suggest that object features such as shape may build on the process of discriminating target distances from echo delays (Simmons 2012).

In vision, precise determination of distance is a complex process involving not only binocular disparity but also monocular cues, such as occlusion. Memory and experience also play a role, as demonstrated by the "Ames Room" optical illusion. Experimental data from humans show a size-distance invariance in determining egocentric distance. This means that relative accuracy in range estimation remains constant, and absolute accuracy depends on absolute distance. The binocular cues for determining depth in a scene (convergence, accommodation, stereopsis) are limited to distances of only a few tens of meters (Philbeck and Loomis 1997). Accuracy in determining distance by vision has been revealed by various methods. These include verbal reports, walking to a goal, or judging which of two objects is nearer. Results vary with methods, but even with all distance cues available, the accuracy is limited to about 10 cm (±20 %) at 1 m absolute distance (Kunnapas 1968; Silva 1985), which means that nearby distance discrimination by vision in humans is, in fact, somewhat less accurate than that by echolocation in bats.

Behavioral studies on perception of horizontal position show that localization accuracy in bats is about 1–2°, either for tracking targets (Griffin et al. 1965; Masters et al. 1985; Trappe and Schnitzler 1982) or for perceiving small changes in the spacing between vertical rods (Simmons et al. 1983). Vertical resolution is about 3°

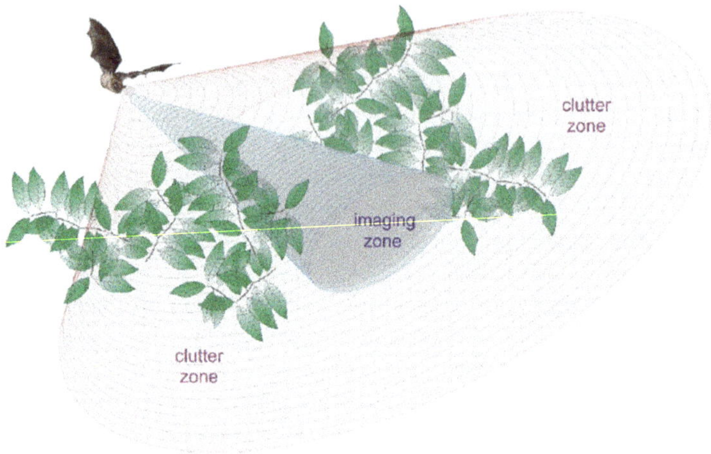

Fig. 10.4 Diagram illustrating the forward-pointing region of high-resolution sonar imaging and the side-looking peripheral regions of clutter suppression, where information about offside objects is available for guiding flight from the acoustic flow field while not interfering with forward-looking imaging

(Lawrence and Simmons 1982a). What about spatial acuity in vision? Visual resolution of high-contrast stimuli in humans is on the order of 1 min of arc (1/60°), far greater than horizontal localization accuracy by echolocation in bats. Visual resolution in animals varies greatly, from birds of prey that show visual acuity about two times higher than humans to rodents that show acuity about an order of magnitude below that of humans, thus comparable to bats (Fox et al. 1976; Prusky et al. 2000). However, visual acuity may not be directly comparable to horizontal localization accuracy in bats. Instead, it may be more appropriate to compare to the accuracy of pointing to a visual target or to the precision of visual gaze control. Both measures of visual aim accuracy are of the order of 1–5° in humans (Biguer et al. 1984; Land and Tatler 2009; Tatler and Land 2011), which is comparable to acoustic gaze accuracy in echolocating bats (Ghose and Moss 2003).

Echolocating bats are best known for their ability to find and intercept flying insects. But bats are also very adept at flying through vegetation, finding open paths through arrays of obstacles and even finding targets embedded in clutter, such as leaves and branches (Siemers and Schnitzler 2000). Understanding the means by which flying bats perceive targets and immediate obstacles without suffering interference from objects located to the sides is important in the context of spatial perception. The high level of sonar performance achieved by bats flying in clutter depends on auditory computations that create regions of sensitivity for features of objects in the immediate surroundings (Figure 10.4). The central perceptual imaging zone is estimated to be narrow, about 20° (±10°) across (Sümer et al. 2009). When a big brown bat tracks a flying insect, it keeps the target centered in this imaging zone (Figure 10.4) so the full spectrum of broadcasts impinges on the target (Ghose and Moss 2006). As a result, changes in the spectrum of echoes can be attributed to the shape of a target more than its location. Surrounding the central imaging zone (Figure 10.4) is a wider zone that extends out to about 70–90° off to the bat's sides (Jakobsen et al. 2013). Echoes from objects in this peripheral zone are low-pass filtered because of their off-axis position (Bates and Simmons 2011). This would be expected to carry low-resolution sonar information, comparable to visual acuity in the periphery. The low-pass filtering of peripheral echoes may provide a mechanism to minimize perceptual interference of full-spectrum echoes returned by objects located straight ahead. Such a mechanism of central imaging and peripheral blurring may give the bat a "tunnel vision" effect comparable to what humans experience when fixating the center of a photograph or a natural scene. The operation of these central and peripheral zones is difficult to discern because they are manifested in perception and closely related to "attention" both in echolocation and in vision (Simmons 2012, 2014).

Interestingly, while clutter has to be sidelined in perception to prevent perceptual interference, representation of clutter echoes is still necessary to guide flight through clutter. However, a broad sonar beam close to objects is not universal. Some bats that emit echolocation signals through their nostrils produce very narrow beams in cluttered space (Brinkløv et al. 2011; Surlykke et al. 2013). The wide and narrow acoustic fields of ensonification in clutter by vespertilionids (Jakobsen et al. 2013) and phyllostomids (Surlykke et al. 2013), respectively, suggest quite different sonar approaches to clutter echo processing in these different bat families.

10.5 Stroboscopic Nature of Echolocation

Few temporal changes in the sonar scene happen at a timescale fast enough to be encoded in a single echo only a few milliseconds long. Only very fast movements, such as the wingbeats of insects, might occur within the duration of a single long call of a CF bat and, perhaps, the longest search calls of FM bats. Most insects beat their wings at rates below 50 Hz, which requires a call duration of at least 20 ms to capture one full wingbeat cycle (Schnitzler et al. 1983). In general, the changes in the scene are encoded by the changes in echo composition from call to call. Thus, the update rate is set by the bat's own call rate. Bats produce their FM echolocation sounds intermittently, with silent intervals between sounds that vary, depending on the task (Kothari et al. 2014). This raises the question of how bats may experience a dynamic echo scene and track moving objects as humans do through vision when they sample the scene intermittently (Moss and Surlykke 2001).

In open spaces, a long silent period of up to 200 ms between calls allows undisturbed time for echoes to return from large reflecting surfaces, such as the ground, as far away as 20–30 m (Stilz and Schnitzler 2012). For big brown bats in more complex spaces, intervals of 20–40 ms give time for echoes to arrive from objects at distances up to approximately 3–7 m. If the task involves maneuvering in close proximity to dense clutter, the intervals become shorter, and the sounds are grouped together into pairs or triplets, referred to as sound groups or "strobe groups," as described above (Figure 10.2) (Petrites et al. 2009).

In the terminal stage of interception, the buzz, call intervals are 6–7 ms, allowing echoes to arrive from about 1 m away before the next sound is emitted (see Figure 10.1f for sounds recorded from a big brown bat during a terminal buzz). It makes sense that bats emit their sounds with intervals adjusted to receive and process echo returns. At least acoustically, the echo scene only exists in the fleeting moment following each broadcast, when echoes return to the bat, followed by an interval of silence until the scene is updated the next time the bat emits a sound. This is the background for referring to echolocation as a "stroboscopic" sense. However, the question remains as to whether the scenes perceived by bats also are intermittent.

Under normal, continuous illumination of the visual scene, humans experience a stable visual percept. In part, this stability is illusory, however; it is the product of mechanisms in vision preventing perception of the actual intermittent nature of visual stimulation. When humans scan a scene by serially inspecting different objects, they make high-velocity saccadic eye movements, which are followed by periods of fixation. Visual sensitivity during saccades is suppressed (Volkmann et al. 1968), and yet these interruptions of visual sensory input are not detected by the observer; instead, the perceptual experience is of a continuous and stable visual world built from retinal stimulation during periods of fixation. Echolocation and human vision share similarities in this respect: both operate with intrinsically stroboscopic signal transmission. The fact that humans report a stable percept of the world from interrupted visual snapshots raises the question if bats also perceive a continuous representation of a scene from echoes received at the rate of the broadcasts.

Bats also scan their surroundings in a way that is strikingly similar to visual scanning. The sonar beam of *E. fuscus* is broad enough to simultaneously collect echo information from many objects within a 60–90° cone (Figure 10.3) in the frontal plane (Ghose and Moss 2003). In spite of this broad sonar beam pattern, the bat shifts its attention to sequentially and accurately point the beam axis in the direction of closely spaced objects when it encounters a complex environment with a number of salient objects distributed around the scene. In an experiment where big brown bats were trained to fly through an opening in a mist net to gain access to a tethered insect, the bats sequentially scanned the edges of the opening and more distant prey by pointing the axis of the sonar beam with an accuracy of about 5° (Figure 10.5) (Surlykke et al. 2009a). In another task pipistrelle bats, *Pipistrellus pipistrellus*, also fixated the sonar beam sequentially with high accuracy when approaching an array of microphones (Seibert et al. 2013). Such sequential focusing on salient features in the echo scene shows clear parallels with visual inspection of complex scenes, as revealed by eye-tracking devices. When entering a room (or making tea), humans sequentially, 1–2 times per second, make saccadic eye movements to fixate on objects and obstacles with accuracy of around 1°, in spite of a broad visual field of view (Land and Tatler 2009). Thus, human observers and echolocating bats show similarities in scene scanning.

In vision, another factor that helps to stabilize the perceptual experience is the slow response of photoreceptors and visual neurons to changes in stimulation. This is obvious from the illusion of movement when watching movies even at low frame rates of 25–30 frames per second. In contrast, the temporal precision of the auditory

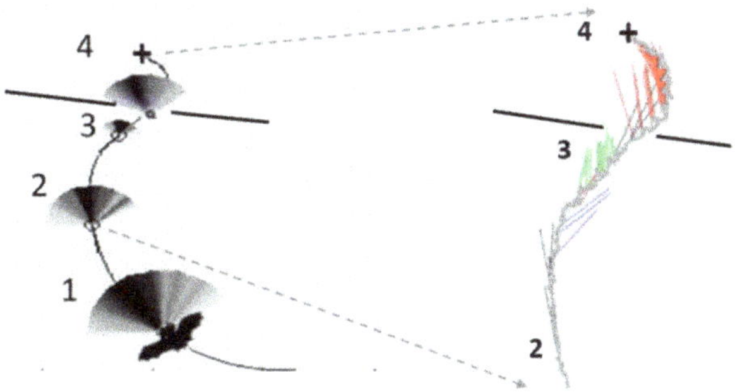

Fig. 10.5 The sonar beam of *E. fuscus* as it approaches a net opening with a reward behind. The *left plot* shows the beam pattern for selected vocalizations. The *right plot* displays the beam axis for each vocalization when the bat begins its approach to the opening and tethered worm (phase 2–4). The directional aim of the sonar beam is color coded: *blue* for the *right* edge of the net opening, *green* for the *left* edge of the net opening, and *red* for the tethered worm. Note the bat's sequential, precise scanning of the closely spaced objects

system is much faster (Rabinowitz and King 2011). In situations requiring fast perceptual processing and behavioral reactions, the echolocating bat's acoustic behavior indicates a need and capability for obtaining more precise information about its surroundings. Call rate increases not only when the bat's own behavior produces changes about information from the environment, as occurs when it approaches targets and background or make turns. A change in echolocation call rate is also observed when the bat's surroundings change, e.g., when flying in groups with conspecifics (Brinkløv et al. 2009) or when a prey item suddenly appears (Kothari et al. 2014). In the terminal phase of prey pursuit, the big brown bat's call rate, up to 150 sounds per second (Surlykke and Moss 2000), provides a much faster refresh rate compared with vision. This indicates high temporal resolution, even if bats cannot respond to each echo separately.

Temporal resolution in bat hearing also depends on stimulus level. At near-threshold levels, sensory receptors operate as energy detectors, integrating energy over a long time to facilitate detection. Energy integration times of around 60 ms were measured in a passive-listening detection task in Brazilian free-tailed bats (*Tadarida brasiliensis*) (Schmidt and Thaller 1994). In actively echolocating vespertilionids, much shorter integration times of around 2.5 ms (the duration of the call) were measured (Surlykke and Bojesen 1996). These results, along with range-dependent adjustments in call duration, suggest an adaptable integration time in the bat auditory system, changing from long to short, from passive listening to active echolocation during the course of prey pursuit. The changes in auditory integration time reflect the dynamic task at hand. In the foraging search phase, target detection is the problem. To take advantage of the long search calls, bats must integrate energy over the full duration of the call (up to 20–30 ms in FM bats). In the terminal phase of insect pursuit, precise localization for prey interception requires a short integration time to record a time stamp for echo arrival time, the bat's cue for target distance.

The integration time of the visual system also depends on energy level but operates on longer timescales than the auditory system. Under limited light level conditions, where rod vision dominates (scotopic vision), time resolution is very slow. When light levels and viewing conditions are controlled in a laboratory setting, maximum critical flicker fusion frequency (CFF) in humans approaches 50 Hz (Hecht and Smith 1936), but under natural viewing situations, CFF is under 20 Hz. Humans, for example, cannot resolve the blades of a fan when the motion is fast. Thus, in terms of temporal resolution, bats appear to represent the world at a faster rate through echolocation than visual-dominated animals like humans do through their optical imaging systems.

10.6 Scene Analysis by Vision and Echolocation

An animal's surroundings deliver stimuli arriving from multiple objects located in different directions that comprise the "scene." Objects emerge in perception as a consequence of bottom-up and top-down processing. In much-simplified terms, a scene typically consists of a number of objects that might have biological relevance, for instance, food. Also present is a plethora of other objects that constitute the bulk of the scene. This is the background that must be segregated to minimize interference with perception of relevant objects. The animal analyzes the scene to perceive the locations and identities of targets and obstacles, which guide its goal-directed locomotion. In particular, individual objects are perceived as discrete entities, segregated from their surroundings, so they can be located, classified, and acted upon (Lewicki et al. 2014). If individual, relevant objects are not effectively segregated from the background, important objects could go unrecognized. In vision, objects can have different physical features that distinguish them from the background (e.g., brightness, color, motion, texture), and the stereo view provided by the two eyes can be used to locate individual objects in depth, so segregation is enhanced. Only when one object is partially hidden behind another, occlusion, does scene-derived interference with visual perception occur. Moreover, even a hidden object can be discerned if one or both eyes capture part of it, or the hidden object may come into view simply by changing the viewer's location in the scene.

In contrast to passive listening, echolocation carries information about target distance in the arrival time of sonar returns. Echoes are reflected from objects in "line of sight" of the sonar emission. Thus, sonar reflections from objects can be masked or completely hidden by echoes from large structures in the foreground, a kind of "sonar occlusion." In vision, the 2-D representation on the retina already segregates objects spatially, whereas, in echolocation, objects in a sonar scene at about the same distance as the target, just in different directions, may create a clutter problem because these echoes arrive at about the same delay after broadcasts, overlapping in time. These near-simultaneous echoes have to be separated to discern the target in the presence of the clutter. In contrast to vision, where one object may entirely occlude a more distant one, in echolocation, it is possible for an animal to detect a distant object that is directly behind another from echo returns separated in arrival time.

Examining how bats separate targets and clutter reveals yet another similarity between echolocation and vision, specifically, what information from the environment is actually perceived. Fundamental to this process is *attention*, the selection of objects from the surrounding scene for scrutiny. In vision, when a human looks at an object, perception is momentarily concentrated on that object and its features. Surrounding elements of the scene are outside the region of sharp focus but not out of sight (Tatler and Land 2011). If something in the periphery attracts attention, gaze is shifted to examine it. In echolocation, too, bats attend to selected objects (indicated by where the sonar beam is aimed) and monitor surrounding objects, which constitute clutter (Simmons 2014). Here, echolocation diverges from vision. In vision, objects located off to the sides of the point of fixation are registered outside the fovea of the retina, and there is little problem keeping them segregated from the object registered at the fovea. In echolocation, objects located off to the side might return echoes that overlap echoes received from the object being fixated by the bat's head-aim gaze. In this case, the echoes from surrounding objects constitute a source of perceptual interference that could mask the presence of echoes from the object of immediate interest. This effect is *clutter interference*, and it has no direct parallel in vision. Bats solve the problem of clutter interference using neural computations that spatially blur off axis echoes (Bates et al. 2011; Bates and Simmons 2011). However, whether for vision or for echolocation, once an individual object is selected and its features expressed in a central perceptual display, the object can be discriminated from background (Lewicki et al. 2014).

Comparing a photograph of a visual scene and abstractions of an echo representation of the same scene illustrates how images are composed through echolocation and vision. The image in Figure 10.6a shows a photographic representation, assuming color vision, with brightness and color contrast highlighting dark tree trunks, branches, and green leaves. Light sends rays through the forest and casts shadows from trees across the ground. The location of trees defines open space along a roadway, changes in terrain, and texture. The distance of objects is conveyed by relative size and occlusion. Occluded objects and those further away, deeper in the forest, are not present in the photograph. The image in Figure 10.6b represents an attempt to capture some aspects of an echo scene from the same environment. Bright images are intended to convey objects that would return the strongest most broadband echoes, such as trees and branches closest to the echolocating bat. More distant objects return weak echoes, illustrated by blur. A key feature of a sonar image that cannot be conveyed in the photographic image in Figure 10.6b is that the distance to the different tree trunks and branches would be conveyed through a cascade of echo delays.

The images in Figure 10.6a, b illustrate that both vision and echolocation support spatial perception, while at the same time building on totally different physical stimuli and sensory receptors. In one crucial way, looking at the scene in a picture is much like looking at a natural scene: a viewer is free to scan the photograph focusing gaze at different aspects, just as when scanning objects in the environment. When a viewer fixates the center of the picture in Figure 10.6a, this region of the scene is the focus of attention and is perceived at high spatial resolution because information is captured by the fovea of the retina, with its tightly packed photoreceptors. The parts of the scene in the photograph located off to the sides are rendered comparatively blurred with progressively lower spatial resolution because the retina has lower receptor density. This has functional significance. Looking at the road in the picture, for example, provides information

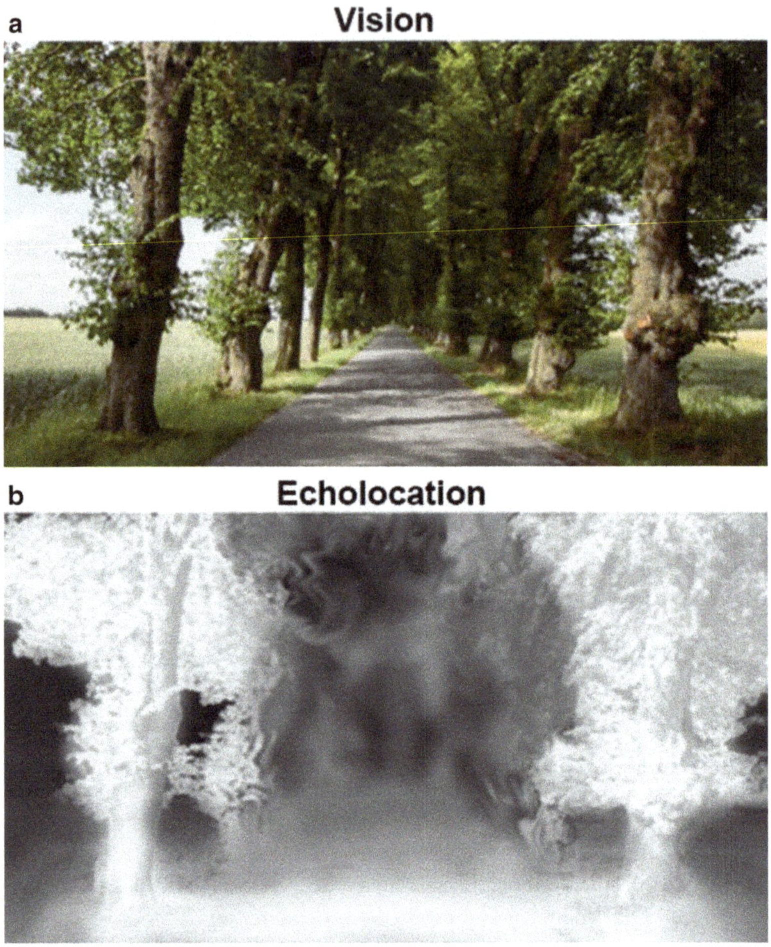

Fig. 10.6 Information carried in a visual scene and an echo scene. (**a**) Visual scene: Brightness and color contrast highlight dark tree trunks, branches, and green leaves. Light source to the left sends rays through the forest and casts shadows from trees across the ground. Location of trees defines open space along roadway, changes in terrain, and texture. Distance of objects is conveyed by relative size and occlusion. Occluded objects far away are invisible. (**b**) Echo scene: Object direction, size, and texture conveyed by echo amplitude, timing, and spectrum. The bat focuses its beam sequentially on the two close trees to the left and right. The closer and the more on-axis objects are, the broader (*whiter*) the spectrum, including more high frequency. Distance of objects is conveyed by echo delay and amplitude. Location of trees, branches, leaves, and ground delineate clutter and open space (Photos taken and edited by Peter Berger)

about the way to walk forward without colliding with trees. Shifting to focus on one of the trees in the picture provides full details of its leaves and branches but with a loss of detailed information about the direction to walk forward along the road. Thus, in the natural environment, an observer's gaze would follow the road while walking forward, only shifting away from the road if attention is redirected. The observer assembles a series of high-resolution foveal snapshots of parts of the scene and would not be aware of the blurred peripheral elements or interruptions in visual signal transmission due to saccadic suppression.

The distinction in vision between the attended region of high resolution and the defocused periphery can be conceptualized as a perceptual center and surround (Figure 10.4), for which there are parallels in echolocation (Simmons 2014).

The echo scene is by nature more difficult to illustrate in a 2-D photograph. Imagine that the sonar beam is first directed at the trees lining the left side of the road and subsequently directed to the right. The accumulated image of this figure conveys the composite echo returns from several sonar emissions. In the illustration of this figure, the whiteness of the portrayed image represents the intensity and bandwidth of spectral energy in each echo return; thus, the closer an object and the more on-axis the sonar emission, the more high-frequency energy is returned. For this reason, at increasing distance, the display becomes darker because echoes become progressively weaker due to spherical spreading losses and excess attenuation of high-frequency sounds. When the bat directs its sonar beam toward trees lining the left side of the road, for example, close trees would return strong broadband echoes, and echoes from objects in the bat's periphery would be low-pass filtered, as described above.

10.7 Summary and Conclusions

Echolocation and vision are distal sensing systems that support 3-D spatial perception of the environment. In spite of distinct differences between the physical stimuli, receptor organs and early neural processing of the two systems, a comparison of vision and echolocation reveals some striking parallels that can guide a more informed understanding of higher level perceptual mechanisms across modalities. For example, high resolution along the central axis is common to vision and echolocation. This suggests that distal sensing systems are organized to extract sharp spatial information in the direction of gaze. This may reflect a common strategy to efficiently allocate computational resources to guide action.

Both echolocation and vision support target tracking and interception, which requires comparable, extremely fast sensorimotor coordination. An entire pursuit sequence for an insectivorous bat typically lasts a fraction of a second to about a second, and, in baseball, there is often less than half a second for a batter to judge the trajectory of the baseball pitch, which gives time for one saccade, or at the most two saccades, to accurately judge the ball's changing position. A player's performance improves greatly with training, and eye tracking devices have shown that experienced baseball players predict the ball's movement; instead of following the ball, they fixate the predicted point in space where it will make contact with the baseball bat. This behavior requires the construction of an internal, neural dynamic model of the ball's movement. Data also show that the model can be rapidly updated (Land and McLeod 2000; Hayhoe and Ballard 2005). These findings lead to the speculation that echolocating bats construct and update internal models of insect trajectories to plan their interception.

In fast ball games, input to the visual system far exceeds the temporal visual resolution and even more so the minimum reaction time, but experience and predictive models result in accurately coordinated behavior. The same may hold true for bats relying on echolocation. The input rate, in particular in the sonar buzz, is faster than auditory processing time, but like a ball player, the combination of detailed perceptual processing and practice of motor skills allow bats to perform impressively swift maneuvers for prey interception. These rapid and agile sonar-guided behaviors can be observed any summer night in insect- and bat-rich areas. In the lab, trained bats often perform in a stereotypical way, developing individual strategies to solve tasks (Moss et al. 2006; Barchi et al. 2013). Lab studies of capture maneuvers indicate a predictive dynamic model of the bat's position relative to obstacles and prey (Miller and Olesen 1979; Ghose et al., 2006). However, in the wild, the situation is much less predictable and, like ball players, bats must adapt and update their model of insect trajectory to cope with the natural variation in prey size, flight speed, and behavior.

Lab and field studies of bat echolocation demonstrate how the animal's adjustments in sonar call features and motor behaviors contribute directly to solving the scene analysis problem of sorting/tracking sounds arriving from different directions and distances. Adjustments in source level, duration, and direction of calls provide reliable indicators of the bat's attention to objects at different locations, like saccadic eye movements, and accommodation reveals the focal point of visual attention.

Observations on visual perception can also lead to insights about echolocation. For example, it is noteworthy that human observers experience a continuous and stable percept of the world, although visual information is suppressed during eye movements. This finding motivates the hypothesis that bats also experience a continuous representation of the world through interrupted echo snapshots of the environment. Importantly, comparisons between echolocation and vision can guide broader insights to the common functions of perceptual systems in interpreting raw sensory input and enabling complex spatial behaviors in dynamic environments.

Acknowledgements We are grateful to Wei Xian for analyses of the data sets and careful preparation of figures. This work was supported by the Danish Council for Natural Sciences, FNU (0602-02529B) to AS; the Human Frontiers Science Program Organization, HFSP (RGP0040/213), the National Science Foundation, NSF (IOS-1010193), the Air Force Office of Sponsored Research (FA9550-14-1-0398), and the Office of Naval Research (N00014-12-1-0339) to CFM; and the Office of Naval Research (N00014-14-1-0588) to JAS. We thank Peter Berger for the images in Figure 10.6 of this chapter.

References

Aytekin, M., Grassi, E., Sahota, M., & Moss, C. F. (2004). The bat head-related transfer function reveals binaural cues for sound localization in azimuth and elevation. *Journal of the Acoustical Society of America*, 116, 3594–3605.

Aytekin, M., Mao, B., & Moss, C. F. (2011). Sparial perception and adaptive sonar behavior. *Journal of the Acoustical Society of America*, 128, 3788–3798.

Barchi, J. R., Knowles, J. M., & Simmons, J. A. (2013). Spatial memory and stereotypy of flight paths by big brown bats in cluttered surroundings. *Journal of Experimental Biology*, 216(6), 1053–1063. doi: 10.1242/jeb.073197

Bates, M. E., & Simmons, J. A. (2011). Perception of echo delay is disrupted by small temporal misalignment of echo harmonics in bat sonar. *Journal of Experimental Biology*, 214, 394–401.

Bates, M. E., Simmons, J. A., & Zorikov, T. V. (2011). Clutter rejection in bat sonar is achieved by image defocusing from harmonic beamforming. *Science*, 333, 627–630.

Batra, R., Kuwada, S., & Fitzpatrick, D. C. (1997). Sensitivity to interaural temporal disparities of low- and high-frequency neurons in the superior olivary complex. I. Heterogeneity of responses. *Journal of Neurophysiology*, 78, 1222–1236.

Biguer, B., Prablanc, C., & Jeannerod, M. (1984). The contribution of coordinated eye and head movements in hand pointing accuracy. *Experimental Brain Research*, 55, 462–469.

Bregman, A. S. (1990). *Auditory scene analysis*. Cambridge, MA: MIT Press. (2nd Edition, 1994)

Brinkløv, S., Kalko, E. K. V., & Surlykke, A. (2009). Intense echolocation calls from two 'whispering' bats, *Artibeus jamaicensis* and *Macrophyllum macrophyllum* (Phyllostomidae). *Journal of Experimental Biology*, 212, 11–20.

Brinkløv, S., Kalko, E. K. V., & Surlykke, A. (2010). Dynamic adjustment of biosonar intensity to habitat clutter in the bat *Macrophyllum macrophyllum* (Phyllostomidae). *Behavioral Ecology and Sociobiology*, 64, 1867–1874.

Brinkløv, S., Jakobsen, L., Ratcliffe, J. M., Kalko, E. K. V., & Surlykke, A. (2011). Echolocation call intensity and directionality in flying short-tailed fruit bats, *Carollia perspicillata* (Phyllostomidae). *Journal of the Acoustical Society of America*, 129, 427–435.

Dear, S. P., Simmons, J. A., & Fritz, J. (1993). A possible neuronal basis for representation of acoustic scenes in auditory cortex of the big brown bat. *Nature*, 364, 620–623.

Falk, B., Williams, T., Aytekin, M., & Moss, C.F. (2011). Adaptive behavior for texture discrimination by the free-flying big brown bat, *E. fuscus*. *Journal of Comparative Physiology A: Neuroethology, Sensory, Neural, and Behavioral Physiology*, 197, 491–503.

Fenton, M. B., Faure, P. A., & Ratcliffe, J. M. (2012). Evolution of high duty cycle echolocation in bats. *Journal of Experimental Biology*, 215, 2935–2944.

Fenton, B. M., Jensen, F. H., Kalko, E. K., & Tyack, P. L. (2014). Sonar signals of bats and toothed whales. In A. Surlykke & P. E. Nachtigall (Eds.). *Biosonar* (pp. 11–59). NewYork, Berlin, Heidelberg: Springer.

Fox, R., Lehmkuhle, S. W., & Westendorf, D. H. (1976). Falcon visual acuity. *Science*, 192, 263–265.

Ghose, K., & Moss, C. F. (2003). The sonar beam pattern of a flying bat as it tracks tethered insects. *Journal of the Acoustical Society of America*, 114, 1120–1131.

Ghose, K., & Moss, C. F. (2006). Steering by hearing: A bat's acoustic gaze is linked to its flight motor output by a delayed, adaptive linear law. *Journal of Neuroscience*, 26, 1704–1710.

Ghose, K. Horiuchi, T.K., Krishnaprasad, P.S. & Moss, C.F. (2006). Echolocating bats use a nearly time-optimal strategy to intercept prey. *Public Library of Science Biology*, 4(5): 865–873.

Griffin, D. R. (1958). *Listening in the dark*. New York: Yale University Press. (2nd edition, 1986, Ithaca, NY: Cornell University)

Griffin, D. R., Friend, J. H., & Webster, F. A. (1965). Target discrimination by the echolocation of bats. *Journal of Experimental Zoology*, 158, 155–168.

Hayhoe, M., & Ballard, D. (2005). Eye movements in natural behavior. *Trends in Cognitive Science*, 9, 188–194. doi: 10.1016/j.tics.2005.02.009

Hecht S, & Smith EL (1936). Intermittent stimulation by light : VI. Area and the relation between critical frequency and intensity. *The Journal of General Physiology*, 19 (6), 979–989. PMID: 19872977.

Holderied, M. W., & von Helversen, O. (2003). Echolocation range and wingbeat period match in aerial-hawking bats. *Proceedings of the Royal Society of London B: Biological Sciences*, 270, 2293–2299.

Holderied, M. W., & von Helversen, O. (2006). 'Binaural echo disparity' as a potential indicator of object orientation and cue for object recognition in echolocating nectar-feeding bats. *Journal of Experimental Biology*, 209, 3457–3468.

Jakobsen, L., Ratcliffe, J. M., & Surlykke, A. (2013). Convergent acoustic field of view in echolocating bats. *Nature*, 493, 93–96. doi: 10.1038/nature11664

Kössl, M., & Vater, M. (1995). Cochlear structure and function in bats. In A. N. Popper & R. R. Fay (Eds.), *Hearing by bats* (pp. 191–234). New York: Springer.

Kothari, N., Wohlgemuth, M., Hulgard, K., Surlykke, A., & Moss, C. F. (2014). Timing matters: Sonar call groups facilitate target localization in bats. *Frontiers in Physiology*, doi: 10.3389/fphys.2014.00168

Kunnapas, T. (1968). Distance perception as a function of available visual cues. *Journal of Experimental Psychology*, 77(4), 523–529. doi: 10.1037/h0026050

Land, M. F., & McLeod, P. (2000). From eye movements to actions: how batsmen hit the ball. *Nature Neuroscience*, 3(12), 1340–1345.

Land, M. F., & Tatler, B. W. (2009). *Looking and acting: Vision and eye movements in natural behaviour*. Oxford, New York: Oxford University Press.

Lawrence, B. D., & Simmons, J. A. (1982a). Echolocation in bats: the external ear and perception of the vertical position of targets. *Science*, 218, 481–483.

Lawrence, B. D., & Simmons, J. A. (1982b). Measurements of atmospheric attenuation at ultrasonic frequencies and the significance for echolocation by bats. *Journal of the Acoustical Society of America*, 71, 585–590.

Lewicki, M. S., Olshausen, B. A., Surlykke, A., & Moss, C. F. (2014). Scene analysis in the natural environment. *Frontiers in Psychology*, doi: 10.3389/fpsyg.2014.00199

Masters, W. M., Moffat, A. J. M., & Simmons, J. A. (1985). Sonar tracking of horizontally moving targets by the big brown bat *Eptesicus fuscus*. *Science*, 228, 1331–1333.

Miller, L. A., & Olesen, J. (1979). Avoidance behavior in green lacewings. *Journal of Comparative Physiology A: Neuroethology, Sensory, Neural, and Behavioral Physiology*, 131, 113–120.

Moss, C. F., & Surlykke, A. (2001). Auditory scene analysis by echolocation in bats. *Journal of the Acoustical Society of America*, 110, 2207–2226.

Moss, C. F., Bohn, K., Gilkenson, H., & Surlykke, A. (2006). Active listening for spatial orientation in a complex auditory scene. *PLoS Biology*, 4, 615–626.

Moss, C. F., Chiu, C., & Surlykke, A. (2011). Adaptive vocal behavior drives perception by echolocation in bats. *Current Opinion in Neurobiology*, 21, 645–652.

Neuweiler, G. (1989). Foraging ecology and audition in echolocating bats. *Trends in Ecology and Evolution*, 4, 160–166.

O'Neill, W. E., & Suga, N. (1979). Target range-sensitive neurons in the auditory cortex of the mustache bat. *Science*, 203, 69–73.

Olsen, J. F., & Suga, N. (1991). Combination-sensitive neurons in the medial geniculate body of the mustached bat: Encoding of target range information. *Journal of Neurophysiology*, 65, 1275–1296.

Petrites, A. E., Eng, O. S., Mowlds, D. S., Simmons, J. A., & DeLong, C. M. (2009). Interpulse interval modulation by echolocating big brown bats (*Eptesicus fuscus*) in different densities of obstacle clutter. *Journal of Comparative Physiology A: Neuroethology, Sensory, Neural, and Behavioral Physiology*, 195, 603–617.

Philbeck, J. W., & Loomis, J. M. (1997). Comparison of two indicators of perceived egocentric distance under full-cue and reduced-cue conditions. *Journal of Experimental Psychology: Human Perception and Performance*, 23, 72–85.

Portfors, C. V., & Wenstrup, J. J. (1999). Delay-tuned neurons in the inferior colliculus of the mustached bat: implications for analyses of target distance. *Journal of Neurophysiology*, 82(3), 1326–1338.

Prusky, G. T., West, P. W., & Douglas, R. M. (2000). Behavioral assessment of visual acuity in mice and rats. *Vision Research*, 40, 2201–2209.

Rabinowitz, N. C., & King, A. J. (2011). Auditory perception: Hearing the texture of sounds. *Current Biology*, 21(23), R967–R968.

Ratcliffe, J. M., Elemans, C. P. H., Jakobsen, L., & Surlykke, A. (2013). How the bat got its buzz. *Biology Letters*, 9, 20121031. doi: 10.1098/rsbl.2012.1031

Sändig, S., Schnitzler, H.-U., & Denzinger, A. (2014). Echolocation behaviour of the big brown bat (*Eptesicus fuscus*) in an obstacle avoidance task of increasing difficulty. *Journal of Experimental Biology*, doi: 10.1242/jeb.099614

Schmidt, S., & Thaller, J. (1994). Temporal auditory summation in the echolocating bat, *Tadarida brasiliensis*. *Hearing Research*, 77, 125–134.

Schnitzler, H.-U., & Kalko, E. K. V. (2001). Echolocation by insect-eating bats. *BioScience*, 51, 557–569.

Schnitzler, H.-U., Menne, D., Kober, R., & Heblich, K. (1983). The acoustical image of fluttering insects in echolocating bats. In F. Huber & H. Markl (Eds.), *Neuroethology and behavioral physiology* (pp. 235–250). New York: Springer.

Seibert, A.-M., Koblitz, J. C., Denzinger, A., & Schnitzler, H.-U. (2013). Scanning behavior in echolocating common pipistrelle bats (*Pipistrellus pipistrellus*). *PLoS ONE*, 8, e60752. doi: 10.1371/journal.pone.0060752

Siemers, B. M., & Schnitzler, H.-U. (2000). Natterer's bat (*Myotis nattereri* Kuhl, 1818) hawks for prey close to vegetation using echolocation signals of very broad bandwidth. *Behavioral Ecology and Sociobiology*, 47, 400–412.

Silva, J. A. D. (1985). Scales for perceived egocentric distance in a large open field: comparison of three psychophysical methods. *The American Journal of Psychology*, 98(1), 119–144. doi: 10.2307/1422771

Simmons, J. A. (1971). Echolocation in bats: signal processing of echoes for target range. *Science*, 171, 925–928.

Simmons, J. A. (1973). The resolution of target range by echolocating bats. *Journal of the Acoustical Society of America*, 54, 157–173.

Simmons, J. A. (2012). Bats use a neuronally implemented computational acoustic model to form sonar images. *Current Opinion in Neurobiology*, 22, 311–319. doi: 10.1016/j.conb.2012.02.007

Simmons, J. A. (2014). Temporal binding of neural responses for focused attention in biosonar. *Journal of Experimental Biology*, 217, 2834–2843.

Simmons, J. A. & Grinnell, A. D. (1988). The performance of echolocation: Acoustic images perceived by echolocating bats. In P. Nachtigall & P. W. B. Moore (Eds.), *Animal sonar systems: Processing and performance*. (pp. 353–385). New York: Plenum Press.

Simmons, J. A., & Stein, R. A. (1980). Acoustic imaging in bat sonar: Echolocation signals and the evolution of echolocation. *Journal of Comparative Physiology A: Neuroethology, Sensory, Neural, and Behavioral Physiology*, 135, 61–84.

Simmons, J. A., Kick, S. A., Lawrence, B. D., Hale, C., Bard, C., & Escudié, B. (1983). Acuity of horizontal angle discrimination by the echolocating bat, *Eptesicus fuscus*. *Journal of Comparative Physiology A: Neuroethology, Sensory, Neural, and Behavioral Physiology*, 153, 321–330.

Simmons, J. A., Freedman, E. G., Stevenson, S. B., Chen, L., & Wohlgenant, T. J. (1989). Clutter interference and the integration time of echoes in the echolocating bat, *Eptesicus fuscus*. *Journal of the Acoustical Society of America*, 86, 1318–1332.

Stilz, W.-P., & Schnitzler, H.-U. (2012). Estimation of the acoustic range of bat echolocation for extended targets. *Journal of the Acoustical Society of America*, 132(3), 1765–1775.

Sümer, S., Denzinger, A., & Schnitzler, H.-U. (2009). Spatial unmasking in the echolocating Big Brown Bat, *Eptesicus fuscus*. *Journal of Comparative Physiology A: Neuroethology, Sensory, Neural, and Behavioral Physiology*, 195, 463–472.

Surlykke, A. (1992). Target ranging and the role of time-frequency structure of synthetic echoes in big brown bats, *Eptesicus fuscus*. *Journal of Comparative Physiology A: Neuroethology, Sensory, Neural, and Behavioral Physiology*, 170, 83–92.

Surlykke, A., & Bojesen, O. (1996). Integration time for short broad band clicks in echolocating FM-bats (*Eptesicus fuscus*). *Journal of Comparative Physiology A: Neuroethology, Sensory, Neural, and Behavioral Physiology*, 178, 235–241.

Surlykke, A., & Kalko, E. K. V. (2008). Echolocating bats cry out loud to detect their prey. *PLoS ONE*, 3, e2036(2031)–e2036(2010).

Surlykke, A., & Moss, C. F. (2000). Echolocation behavior of big brown bats, *Eptesicus fuscus*, in the field and the laboratory. *Journal of the Acoustical Society of America*, 108, 2419–2429.

Surlykke, A., Ghose, K., & Moss, C. F. (2009a). Acoustic scanning of natural scenes by echolocation in the big brown bat, *Eptesicus fuscus*. *Journal of Experimental Biology*, 212, 1011–1020.

Surlykke, A., Pedersen, S. B., & Jakobsen, L. (2009b). Echolocating bats emit a highly directional sonar sound beam in the field. *Proceedings of the Royal Society of London B: Biological Sciences*, 276, 853–860.

Surlykke, A., Jakobsen, L., Kalko, E. K. V., & Page, R. A. (2013). Echolocation intensity and directionality of perching and flying fringe-lipped bats, *Trachops cirrhosus* (Phyllostomidae). *Frontiers in Physiology*, doi: 10.3389/fphys.2013.00143

Tatler, B. W., & Land, M. F. (2011). Vision and the representation of the surroundings in spatial memory. *Philosophical Transactions of the Royal Society B: Biological Sciences*, 366, 596–610.

Thomas, J. A., Moss, C. F., & Vater, M. (2004). *Echolocation in bat and dolphins*. Chicago, IL: The Chicago University Press.

Trappe, M., & Schnitzler, H.-U. (1982). Doppler-shift compensation in insect-catching horseshoe bats. *Naturwissenschaften*, 69(4), 193–194.

Triblehorn, J. D., Ghose, K., Bohn, K., Moss, C. F., & Yager, D. D. (2008). Free-flight encounters between praying mantids (*Parasphendale agrionina*) and bats (*Eptesicus fuscus*). *Journal of Experimental Biology*, 211, 555–562.

Valentine, D. E., & Moss, C. F. (1997). Spatially selective auditory responses in the superior colliculus of the echolocating bat. *Journal of Neuroscience*, 17(5), 1720–1733.

Volkmann, F. C., Schick, A., & Riggs, L. A. (1968). Time course of visual inhibition during voluntary saccades. *Journal of the Optical Society of America*, *58*(4), 562–569.

von der Emde, G., & Schnitzler, H.-U. (1990). Classification of insects by echolocating greater horseshoe bats. *Journal of Comparative Physiology A: Neuroethology, Sensory, Neural, and Behavioral Physiology*, 167, 423–430.

Wahlberg, M., & Surlykke, A. (2014). Sound intensities of biosonar signals from bats and toothed whales. In A. Surlykke & P. E. Nachtigall (Eds.), *Biosonar.* (pp. 107–141). NewYork: Springer.

Wotton, J. M., & Simmons, J. A. (2000). Spectral cues and perception of the vertical position of targets by the big brown bat, *Eptesicus fuscus*. *Journal of the Acoustical Society of America*, 107, 1034–1041.

Chapter 11
Perspectives and Challenges for Future Research in Bat Hearing

Lutz Wiegrebe, Alan D. Grinnell, and M. Brock Fenton

Keywords Bat communication • Bat genetics • Bat guilds • Bat navigation • Bat socialization • Echo imaging • Echolocation • Neurobiology of biosonar • Predator-prey interaction

11.1 Introduction

The previous chapters provide an overview of advances in our understanding of bat hearing and echolocation. Here major challenges are identified that continue to face those who study hearing in bats. These challenges are presented under the following headings: Phylogeny and evolution of echolocation; Prey detection, Call production: measuring azimuth and elevation; Lifestyle and echolocation in laryngeally echolocating bats; Correlation of echolocation strategies with niches: Guilds of bats; Echolocation and social signals; Neural basis of biosonar; Doppler shift compensation; Acoustic versus visual images and auditory scene analysis; and How do bats navigate?

L. Wiegrebe (✉)
Division of Neurobiology, Department Biology II, Ludwig-Maximilians-Universität München, Großhaderner Straße 2, Planegg-Martinsried D-82152, Germany
e-mail: lutzw@lmu.de

A.D. Grinnell
Department of Physiology, David Geffen School of Medicine, University of California, Los Angeles, CA 90095, USA
e-mail: adg@ucla.edu

M.B. Fenton
Department of Biology, University of Western Ontario, London, ON, Canada N6A 5B7
e-mail: bfenton@uwo.ca

M.B. Fenton et al. (eds.), *Bat Bioacoustics*, Springer Handbook of Auditory Research 54, DOI 10.1007/978-1-4939-3527-7_11

11.2 Phylogeny and Evolution of Echolocation

Important advances in modern molecular biology have changed the view of phylogenetic relationships among bats, which are discussed in Chapter 2 (Teeling, Jones, and Rossiter). Future studies of the molecular basis of echolocation can now build upon a rich resource of published genome data sets for bats (Seim et al. 2013; Tsagkogeorga et al. 2013; Zhang et al. 2013), and cetaceans (Gui et al. 2013; Zhou et al. 2013; Yim et al. 2014) with more genomes to be published by the Genome 10K project (a proposal to obtain whole-genome sequences for 10,000 vertebrate species; Genome 10K 2009).

An additional application of high-throughput sequencing can provide strong verification of a role for organs or structures associated with echolocation. In general, sequencing RNA transcripts offers a cost-effective means of obtaining coding gene sequences, especially for specific tissues, and has already been used in several studies of bats (Francischetti et al. 2013; Phillips et al. 2014). However, obtaining sufficient yields of non-degraded RNA from the cochleae of bats is technically difficult because of the high degree of mineralization of the cochlea and the small amounts of starting material. Such problems might partially be addressed by the emerging field of single-cell transcriptomics. Obtaining transcriptome data from the ears of cetaceans is arguably even more challenging because RNA degrades rapidly postmortem, precluding the use of stranded animals. Currently there is only one published study of cochlear transcriptomes of bats, in which the echolocating Rickett's big-footed bat (*Myotis ricketti*) and the Old World greater short-nosed fruit bat (*Cynopterus sphinx*) were examined (Dong et al. 2013). Comparisons of expression profiles revealed 987 genes that were significantly upregulated in the echolocating species, including 18 known hearing genes. *FoxP2*, a gene associated with fine motor control and speech in humans and songbirds, shows unusually high variation among bats compared to other mammalian orders (Li et al. 2007).

Central to the understanding of the evolution of echolocation in bats is a better appreciation of how widespread echolocation is among species of Pteropodidae. Boonman et al. (2014) demonstrated that three pteropodid species other than rousette bats (*Rousettus* spp.) echolocate. Boonman et al. (2014) demonstrated that these other pteropodids (dawn bats, *Eonycteris spelaea*; lesser long-tongued fruit bats, *Macroglossus minimus*; and short-nosed fruit bats) had a rudimentary capacity for echolocation compared to Egyptian rousettes. Although Gould (1988) proposed that *E. spelaea* generated clicks by wing claps, the data collected by Boonman et al. (2014) did not support this proposal.

11.3 Prey Detection, Call Production: Measuring Azimuth and Elevation

The ability of echolocating bats to accurately determine the range of targets (the *z*-dimension) is remarkable and well understood. However, the neural basis of localization in the other two dimensions, azimuth and elevation, are still somewhat unclear. The ability of some echolocating bats to locate targets in the vertical dimension (~3°)

is about half as accurate as the ability to locate a target in the horizontal dimension (~1.5°) (Lawrence and Simmons 1982a; Simmons et al. 1983). Either is much better than in most mammals. Accuracy in the vertical axis is critical for localizing targets in three dimensions.

The external ears (pinnae) are unquestionably important to the directionality of hearing. Species with large and complex-shaped pinnae and tragi may be key to understanding target localization. Removal of the tragus reduces vertical localization accuracy, suggesting that sound reflected into the ears by the tragi creates angle of incidence-specific spectral peaks and notches in echoes and accounts for the accuracy of vertical localization (Lawrence and Simmons 1982b). The sharp frequency-dependent directionality of hearing by each ear, determined electrophysiologically, suggests that directionality patterns are central to target localization. Moreover, electrophysiological recordings show binaural inhibitory sharpening of directional responses and exceedingly sharp response patterns of single units in the bat inferior colliculus. These data make it likely that accurate vertical sound localization involves binaural comparison of intensities at many different frequencies (Grinnell and Grinnell 1965; Fuzessery and Pollak 1984).

It is unknown how bats determine echo azimuth. Echolocating bats have phenomenal temporal acuity, but the 15–16 µs/dB time-intensity trade-off makes it unlikely that localization of target angle in azimuth is based on differences in arrival times. A recent study on the neural sensitivity to differences in the envelopes of complex echoes, however, indicates that interaural timing can contribute to azimuthal target localization (Borina et al. 2011). Indeed, there may be no fundamental difference between localization in the horizontal and vertical dimensions. Some insight into these questions might be obtained by behavioral experiments in which vertical localization is tested with only one ear, and both vertical and horizontal localization are tested with echoes containing different fractions of the echo spectrum.

A central and open question concerns the spatial resolution of bat sonar: localizing a single reflector, as described above, does not require high resolution, just high acuity. Spatial resolution (as defined in astronomy; Rayleigh 1879) quantifies the extent to which an imaging system can detect the difference between two point reflectors and a single, extended reflector, for example. Thus, the concept of spatial resolution addresses to what extent bats can perceptually separate simultaneous reflections from multiple, equidistant objects. Based on what is known about mammalian auditory binaural processing (Grothe et al. 2010), it seems unlikely that bats can spatially resolve objects as well as our visual system can. Thus, comparisons of bat biosonar and stroboscopic vision may be misleading because for every flash of light, a visual system immediately benefits from the extraordinary spatial resolution of a retina. While auditory spatial receptive fields in mammals can be sharp, especially in bats (Hoffmann et al. 2010, 2013; Bartenstein et al. 2014) and, as shown for cats, can be sharpened further, dependent on the perceptual task (Lee and Middlebrooks 2011), there exists no psychophysical or neurophysiological evidence that space is represented independently in each of these sharp fields. This independence, however, would be required to resolve multiple equidistant objects through biosonar.

Directional sensitivity of hearing is undoubtedly affected by the ability of most bats to move their pinnae forward and back and to rotate them from side to side.

Perhaps most interesting is the special case of high duty cycle echolocating bats: rhinolophids, hipposiderids, and several species of mormoopids in the complex, including Parnell's mustached bats (*Pteronotus parnellii*) (Clare et al. 2011). These bats separate pulse and echo in frequency by exploiting Doppler-shifted echoes (Chapter 9, Hiryu, Mora, and Riquimaroux) and alternately flick their ears forward and back, apparently in synchrony with pulse emission; as one ear is moved forward and down, the other is rotated back and up (Gao et al. 2011). Nasal structures associated with pulse emission move at the same time. Hipposiderids and rhinolophids are closely related to one another but not to the mormoopids (Simmons 2005; Teeling 2009).

Recent advances in digital signal throughput have allowed recording bat sonar emissions with arrays of ultrasonic microphones and thus measure and reconstruct the directionality of the sonar emissions. These data serve to verify the modeled emission directionality but, importantly, microphone arrays allow investigating movement, the second source of information. Compared to the visual system, which shuts down during every fast eye movement (Thiele et al. 2002), the exceptional temporal resolution of the auditory system allows a systematic analysis of how echoes change when the directional properties of a biosonar system change. Well-controlled behavioral studies with microphone-array recordings have shown how bats aim their sonar beam during both obstacle avoidance and hunting (Ghose and Moss 2003, 2006; Moss et al. 2006). However, the overall directionality of a sonar system is the product of emission and reception directionality, so analyses of time-variant sonar-beam directionality are just half of the story.

The second and likely more important part of the story is in the directionality of the receivers, namely the pinnae. Large pinnae are more directional than the typically smaller sound-emitting structures (Vanderelst et al. 2010) and thus more effectively confine the acquired space. As the pinnae do not emit a signal, their time-variant directionality cannot be tracked like the emission directionality. A major challenge for the coming years is assessing ear movements from bats engaged in a clearly defined sonar task and combining time-variant reception directionality from these ear movements with the time-variant emission directionality. Only through this combination of directionalities can we determine the point in space onto which a bat focuses its sonar system, and possibly quantifies its distance at each point in time.

The following illustrates the potential importance of such a reconstruction. In a behavioral experiment with rousette bats, Yovel et al. (2010) showed that these animals alternate the azimuth of their sonar-beam axis left and right relative to the attended object. Studies in laryngeally echolocating bats have never found this left-right alternation. However, it remains possible that some bats may follow a similar localization strategy as rousette bats without varying the directionality of the emission but by varying the directionality of reception. Specifically, bats could use alternating ear movements synchronized with sonar emissions (Gao et al. 2011). Due to the multiplicative nature of the emission and reception directionalities, emission-synchronized, alternating ear movements would be an equally effective localization strategy as alternating emission directionality. But alternating ear movements would not leave a trace on a microphone array recording the emissions.

Major challenges also remain for those trying to obtain comprehensive and precise data from free-flying bats. Even with the fastest high-resolution cameras and/or latest optical-tracking techniques, it will be difficult to reconstruct fast ear movements of a flying bat. When the non-rigid movements and possibly flight-induced deformations of the pinnae (see Metzner and Mueller, Chapter 3) are also taken into account, reconstructing ear movements appears almost impossible. It may, however, be possible to record myogenic signals from the ear muscles and from there reconstruct ear movements in flight.

As more behavioral and electrophysiological data accumulate about species-specific specializations in echolocation, one can better appreciate specializations at different neural levels. Such specializations may be reflected in neurons sharply tuned in azimuth, elevation, and echo delay. It is unclear how sharp tuning in 3-D space is neurally generated and read out. Short CF-FM bats might be an interesting place to start, for example, because the function of the CF component in their calls is unclear (but see Roverud and Grinnell 1985a, b). Some very sophisticated experimentation is now being directed at these problems, but more accurate measures of what bats can do would facilitate the development of good models for how this is achieved by the bat nervous system.

11.4 Life Style and Echolocation in Laryngeally Echolocating Bats

Each investigation of the lifestyle and foraging behavior of another species of bat produces another fascinating discovery. Detailed study of fringe-lipped bats (*Trachops cirrhosus*), for example, led to revelations about passive-acoustic prey detection and gleaning behavior on the part of the bat, as well as adaptations of frogs to avoid predation. Studies of rousette bats by Yovel et al. (2010) revealed a previously unrecognized behavioral mechanism for enhancing target localization, as mentioned above. Another elegant example is the studies of the specialized lifestyle of the common big-eared bats (*Micronycteris microtis*). Geipel et al. (2013) showed that common big-eared bats detect and identify silent, motionless dragonflies and other insect prey from the surfaces of leaves in the dense rainforest understory at night. Whether or not other gleaning bat species have similar echo-imaging capabilities remains to be seen. To date it is also unclear to what extent gleaning bats possess the more modest ability to track prey as it moves on a surface by detecting changes in echoes during a sequence of echolocation calls (Ratcliffe et al. 2005; Ratcliffe 2009).

Many species in several orders of insects (at least Lepidoptera, Coleoptera, Orthoptera, Mantoidea, Neuroptera) have ears sensitive to the echolocation calls of bats. And yet, not surprisingly, when details about species consumed are available, usually through analysis of insect remains by DNA barcoding, we know that some bats eat insects that are sensitive to their calls. The examples range from eastern red bats (*Lasiurus boreali*) (Clare et al. 2009) to little brown bats (*Myotis lucifugus*) (Clare et al. 2011, 2014a) and big brown bats (*Eptesicus fuscus*) (Clare et al. 2014b).

In other cases, bats such as barbastelles (*Barbastella barbastellus*) adjust their echolocation calls to reduce the conspicuousness of their calls and thwart hearing-based defenses (Goerlitz et al. 2010). Given the millions of years of co-evolution of bats and insects, this is potentially a rich field for further study.

11.5 Correlation of Echolocation Strategies with Niches: Guilds of Bats

In Chapter 6 (by Denzinger, Kalko, Tschapka, Grinnell, and Schnitzler), bats are assigned to seven guilds based on their preferred habitat type and foraging mode. This approach allows tentative identification of three types of niche spaces used by bats according to similarities in the combinations of niche dimensions. Such an assignment can be difficult, especially in behaviorally flexible bats that switch among habitats and foraging modes and bats that vary their diets. Nevertheless, this approach allows tentative assignment of a species to a guild that indicates its ecological position within the local ecosystems. The guild concept identifies bats that live under similar ecological conditions and share similar sensory and motor adaptations. At the same time, there is little or no evidence of competition among bats within a guild for food or foraging areas and there can be considerable overlap in diet.

Similarly, the segregation of bat species into guilds is based largely on foraging behavior. While this is a major step forward in understanding bat/echolocation specializations, this segregation needs to be complemented by more thorough analyses of the versatility of echolocation strategies when the same species, in other aspects of its life, navigates through different environments, e.g., open space hunters flying through dense vegetation when commuting from their roost to the foraging area.

At the same time, DNA barcode analysis has permitted more accurate definition of species (lineages), which has enhanced our knowledge about the diversification of bats (e.g., Clare et al. 2011), as well as identification of the species they eat (Clare et al. 2009). People who study bat bioacoustics can use DNA barcode analysis to their advantage, whether the focus is on previously unrecognized biodiversity or on details of the diet. Either development promises to further enrich our knowledge of the bioacoustics of bats.

11.6 Echolocation and Social Signals

We need to learn more about the social structures of bat colonies and communities as well as their vocal and other forms of communication. Producing social signals can be energetically expensive, especially the complex songs and/or displays that males produce to attract females. For example, displaying male hammer-headed bats (*Hypsignathus monstrosus*) produce loud display calls at leks. When a female approaches, males roughly double their rate of display calling, resulting in a buzz-like sound (Bradbury 1977) that must reflect a major increase in energy use. Such high-intensity courtship songs have been documented in other species as well, including non-echolocating pteropodids (Wickler and Seibt 1976). Dechmann et al. (2013) provide a nice example of such a study, examining how the use of echolocation in a communicative, non-foraging context can have a non-trivial metabolic cost. Further work on acoustic signaling in bats, especially high-amplitude social calls and mother-young communication, would provide insight into the cost-benefit trade-off of signaling, potentially allowing us to understand under what conditions such communicative systems would be expected to evolve.

As mating behavior and sexual selection in bats continues to be explored, the focus should move beyond delineating call function and toward understanding the role that such signals play in affecting male reproductive success and fitness. If females can detect differences in male signals and such signals indicate overall male vigor, then females may base much of their decision about who to mate with on information provided in the songs of males.

Increased availability of molecular tools in recent decades has been used to interpret behavioral patterns in a phylogenetic context. Some studies have already begun to look at patterns of echolocation divergence in terms of phylogenetic relatedness (Jones and Teeling 2006; Jung et al. 2014). The potential for using molecular and phylogenetic analyses to understand the evolution of social calls in bats remains, to our knowledge, an untapped area of research.

Bats excel in hearing and echolocation and, like humans and toothed whales, they are mammals that exhibit vocal learning (Janik and Slater 2000; King and Janik 2013). Knörnschild (2014) reviewed the importance of vocal learning in some species, and it is likely that more comprehensive analyses of vocal communication in natural settings will yield many more examples of bat vocal learning. To date, the neural basis of vocal learning has been studied only in songbirds (Amador and Margoliash 2013; Mooney 2014). The early evolutionary divergence between birds and mammals may limit the relevance of bird vocal learning for our understanding of the phylogeny and ontogeny of human speech. The fact that some bat species that exhibit vocal learning can thrive and reproduce in captivity (Knörnschild 2014) paves the way to establishing bats as a tractable mammalian animal model for mammalian vocal learning. This has implications for learning more about the evolution of human speech. This approach is facilitated further by the extensive knowledge available on the neural basis of bat sound production (Chapter 3, Metzner and Mueller) and about the auditory representation of communication calls (Naumann and Kanwal 2011; Kanwal 2012).

11.7 Neural Basis of Biosonar

Since publication of the 1995 SHAR volume *Hearing by Bats*, significant advances have been made on the neural mechanisms underlying hearing and echolocation by bats. Because biosonar is an active imaging process, future experiments should further aim to exploit new concepts and technologies to investigate neural processing of biosonar information.

To give an example, navigation and hunting on the wing in cluttered environments imposes demanding problems (commonly referred to as the "cocktail-party effect") on the auditory system of bats. The bat has to sort out the most important echoes from a wealth of unimportant background echoes to identify relevant objects and to fulfill task-relevant actions. To solve these problems of echo-acoustic scene analysis, several neural mechanisms will have to interact; in addition to hardwired neural circuits, task-specific dynamic processing as well as attention-based mechanisms will come into play. Evidence from non-echolocating mammals shows that task-specific auditory information is specifically processed between non-primary auditory cortex and the frontal cortex (Fritz et al. 2010; Atiani et al. 2014). A corticostriatal pathway has been identified as crucial for the transformation of task-specific auditory information into appropriate action (Xiong et al. 2015). One of the major challenges of the next decade will be to translate these findings and concepts into research on the neural basis of biosonar. For this, future experiments should utilize actively vocalizing bats behaving under both controlled laboratory conditions and natural conditions. This approach would allow more comprehensive studies of the neural code(s) of echolocation in "naturally" behaving bats. The data must be acquired in the context of equally well-controlled behavioral paradigms, as understanding a neural code is more difficult when the animal's intention is unclear. New techniques in wireless electrophysiological recordings from freely behaving bats or techniques that allow storing digitized electrode traces in data loggers carried on animals may bring us closer to realizing some of these goals and will allow combining electrophysiological recordings with behavioral tasks.

An intermediate, established procedure for a well-defined behavioral context in a laboratory setting is to have a bat suspended on a pendulum. This classical setup was used to describe Doppler-shift compensation in horseshoe bats. Very recent work on bat biosonar has seen a revival of the classical pendulum approach (Chapter 9, Hiryu, Mora, and Riquimaroux) in combination with modern, multi-channel neural recording techniques.

Biosonar research will also benefit from molecular tools that typically have been developed in mice but which will become more and more available for "non-standard" animal models such as bats. For example, optogenetic techniques could help to dissect the neural circuitry underlying biosonar processing and task-dependent decisions.

Also concerning the neural coding of stimulus duration and complex acoustic objects, neurophysiological approaches should be extended to include actively vocalizing bats behaving under both controlled conditions. This will be especially important for studies on bat vocal communication given that behavioral context is crucial for inferring signal function(s). Again, advances in telemetric electrophysiological recordings from freely behaving bats may bring us closer to realizing some of these goals, as will playback discrimination experiments with virtual acoustic targets.

Examples for the enhancement of neural-processing features in the context of echolocation are the coding of FM directional selectivity (Suga 1965), combination sensitivity (Mittmann and Wenstrup 1995), delay tuning (Suga 1970; O'Neill and Suga 1979), and duration tuning (Casseday et al. 1994), features that were subsequently seen in the auditory systems of other mammals (Brand et al. 2000; Zhang et al. 2003; Portfors and Felix 2005). While all of these features are also seen in the auditory systems of other mammals, they are much less prominent than in bats. It is these differences that enable bats to capture small flying prey in the dark. For example, there is a pronounced over-representation of neural selectivity for downward FM sweeps in the auditory systems of bats (Fuzessery et al. 2006; Andoni and Pollak 2011), whereas in rats upward and downward FM-selective neurons are about equally prevalent (Zhang et al. 2003; Kuo and Wu 2012). Similarly, combination sensitivity is seen in about half of the auditory neurons in the mustache bat's IC (Wenstrup et al. 2012), whereas combination sensitivity occurs in only about 15 % of neurons in the mouse IC (Portfors and Felix 2005). In short, the adaptations in the brain stem auditory nuclei are primarily, although not exclusively, a matter of degree. Among the important possible qualitative differences between bats and other mammals are (a) the extraordinary temporal resolution and exaggeration of responses to echoes (Grinnell 1963), whereas the nervous system in most other mammals suppresses responses to echoes; and (b) the neural circuitry that enables bats to interpret echo delay as target range. That they appear to be able to do this for multiple targets nearly simultaneously may be more a quantitative than qualitative difference, but it is a critical and highly impressive difference.

Our concept of neural sonar processing may be biased toward an unnatural steady state. Bartenstein et al. (2014) have shown, however, that the chronotopic organization found in the bat auditory cortex is adjusted according to the absolute lateral displacement of the passing reflector. Thus, the chronotopic cortex does not encode echo delay per se but a much more relevant perceptual parameter, namely, the likelihood of colliding with the reflector. These findings are also supported by the description of blurry chronotopic maps (Hechavarria et al. 2013). While these studies were still done under anesthesia and without any animal motion or vocalization, these data indicate what we miss in a biological, highly non-linear system by trying to understand the dynamics of the system from simply concatenating steady states over time.

11.8 Doppler Shift Compensation

High duty cycle (HDC) echolocators use narrowband signal components to detect the Doppler shifts caused by the moving wings of prey among clutter echoes reflected from the environment. The entire suite of essentially identical adaptations for the detection of Doppler shifts in the biosonar system of these unrelated bat groups is a case of convergent evolution driven by a highly advantageous solution to a sonar sensing problem (Chapter 9, Hiryu, Mora, and Riquimaroux).

High duty cycle echolocators are interesting from this perspective. How diverse are their hunting strategies and niches? Do they all hunt in highly cluttered environments? Are Parnell's mustached bats foraging as they fly along paths in the rain forest, or are they simply traveling from place to place? How does Doppler shift detection function in a cluttered environment, where faint periodically Doppler-shifted echoes from small moving prey must be distinguished from unidirectionally shifted echoes from multiple stationary objects? What are the behavioral benefits of Doppler shift detection; what are the liabilities?

The behavioral and physiological ontogeny of Doppler shift detection has not been well studied nor has its evolution. Structured comparative studies of the mormoopids would help to reveal the evolutionary history of Doppler shift detection. Furthermore, a more interesting question is why HDC bats diversified in the Old World but not in the New World. Consideration of that question can also give new leads in attempting to elucidate the evolution of Doppler shift detection. Further investigation is also needed to understand how these bats adapt their echolocation both behaviorally and physiologically to overcome unexpected jamming sounds while flying with conspecifics. Doppler shift detection is a unique strategy for echolocation that will provide new perspectives not only for animal neuroethology but potentially also for various design concepts in the technology and engineering fields.

11.9 Acoustic Versus Visual Images and Auditory Scene Analysis

In spite of distinct differences between the physical stimuli, receptor organs, and early neural processing for the two systems, a comparison of vision and echolocation reveals striking parallels that can guide a more informed understanding of higher level perceptual mechanisms across modalities. For example, high resolution along the direction of gaze or the central axis is common to vision and echolocation. In both systems, spatial information is sharpest in the direction of gaze. This may reflect a common strategy to efficiently allocate computational resources to guide action.

Observations on visual perception can lead to insights about echolocation. For example, human observers experience a continuous and stable percept of the world, although visual information is suppressed during eye movements. Bats may also experience a continuous representation of the world through interrupted echo snapshots of the environment. We need to know more about how bats segregate objects through echo-acoustic imaging. In vision, object segregation is achieved, in large part, through an analysis of spatial frequency. When objects are segregated by the hard edges (= high spatial frequency) that outline them against their background (Klymenko and Weisstein 1986), bats may be capable of a similar spatial-frequency analysis.

Some trawling bats detect objects on the water's surface where the object is surrounded by a sharp edge, much sharper than surface waves on the water itself. Trawling bats successfully detect prey even from flowing or disturbed water with pronounced ripples. The spatial frequencies of ripples on water are still lower than that of the edge between the water surface and the object. Thus, spatial high-pass filtering, a classical way to segregate objects in vision, would be equally effective for finding an object moving on flowing water (Zsebok et al. 2013). However, it remains unclear how bats accomplish this, given the lack of spatial information at the level of either ear in audition and echolocation.

Despite the fundamentally different peripheral-sensory representations of an object in vision and echolocation, the object remains the same, so vision and echolocation might be expected to present a recognizable, modality-independent representation of the object. The extent to which such a modality-independent object representation is achieved has been documented for echolocating dolphins, which can match the shape of objects across vision and echolocation (Harley et al. 1996, 2003). While cross-modal object recognition through echolocation in bats has been well documented, evidence for a modality-independent object representation and cross-modal matching of 3-D shape remains unavailable from bats. Because auditory and visual systems of bats serve the same task, namely imaging the 3-D shape in the field of view, bats are ideal models to study cross-modal integration. Several studies have verified audio and visual inputs to the superior colliculus (SC) of echolocating bats (Covey et al. 1987; Valentine and Moss 1997; Scalia et al. 2014). The SC is a sensorimotor interface in echolocating bats (Valentine et al. 2002). Compared to the barn owl (*Tyto alba*) that currently dominates the topic of audiovisual integration (Winkowski and Knudsen 2007), bats exhibit both experimental disadvantages and possible behavioral/neural advantages.

First, like many mammals, bats have movable ears and eyes. This greatly complicates the situation relative to the barn owl because the bat would have to know at every instant the position and directionality of its ears at each frequency relative to its head and eye direction. Barn owls do not have this problem because they cannot move the eyes, ears, or facial ruff independent of the movement of their heads. The exceptionally high directionality of bat hearing (Grothe et al. 1996; Hoffmann et al. 2010; Hoffmann et al. 2013) makes bat audition somewhat more similar to vision. Very recent work in the bat inferior colliculus indicates that bats have space-specific cross-modal facilitation between vision and echolocation. Interestingly, there is no cross-modal facilitation between vision and lower frequency acoustic stimulation (Hoffmann and Luksch, unpublished observations).

A related question is the degree of accuracy of an acoustic image compared with a visual image. Echolocating bats with moderately good vision should be tested for their ability to recognize, acoustically and seamlessly, learned visual discriminations of increasing depth, breadth, and complexity (and vice versa).

11.10 How Do Bats Navigate?

Bats clearly can navigate in familiar space by using spatial memory that is fed from its biosonar. Exciting progress is being made in analyzing 3-D grid and place cells in the bat brain (Yartsev and Ulanovsky 2013; Finkelstein et al. 2015). Possibly this can be expanded to get some idea of what kind of biosonar map a bat has of its "home territory." It seems highly unlikely that a bat can have a biosonar map of an area several square kilometers in dimensions. How is home area learned? What are the differences in home area and behavior in a tagged, blindfolded bat compared with others in the colony or itself before or after blindfolding? And what about homing? Are magnetic fields involved and over what distances? This is a fundamental neuroethological question, and it may turn out that bats would be an ideal model in which to address it. Tsoar et al. (2011) demonstrated that rousette bats have an internal map they use in orientation.

11.11 Summary

The current chapter is intended to identify some of the major challenges that only now emerge as technically feasible when studying hearing in bats. These technical advances include all aspects of biological research, from novel genetic tools to novel recording techniques of ultrasonic audio, high-speed video, and global positioning data. With these new techniques we will be able to address specializations in the bat vocalization and auditory systems, from its genetic foundations to its behavioral dynamics. We hope that the current chapter is received as an 'appetizer' for young biologists across disciplines to focus their scientific efforts on the fascinating topic of how it is to be a bat that has made the nocturnal airspace its home.

Acknowledgements The authors thank Uwe Firzlaff for helpful comments on an earlier version. And we extend special thanks to the >1,200 species of bats, whose extraordinary capabilities are the inspiration for this book.

References

Amador, A., & Margoliash, D. (2013). A mechanism for frequency modulation in songbirds shared with humans. *Journal of Neuroscience*, 33(27), 11136–11144. doi:10.1523/jneurosci.5906-12.2013

Andoni, S., & Pollak, G. D. (2011). Selectivity for spectral motion as a neural computation for encoding natural communication signals in bat inferior colliculus. *Journal of Neuroscience*, 31(46), 16529–16540. doi:10.1523/jneurosci.1306-11.2011

Atiani, S., David, S. V., Elgueda, D., Locastro, M., Radtke-Schuller, S., Shamma, S. A., & Fritz, J. B. (2014). Emergent selectivity for task-relevant stimuli in higher-order auditory cortex. *Neuron*, 82(2), 486–499. doi:10.1016/j.neuron.2014.02.029

Bartenstein, S. K., Gerstenberg, N., Vanderelst, D., Peremans, H., & Firzlaff, U. (2014). Echo-acoustic flow dynamically modifies the cortical map of target range in bats. *Nature Communications*, 5, 4668. doi:10.1038/ncomms5668

Boonman, A., Bumrungsri, S., & Yovel, Y. (2014). Nonecholocating fruit bats produce biosonar clicks with their wings. *Current Biology*, 24(24), 2962–2967. doi:10.1016/j.cub.2014.10.077

Borina, F., Firzlaff, U., & Wiegrebe, L. (2011). Neural coding of echo-envelope disparities in echolocating bats. *Journal of Comparative Physiology A: Neuroethology, Sensory, Neural, and Behavioral Physiology*, 197(5), 561–569.

Bradbury, J. W. (1977). Lek mating behavior in the hammer-headed bat. *Zeitschrift für Tierpsychologie*, 45, 30.

Brand, A., Urban, R., & Grothe, B. (2000). Duration tuning in the mouse auditory midbrain. *Journal of Neurophysiology*, 84(4), 1790–1799.

Casseday, J. H., Ehrlich, D., & Covey, E. (1994). Neural tuning for sound duration: Role of inhibitory mechanisms in the inferior colliculus. *Science*, 264, 847–850.

Clare, E. L., Fraser, E. E., Braid, H. E., Fenton, M. B., & Hebert, P. D. (2009). Species on the menu of a generalist predator, the eastern red bat (*Lasiurus borealis*): Using a molecular approach to detect arthropod prey. *Molecular Ecology*, 18(11), 2532–2542. doi:10.1111/j.1365-294X.2009.04184.x

Clare, E. L., Lim, B. K., Fenton, M. B., & Hebert, P. D. (2011). Neotropical bats: Estimating species diversity with DNA barcodes. *PLoS ONE*, 6(7), e22648. doi:10.1371/journal.pone.0022648
Clare, E. L., Symondson, W. O., Broders, H., Fabianek, F., Fraser, E. E., MacKenzie, A., Boughen, A., Hamilton, R., Willis, C.K., Martinez-Nunez, F., Menzies, A. K., Norquay, K. J., Brigham, M., Poissant, J., Rintoul, J., Barklay, R. M., & Reimer, J. P. (2014a). The diet of *Myotis lucifugus* across Canada: Assessing foraging quality and diet variability. *Molecular Ecology*, 23(15), 3618–3632. doi:10.1111/mec.12542
Clare, E. L., Symondson, W. O., & Fenton, M. B. (2014b). An inordinate fondness for beetles? Variation in seasonal dietary preferences of night-roosting big brown bats (*Eptesicus fuscus*). *Molecular Ecology*, 23(15), 3633–3647.
Covey, E., Hall, W. C., & Kobler, J. B. (1987). Subcortical connections of the superior colliculus in the mustache bat, *Pteronotus parnellii*. *Journal of Comparative Neurology*, 263(2), 179–197. doi:10.1002/cne.902630203
Dechmann, D. K., Wikelski, M., van Noordwijk, H. J., Voigt, C. C., & Voigt-Heucke, S. L. (2013). Metabolic costs of bat echolocation in a non-foraging context support a role in communication. *Frontiers in Physiology*, 4, 66. doi:10.3389/fphys.2013.00066
Dong, D., Lei, M., Liu, Y., & Zhang, S. (2013). Comparative inner ear transcriptome analysis between the Rickett's big-footed bats (*Myotis ricketti*) and the greater short-nosed fruit bats (*Cynopterus sphinx*). *BMC Genomics*, 14(1), 916.
Finkelstein, A., Derdikman, D., Rubin, A., Foerster, J. N., Las, L., & Ulanovsky, N. (2015). Three-dimensional head-direction coding in the bat brain. *Nature*, 517(7533), 159–164. doi: 10.1038/nature14031
Francischetti, I. M., Assumpcao, T. C., Ma, D., Li, Y., Vicente, E. C., Uieda, W., & Ribeiro, J. M. (2013). The "Vampirome": Transcriptome and proteome analysis of the principal and accessory submaxillary glands of the vampire bat *Desmodus rotundus*, a vector of human rabies. *Journal of Proteomics*, 82, 288–319. doi:10.1016/j.jprot.2013.01.009
Fritz, J. B., David, S. V., Radtke-Schuller, S., Yin, P., & Shamma, S. A. (2010). Adaptive, behaviorally gated, persistent encoding of task-relevant auditory information in ferret frontal cortex. *Nature Neuroscience*, 13(8), 1011–1019. doi:10.1038/nn.2598
Fuzessery, Z. M., & Pollak, G. D. (1984). Neural mechanisms of sound localization in an echolocating bat. *Science*, 225(4663), 725–728.
Fuzessery, Z. M., Richardson, M. D., & Coburn, M. S. (2006). Neural mechanisms underlying selectivity for the rate and direction of frequency-modulated sweeps in the inferior colliculus of the pallid bat. *Journal of Neurophysiology*, 96(3), 1320–1336. doi: 10.1152/jn.00021.2006
Gao, L., Balakrishnan, S., He, W., Yan, Z., & Muller, R. (2011). Ear deformations give bats a physical mechanism for fast adaptation of ultrasonic beam patterns. *Physical Review Letters*, 107(21), 214301.
Geipel, I., Jung, K., & Kalko, E. K. (2013). Perception of silent and motionless prey on vegetation by echolocation in the gleaning bat *Micronycteris microtis*. *Proceedings in Biological Sciences*, 280(1754), 20122830. doi: 10.1098/rspb.2012.2830
Genome 10K. (2009). A proposal to obtain whole-genome sequence for 10,000 vertebrate species. *Journal of Heredity*, 100(6), 659–674. doi:10.1093/jhered/esp086
Ghose, K., & Moss, C. F. (2003). The sonar beam pattern of a flying bat as it tracks tethered insects. *Journal of the Acoustical Society of America*, 114(2), 1120–1131.
Ghose, K., & Moss, C. F. (2006). Steering by hearing: A bat's acoustic gaze is linked to its flight motor output by a delayed, adaptive linear law. *Journal of Neuroscience*, 26(6), 1704–1710.
Goerlitz, H. R., ter Hofstede, H. M., Zeale, M. R., Jones, G., & Holderied, M. W. (2010). An aerial-hawking bat uses stealth echolocation to counter moth hearing. *Current Biology*, 20(17), 1568–1572. doi: 10.1016/j.cub.2010.07.046
Gould, E. (1988). Wing-clapping sounds of *Eonycteris spelaea* (Pteropodidae) in Malaysia. *Journal of Mammalogy*, 69, 378–379.
Grinnell, A. D. (1963). The neurophysiology of audition in bats: Temporal parameters. *Journal of Physiology*, 167, 67–96.
Grinnell, A. D., & Grinnell, V. S. (1965). Neural correlates of vertical localization by echo-locating bats. *Journal of Physiology*, 181(4), 830–851.

Grothe, B., Covey, E., & Casseday, J. H. (1996). Spatial tuning of neurons in the inferior colliculus of the big brown bat: Effects of sound level, stimulus type and multiple sound sources. *Journal of Comparative Physiology A: Neuroethology, Sensory, Neural, and Behavioral Physiology*, 179(1), 89–102.

Grothe, B., Pecka, M., & McAlpine, D. (2010). Mechanisms of sound localization in mammals. *Physiological Reviews*, 90(3), 983–1012. doi: 10.1152/physrev.00026.2009

Gui, D., Jia, K., Xia, J., Yang, L., Chen, J., Wu, Y., & Meisheng, Y. (2013). De novo assembly of the Indo-Pacific humpback dolphin leucocyte transcriptome to identify putative genes involved in the aquatic adaptation and immune response. *PLoS ONE*, 8(8), e72417. doi: 10.1371/journal.pone.0072417

Harley, H. E., Roitblat, H. L., & Nachtigall, P. E. (1996). Object representation in the bottlenose dolphin (*Tursiops truncatus*): Integration of visual and echoic information. *Journal of Experimental Psychology: Animal Learning and Cognition*, 22(2), 164–174.

Harley, H. E., Putman, E. A., & Roitblat, H. L. (2003). Bottlenose dolphins perceive object features through echolocation. *Nature*, 424(6949), 667–669.

Hechavarria, J. C., Macias, S., Vater, M., Voss, C., Mora, E. C., & Kossl, M. (2013). Blurry topography for precise target-distance computations in the auditory cortex of echolocating bats. *Nature Communications*, 4, 2587. doi:10.1038/ncomms3587

Hoffmann, S., Schuller, G., & Firzlaff, U. (2010). Dynamic stimulation evokes spatially focused receptive fields in bat auditory cortex. *European Journal of Neuroscience*, 31(2), 371–385. doi:10.1111/j.1460-9568.2009.07051.x

Hoffmann, S., Warmbold, A., Wiegrebe, L., & Firzlaff, U. (2013). Spatiotemporal contrast enhancement and feature extraction in the bat auditory midbrain and cortex. *Journal of Neurophysiology*, 110(6), 1257–1268. doi: 10.1152/jn.00226.2013

Janik, V. M., & Slater, P. J. (2000). The different roles of social learning in vocal communication. *Animal Behaviour*, 60(1), 1–11. doi: 10.1006/anbe.2000.1410

Jones, G., & Teeling, E. C. (2006). The evolution of echolocation in bats. *Trends in Ecology and Evolution*, 21(3), 149–156. doi:10.1016/j.tree.2006.01.001

Jung, K., Molinari, J., & Kalko, E. K. (2014). Driving factors for the evolution of species-specific echolocation call design in new world free-tailed bats (Molossidae). *PLoS ONE*, 9(1), e85279. doi:10.1371/journal.pone.0085279

Kanwal, J. S. (2012). Right-left asymmetry in the cortical processing of sounds for social communication vs. navigation in mustached bats. *European Journal of Neuroscience*, 35(2), 257–270. doi:10.1111/j.1460-9568.2011.07951.x

King, S. L., & Janik, V. M. (2013). Bottlenose dolphins can use learned vocal labels to address each other. *Proceedings of the National Academy of Sciences of the USA*, 110(32), 13216–13221. doi:10.1073/pnas.1304459110

Klymenko, V., & Weisstein, N. (1986). Spatial frequency differences can determine figure-ground organization. *Journal of Experimental Psychology: Human Perception and Performance*, 12(3), 324–330.

Knörnschild, M. (2014). Vocal production learning in bats. *Current Opinion in Neurobiology*, 28, 80–85. doi:10.1016/j.conb.2014.06.014

Kuo, R. I., & Wu, G. K. (2012). Generation of direction selectivity in the auditory system. *Neuron*, 73, 1016–1027.

Lawrence, B. D., & Simmons, J. A. (1982a). Echolocation in bats: The external ear and perception of the vertical positions of targets. *Science*, 218(4571), 481–483.

Lawrence, B. D., & Simmons, J. A. (1982b). Measurements of atmospheric attenuation at ultrasonic frequencies and the significance for echolocation by bats. *Journal of the Acoustical Society of America*, 71(3), 585–590.

Lee, C. C., & Middlebrooks, J. C. (2011). Auditory cortex spatial sensitivity sharpens during task performance. *Nature Neuroscience*, 14(1), 108–114. doi:10.1038/nn.2713

Li, G., Wang, J., Rossiter, S. J., Jones, G., & Zhang, S. (2007). Accelerated FoxP2 evolution in echolocating bats. *PLoS ONE*, 2(9), e900. doi: 10.1371/journal.pone.0000900

Mittmann, D. H., & Wenstrup, J. J. (1995). Combination-sensitive neurons in the inferior colliculus. *Hearing Research*, 90(1–2), 185–191.
Mooney, R. (2014). Auditory-vocal mirroring in songbirds. *Philosophical Transactions of the Royal Society B*, 369(1644), 20130179. doi:10.1098/rstb.2013.0179
Moss, C. F., Bohn, K., Gilkenson, H., & Surlykke, A. (2006). Active listening for spatial orientation in a complex auditory scene. *PLoS Biology*, 4(4), e79.
Naumann, R. T., & Kanwal, J. S. (2011). Basolateral amygdala responds robustly to social calls: Spiking characteristics of single unit activity. *Journal of Neurophysiology*, 105(5), 2389–2404. doi:10.1152/jn.00580.2010
O'Neill, W. E., & Suga, N. (1979). Target range-sensitive neurons in the auditory cortex of the mustache bat. *Science*, 203(4375), 69–73.
Phillips, C. J., Phillips, C. D., Goecks, J., Lessa, E. P., Sotero-Caio, C. G., Tandler, B., Gannon, M. R., & Baker, R. J. (2014). Dietary and flight energetic adaptations in a salivary gland transcriptome of an insectivorous bat. *PLoS ONE*, 9(1), e83512. doi:10.1371/journal.pone.0083512
Portfors, C. V., & Felix, R. A., II. (2005). Spectral integration in the inferior colliculus of the CBA/CaJ mouse. *Neuroscience*, 136(4), 1159–1170. doi: 10.1016/j.neuroscience.2005.08.031
Ratcliffe, J. M. (2009). Neuroecology and diet selection in phyllostomid bats. *Behavioural Processes*, 80(3), 247–251. doi:10.1016/j.beproc.2008.12.010
Ratcliffe, J. M., Raghuram, H., Marimuthu, G., Fullard, J. H., & Fenton, M. B. (2005). Hunting in unfamiliar space: Echolocation in the Indian false vampire bat, *Megaderma lyra*, when gleaning prey. *Behavioral Ecology and Sociobiology*, 58(2), 157–164. doi:10.2307/25063599
Rayleigh, LFRS (1879) Investigations in optics, with special reference to the spectroscope. *Philosophical Magazine Series* 8(49): 261–274. doi:10.1080/14786447908639684
Roverud, R., & Grinnell, A. (1985a). Echolocation sound features processed to provide distance information in the CF/FM bat, *Noctilio albiventris*: Evidence for a gated time window utilizing both CF and FM components. *Journal of Comparative Physiology A: Neuroethology, Sensory, Neural, and Behavioral Physiology*, 156(4), 457–469. doi:10.1007/BF00613970
Roverud, R., & Grinnell, A. (1985b). Discrimination performance and echolocation signal integration requirements for target detection and distance determination in the CF/FM bat, *Noctilio albiventris. Journal of Comparative Physiology A: Neuroethology, Sensory, Neural, and Behavioral Physiology*, 156(4), 447–456. doi:10.1007/BF00613969
Scalia, F., Rasweiler, J. J. T., & Danias, J. (2014). Retinal projections in the short-tailed fruit Bat, *Carollia perspicillata*, as studied using the axonal transport of cholera toxin B subunit: Comparison with mouse. *Journal of Comparative Neurology*, doi:10.1002/cne.23723
Seim, I., Fang, X., Xiong, Z., Lobanov, A. V., Huang, Z., Ma, S., Feng, Y., Turanov, A. A., Zhu, Y., Lenz, T. L., Gerashchenko, M. V., Fan, D., Yim, S. H., Yao, X., Jordan, D., Xiong, Y., Ma, Y., Lyapunov, A. N., Chen, G., Kulakova, O. I., Sun, Y., Lee, S.-G., Bronson, R. T., Moskalev, A. A., Sunyaev, S. R., Zhang, G., Krogh, A., Wang, J., & Gladyshev, V. N. (2013). Genome analysis reveals insights into physiology and longevity of the Brandt's bat *Myotis brandtii. Nature Communications*, 4, 2212. doi: 10.1038/ncomms3212
Simmons, J. A., Kick, S. A., Lawrence, B. D., Hale, C., Bard, C., & Escudie, B. (1983). Acuity of horizontal angle discrimination by the echolocating bat, *Eptesicus fuscus. Journal of Comparative Physiology A: Neuroethology, Sensory, Neural, and Behavioral Physiology*, 153, 321–330.
Simmons, N. B. (2005). Order Chiroptera. In D. E. Wilson & D. M. Reeder (Eds.), *Mammal species of the World: A taxonomic and geographic reference*. 3rd ed. Baltimore, MD: Johns Hopkins University Press.
Suga, N. (1965). Analysis of frequency modulated sounds by auditory neurons of echolocating bats. *Journal of Physiology*, 179, 26–53.
Suga, N. (1970). Echo-ranging neurons in the inferior colliculus of bats. *Science*, 170(3956), 449–452.
Teeling, E. C. (2009). Hear, hear: The convergent evolution of echolocation in bats? *Trends in Ecology and Evolution*, 24(7), 351–354. doi:10.1016/j.tree.2009.02.012
Thiele, A., Henning, P., Kubischik, M., & Hoffmann, K. P. (2002). Neural mechanisms of saccadic suppression. *Science*, 295(5564), 2460–2462. doi:10.1126/science.1068788

Tsagkogeorga, G., Parker, J., Stupka, E., Cotton, J. A., & Rossiter, S. J. (2013). Phylogenomic analyses elucidate the evolutionary relationships of bats. *Current Biology*, 23(22), 2262–2267. doi:10.1016/j.cub.2013.09.014

Tsoar, A., Nathan, R., Bartan, Y., Vyssotski, A., Dell'Omo, G., & Ulanovsky, N. (2011). Large-scale navigational map in a mammal. *Proceedings of the National Academy of Sciences of the USA*, 108(37), E718–E724. doi:10.1073/pnas.1107365108

Valentine, D. E., & Moss, C. F. (1997). Spatially selective auditory responses in the superior colliculus of the echolocating bat. *Journal of Neuroscience*, 17(5), 1720–1733.

Valentine, D. E., Sinha, S. R., & Moss, C. F. (2002). Orienting responses and vocalizations produced by microstimulation in the superior colliculus of the echolocating bat, *Eptesicus fuscus*. *Journal of Comparative Physiology A: Neuroethology, Sensory, Neural, and Behavioral Physiology*, 188(2), 89–108. doi:10.1007/s00359-001-0275-5

Vanderelst, D., De Mey, F., Peremans, H., Geipel, I., Kalko, E., & Firzlaff, U. (2010). What noseleaves do for FM bats depends on their degree of sensorial specialization. *PLoS ONE*, 5(8), e11893. doi:10.1371/journal.pone.0011893

Wenstrup, J. J., Nataraj, K., & Sanchez, J. T. (2012). Mechanisms of spectral and temporal integration in the mustached bat inferior colliculus. *Frontiers in Neural Circuits*, 6, 75. doi:10.3389/fncir.2012.00075

Wickler, W., & Seibt, U. (1976). Field studies on the African fruit bat *Epomophorus wahlbergi* (Sundevall), with special reference to male calling. *Zeitung für Tierpsychologie*, 40(4), 345–376.

Winkowski, D. E., & Knudsen, E. I. (2007). Top-down control of multimodal sensitivity in the barn owl optic tectum. *Journal of Neuroscience*, 27(48), 13279–13291. doi:10.1523/jneurosci.3937-07.2007

Xiong, Q., Znamenskiy, P., & Zador, A. M. (2015). Selective corticostriatal plasticity during acquisition of an auditory discrimination task. *Nature*, 521(7552), 348–351. doi:10.1038/nature14225

Yartsev, M. M., & Ulanovsky, N. (2013). Representation of three-dimensional space in the hippocampus of flying bats. *Science*, 340(6130), 367–372. doi:10.1126/science.1235338

Yim, H.-S., Cho, Y. S., Guang, X., Kang, S. G., Jeong, J.-Y., Cha, S.-S., Oh, H.-M., Lee, J.-H., Yang, E. C., Kwon, K. K., Kim, Y. J., Kim, T. W., Kim, W., Jeon, J. H., Kim, S.-J., Choi, D. H., Jho, S., Kim, H.-M., Ko, J., Kim, H., Shin, Y.-A., Jung, H.-J., Zheng, Y., Wang, Z., Chen, Y., Chen, M., Jiang, A., Li, E., Zhang, S., Hou, H., Kim, T. H., Yu, L., Liu, S., Ahn, K., Cooper, J., Park, S.-G., Hong, C. P., Jin, W., Kim, H.-S., Park, C., Lee, K., Chun, S., Morin, P. A., O'Brien, S. J., Lee, H., Kimura, J., Moon, D. Y., Manica, A., Edwards, J., Kim, B. C., Kim, S., Wang, J., Bhak, J., Lee, H. S., & Lee, J.-H. (2014). Minke whale genome and aquatic adaptation in cetaceans. *Nature Genetics,* 46(1), 88–92. doi:10.1038/ng.2835

Yovel, Y., Falk, B., Moss, C. F., & Ulanovsky, N. (2010). Optimal localization by pointing off axis. *Science*, 327(5966), 701–704. doi:10.1126/science.1183310

Zhang, L. I., Tan, A. Y., Schreiner, C. E., & Merzenich, M. M. (2003). Topography and synaptic shaping of direction selectivity in primary auditory cortex. *Nature*, 424(6945), 201–205. doi:10.1038/nature01796

Zhang, G., Cowled, C., Shi, Z., Huang, Z., Bishop-Lilly, K. A., Zhang, G., Cowled, C., Shi, Z., Huang, Z., Bishop-Lilly, K. A., Fang, X., Wynne, J. W., Xiong, Z., Baker, M. L., Zhao, W., Tachedjian, M., Zhu, Y., Zhou, P., Jiang, X., Ng, J., Yang, L., Wu, L., Xiao, J., Feng, Y., Chen, Y., Sun, X., Zhang, Y., Marsh, G. A., Crameri, G., Broder, C. C., Frey, K. G., Wang, L.-F., & Wang, J. (2013). Comparative analysis of bat genomes provides insight into the evolution of flight and immunity. *Science*, 339(6118), 456–460. doi:10.1126/science.1230835

Zhou, X., Sun, F., Xu, S., Fan, G. Zhu, K., Liu, X., Guo, X., Zhou, J., Fang, X., Li, M., Wei, F., Poulain, J., Zhou, K., Wang, J., Chen, Y., Shi, C., Yang, Y., Huang, Z., Chen, J., Hou, H., Chen, W., Chen, Y., Wang, X., Lv, T., Yang, D., Huang, B., Wang, Z., Tian, R., Xiong, Z., Xu, J., Liang, X., Chen, B., Liu, W., Wang, J., Pan, S., & Yang, G. (2013). Baiji genomes reveal low genetic variability and new insights into secondary aquatic adaptations. *Nature Communications*, 4, 1–6, 2708 doi:10.1038/ncomms3708

Zsebok, S., Kroll, F., Heinrich, M., Genzel, D., Siemers, B. M., & Wiegrebe, L. (2013). Trawling bats exploit an echo-acoustic ground effect. *Frontiers in Physiology*, 4, 65. doi:10.3389/fphys.2013.00065

GPSR Compliance
The European Union's (EU) General Product Safety Regulation (GPSR) is a set of rules that requires consumer products to be safe and our obligations to ensure this.

If you have any concerns about our products, you can contact us on

ProductSafety@springernature.com

In case Publisher is established outside the EU, the EU authorized representative is:

Springer Nature Customer Service Center GmbH
Europaplatz 3
69115 Heidelberg, Germany

www.ingramcontent.com/pod-product-compliance
Ingram Content Group UK Ltd.
Pitfield, Milton Keynes, MK11 3LW, UK
UKHW022214051026
15624UKWH00001B/151
* 9 7 8 1 4 9 3 9 3 5 2 5 3 *